MANAGING WATER RESOURCES IN THE WEST UNDER CONDI[illegible] OF CLIMATE UNCERTAINTY

Proceedings of a Colloquium
November 14-16, 1990
Scottsdale, Arizona

Committee on Climate Uncertainty
and Water Resources Management
Water Science and Technology Board
Commission on Geosciences, Environment,
and Resources

NATIONAL ACADEMY PRESS
Washington, D.C. 1991

NOTICE: The project that is the subject of this report was approved by the Governing Board of the National Research Council, whose members are drawn from the councils of the National Academy of Sciences, the National Academy of Engineering, and the Institute of Medicine. The members of the committee responsible for the report were chosen for their special competences and with regard for appropriate balance.

This report has been reviewed by a group other than the authors according to procedures approved by a Report Review Committee consisting of members of the National Academy of Sciences, the National Academy of Engineering, and the Institute of Medicine.

The National Academy of Sciences is a private, nonprofit, self-perpetuating society of distinguished scholars engaged in scientific and engineering research, dedicated to the furtherance of science and technology and to their use for the general welfare. Upon the authority of the charter granted to it by the Congress in 1863, the Academy has a mandate that requires it to advise the federal government on scientific and technical matters. Dr. Frank Press is president of the National Academy of Sciences.

The National Academy of Engineering was established in 1964, under the charter of the National Academy of Sciences, as a parallel organization of outstanding engineers. It is autonomous in its administration and in the selection of its members, sharing with the National Academy of Sciences the responsibility for advising the federal government. The National Academy of Engineering also sponsors engineering programs aimed at meeting national needs, encourages education and research, and recognizes the superior achievements of engineers. Dr. Robert M. White is president of the National Academy of Engineering.

The Institute of Medicine was established in 1970 by the National Academy of Sciences to secure the services of eminent members of appropriate professions in the examination of policy matters pertaining to the health of the public. The Institute acts under the responsibility given to the National Academy of Sciences by its congressional charter to be an adviser to the federal government and, upon its own initiative, to identify issues of medical care, research, and education. Dr. Stuart Bondurant is acting president of the Institute of Medicine.

The National Research Council was organized by the National Academy of Sciences in 1916 to associate the broad community of science and technology with the Academy's purposes of furthering knowledge and advising the federal government. Functioning in accordance with general policies determined by the Academy, the Council has become the principal operating agency of both the National Academy of Sciences and the National Academy of Engineering in providing services to the government, the public, and the scientific and engineering communities. The Council is administered jointly by both Academies and the Institute of Medicine. Dr. Frank Press and Dr. Robert M. White are chairman and vice chairman, respectively, of the National Research Council.

Support for the project was provided by the Bureau of Reclamation under Cooperative Agreement 1-FC-81-00900.

INTERNATIONAL STANDARD BOOK NUMBER 0-309-04677-7

LIBRARY OF CONGRESS CATALOG CARD NO. 91-67298

Copies of this report are available from:

Water Science and Technology Board
2101 Constitution Avenue, N.W.
Washington, D.C. 20418

National Academy Press
2101 Constitution Avenue, N.W.
Washington, D.C. 20418
S491

Printed in the United States of America

Cover photograph, "Colorado River Delta Meets the Sea of Cortés," used with permission from Annie Griffiths Belt, Silver Spring, Maryland.

Dedication

This volume is dedicated to the memory of Roger R.D. Revelle (1909-1991), renowned oceanographer and student of the earth and its processes. Dr. Revelle worked at the forefront of efforts to understand the possible warming effects of greenhouse gases. A scientist of diverse talents, Dr. Revelle received the National Medal of Science from President Bush in 1990 . . . "for his pioneering work in the areas of carbon dioxide and climate modification, oceanographic exploration presaging plate tectonics, the biological effects of radiation on the marine environment, and studies of human population growth and food supplies." The Water Science and Technology Board and the Committee on Climate Uncertainty and Water Resources Management were honored that Dr. Revelle chose to participate in the colloquium that is the basis for this volume. His insights and humor will be greatly missed.

COLLOQUIUM STEERING COMMITTEE

STEPHEN J. BURGES, *Chair*, University of Washington, Seattle
ROBERT E. DICKINSON, University of Arizona, Tucson
KENNETH D. FREDERICK, Resources for the Future, Washington, D.C.
ROGER E. KASPERSON, Clark University, Worcester, Massachusetts
BRUCE A. KIMBALL, U.S. Water Conservation Laboratory, Phoenix, Arizona
DANIEL P. SHEER, Water Resources Management, Inc., Columbia, Maryland

WATER SCIENCE AND TECHNOLOGY BOARD

DANIEL A. OKUN, *Chair*, University of North Carolina, Chapel Hill
MICHAEL C. KAVANAUGH, *Chair*, James M. Montgomery Consulting Engineers, Walnut Creek, California (through June 1991)
A. DAN TARLOCK, *Vice Chair*, Illinois Institute of Technology, Chicago-Kent College of Law, Chicago, Illinois
NORMAN H. BROOKS, California Institute of Technology, Pasadena
RICHARD A. CONWAY, Union Carbide Corporation, South Charleston, West Virginia
KENNETH D. FREDERICK, Resources for the Future, Washington, D.C.
DAVID L. FREYBERG, Stanford University, Stanford, California
WILFORD R. GARDNER, University of California, Berkeley
DUANE L. GEORGESON, Metropolitan Water District, Los Angeles, California
HOWARD C. KUNREUTHER, University of Pennsylvania, Philadelphia (through June 1991)
JUDY L. MEYER, University of Georgia, Athens
ROBERT R. MEGLEN, University of Colorado, Denver (through June 1991)
DONALD J. O'CONNOR, HydroQual, Inc., Glen Rock, New Jersey
BETTY H. OLSON, University of California, Irvine (through June 1991)

WSTB Colloquium Staff

WSTB Staff

COMMISSION ON GEOSCIENCES, ENVIRONMENT, AND RESOURCES

STAFF

Preface

The question of whether the earth's climate is changing in some significant, human-induced way remains a matter of much debate. But the fact that climate is variable over time is well known. These two elements of climatic uncertainty both affect water resources planning and management in the American West.

Natural variability in hydrologic processes is all-pervasive. Many techniques have been developed to describe components of this variability. Additional tools and strategies are needed to manage water effectively given inherent supply variability and competing demands. With added and improved capabilities and strategies, we will be better prepared to deal with the uncertainties presented by potential climate change.

"Managing Water Resources in the West Under Conditions of Climate Uncertainty," a colloquium held November 14–16, 1990, in Scottsdale, Arizona, was organized by the Water Science and Technology Board (WSTB) at the request of the U.S. Bureau of Reclamation. The colloquium was held to examine the scientific basis for predictions of climate change, the implications of climate uncertainty for water resources management, and the management options available for responding to climate variability and potential climate change. Bureau of Reclamation Commissioner Dennis Underwood, noting the importance of climate variability to his agency's operations in the West, took a personal interest in the colloquium and spoke with participants about his goals for the Bureau of Reclamation as it increases its emphasis on resource management issues.

The colloquium was developed by a steering committee consisting of Robert E. Dickinson, University of Arizona; Kenneth D. Frederick, Resources for the Future; Roger E. Kasperson, Clark

University; Bruce A. Kimball, U.S. Water Conservation Laboratory; Daniel P. Sheer, Water Resources Management, Inc.; and Stephen J. Burges, University of Washington. The proceedings include an overview prepared by the steering committee and the papers presented at the colloquium by individual authors. The introductory remarks made by the session moderators are included to capture the flavor of the colloquium. The overview has been subjected to the report review criteria established by the National Research Council's Report Review Committee; the individual papers and introductory discussions have been reviewed for editorial consistency but have not been peer reviewed.

I thank the steering committee members and WSTB staff, particularly Chris Elfring and Stephen Parker, for their efforts in planning and hosting the event. My great appreciation also goes to the session moderators, the authors of the papers presented, and the members of the discussion panel. These people all contributed generously of their time and expertise. The colloquium participants added much to the richness of the discussion. I thank all who attended.

Stephen J. Burges, Chair
Committee on Managing Water
Resources Under Conditions
of Climate Uncertainty

Contents

SESSION B
THE IMPLICATIONS OF CLIMATIC VARIABILITY FOR WATER IN THE WEST

Papers Presented

Comments

Appendixes

MANAGING WATER RESOURCES IN THE WEST UNDER CONDITIONS OF CLIMATE UNCERTAINTY

1

Overview

Both the natural variability of the hydrologic cycle and potential disruptions of that cycle resulting from possible climate change can affect water supply and thus water management in the western United States. Uncertainty about both types of change poses a challenge for water resource managers. At the request of the Bureau of Reclamation, a committee of the Water Science and Technology Board convened a colloquium on November 14-16, 1990, to draw together material on climate change and climate variability and to explore possible water management responses. This proceedings contains an overview of that colloquium, "Managing Water Resources in the West Under Conditions of Climate Uncertainty," and the individual papers presented there.

As discussed in a related National Research Council report, *Policy Implications of Greenhouse Warming*, increases in atmospheric greenhouse gas concentrations probably will be followed by increases in average temperature. However, we cannot predict how rapidly these changes will occur, how intense they will be, or what regional changes in temperature, precipitation, wind speed, and frost occurrence can be expected. So far, no large or rapid increases in the global average temperature are evident. But if the projections being developed by general circulation models (GCMs) prove to be accurate, the stresses on the planet and its inhabitants would be serious, and substantial responses would be needed (National Research Council, 1991).

Changes in climate—whether brought about by natural variability or by global climate change—would of course have great effects on western water management. Four broad climatic classifications—humid, subhumid, semiarid, and arid—are relevant to western water. Different climatic zones exhibit differences in differences in streamflow, vegetation, evapotranspiration, variability of precipitation, and other factors important to water management. Some of the most noticeable changes in catchment hydrologic fluxes (rainfall amount, evaporation rate and amount, and streamflow rates) and states (soil moisture depth and distribution in time and space) are found in threshold climates—that is, climates where relatively small changes in vegetative state or meteorological inputs have amplified effects.

One important topic, the hydroclimatology and ecosystems of urban areas, many of which are located in hydrologic threshold regions, has not been included explicitly in the colloquium. The relevant issues for urban areas are those associated with massive importation of water and subsequent changes in the natural hydrologic balance. It is not uncommon for imported water to be the equivalent of about twice the annual rainfall volume. Transfers of this magnitude influence the hydrologic cycle locally at the mesoscale (horizontal distances of 20 to 30 km) and influence the hydroclimatological balance at the export locations.

In an editorial written to mark the beginning of the second quarter century of the journal *Water Resources Research*, Charles Howe emphasized the role of technology, institutions, and politics in water resource management. He concluded:

> . . . [I]t remains true, as it was [25 years ago], that socially responsible decisions require broad public participation, channeled through appropriate institutions. Institutions must change in response to changing public values, and institutional change is costly, but vital. Democracy, unfortunately for some, is messy and costly, but we will be better off pursuing the right goals somewhat inefficiently than pursuing the wrong goal efficiently (Howe, 1990).

It was in the spirit of these observations that the colloquium proceeded. All present—participants from academia, industry, and government—were eager to discuss the implications of possible climate change and to find ways to increase the resilience of our water systems in the face of increased uncertainty.

SHARING WATER RESOURCES

Edith Brown Weiss began the colloquium with a keynote address entitled "Sharing Water Resources with Future Generations." She presented a theory of intergenerational equity that relates the present generation to past and future generations and that relates the human species to the natural system of which we are a part. Both relationships can be viewed as trusts, in which the present generation is simultaneously a beneficiary and a trustee.

One significant concern is that traditional economic measures may not be sufficient to ensure that the state of the resources left for future generations reflects present diversity and quality. Her concept of intergenerational equity can be applied to many aspects of society but is particularly relevant to water. "If we are not careful today, we can leave our children a huge bill for cleaning up rivers and lakes we have polluted. We can leave our great-grandchildren a nearly irreversible legacy of eroded watersheds, polluted ground water, and contaminated river bottoms. Water is vital. Simple fairness demands that we conserve it for future generations and that we find ways to consider the interest of future generations in the decisions we make today."

The presentations that followed Dr. Brown Weiss's introduction touched a wide spectrum of issues related to climate and climate change, including measurements of relevant variables and the status of modeling these phenomena at the global scale. The colloquium also explored management responses to climate variability and the affected publics, both present and future.

THE SCIENCE OF CLIMATE CHANGE AND CLIMATE VARIABILITY

The late Roger Revelle opened the first session and commented broadly about factors that affect climate. He made a clear distinction between what is considered "weather" and what is considered "climate." Weather, he noted, is the instantaneous condition of the atmosphere at any particular time and place; climate is the average of weather over some time period. Dr. Revelle expressed his belief that there is good reason to expect that the increase of greenhouse gases in the atmosphere will cause climate warming, but that the degree of warming is very difficult to estimate. Whatever the degree, he predicted profound effects on some aspects of water resources—changes in demand for water, changes in supply, and changes in the general circulation of the atmosphere. Beyond these crude predictions, what we are really faced with is a lack of certainty—a lack of understanding—of what may happen.

Robert Dickinson gave the first formal paper, a primer on climate change. He emphasized the extreme importance of the earth's atmosphere to all life and noted that the portion of the atmosphere that principally generates our climate and weather is relatively thin, extending only about 50,000 feet (16,667 meters) above the earth's surface. (To put this in perspective, if the earth is imagined to be a sphere three feet in diameter, this layer is just four hundredths of an inch thick.) Human activity is changing the infrared radiant transmission properties of this life-sustaining layer. Dr. Dickinson gave a clear indication of the complexity (and relative crudeness) of existing and likely future general circulation models (GCMs). During this presentation and the discussion that followed, he noted the need for better hierarchical links with mesoscale precipitation models. Hillslope flow production, which feeds drainage networks, depends on the spatial and temporal distribution of precipitation and evapotranspiration. We are not yet at the state of development where the output generated by GCMs (where generally the smallest grid is on the order of 200 km by 200 km) can be used as input for hydrologic models for predicting activity (such as streamflow, moisture states, and evapotranspiration) at the catchment scale (where the area included is only ten to a few thousand square km). These links are important areas for research and development.

Kevin Trenberth discussed the recent climatological record and emphasized that many of the data series collected for one set of purposes were measured without uniform techniques. Consequently, few, if any, of the data sets are suitable for detecting global warming or cooling signals. Atmospheric patterns, sea surface temperature anomalies, and land-based hydrology are connected, but the linkages are not readily discernable from existing records or helpful for forecasting. Given observations concerning the heterogeneity of records, renewed examination of existing space-time series of data using some of the tools of modern spatial statistics might show some correlative links.

Malcolm Hughes, David Meko, and Charles Stockton focused on paleo climate information, which scientists use to infer climate states prior to relatively recent direct measurement of climatic variables. The various paleo records indicate that the assumption of weak statistical stationarity (e.g., the average and variance do not change with time) may not be tenable except for relatively short periods. Information may be contained in the various paleo records, but those data may need to be analyzed differently using newer and evolving spatial-temporal statistical tools to reveal any possible significant signals. This is an area where additional research is needed.

Implications of Climate Variability for Management

Stephen Burges and Bruce Kimball chaired the second session, which covered the specific topics of plant-water-atmospheric trace gas relationships, water sources, water resource management, economics, and water law.

Leon Hartwell Allen examined the implications of global warming on plant growth, looking in particular at trace gas enriched and changed thermal environments. The work has implications for future water use in crop production, catchment hydrology and redistribution of water within a catchment, and with linkages with the mesoscale and GCM-scale weather and climate predictions. The growth of plant biomass has been shown to be linearly related to cumulative evapotranspiration. In the case of soybeans, for instance, a doubling of carbon dioxide concentration with no climate change is likely to increase photosynthesis by about 50 percent, growth by about 40 percent, and seed yields by 30 percent. This is consistent with data on other species. Evapotranspiration rates are likely to increase by 5 percent for each 1 °C temperature rise. The work that shows changed biomass and water use has been done at laboratory and relatively small "controlled field" scales. Much work remains to be done to evaluate or estimate such effects at the catchment scale.

Marshall Moss addressed the vexing problem of linkages between catchment-scale hydrology and the larger GCM scale. He emphasized the need for careful use of catchment information to assist in calibrating and validating GCMs. He encouraged everyone to think carefully about the "uncertainty of hydrologic uncertainty." There is a crucial area of research in determining what temporal scales are used to describe hydrologic variability. Decadal average "surrogate streamflow" obtained from paleo records shows that some streamflow patterns derived through GCM modeling are well beyond any historical experience; outputs from hydrologic models that are not in accord with longer term observed patterns in the paleo records are extremely unlikely to happen.

Linda Nash examined alternative possible streamflow scenarios and showed how they could influence storage reservoir states and power production in the Colorado River system. Although there is no way to validate such "what-if" scenarios, they illustrate the range of what might be experienced. This approach is one of few options open for evaluating the influence of climatic variability on water resource system infrastructure. Analysts using this approach should also consider alternative water uses and demands and alternative flow volume scenarios. Many who are concerned about climate change focus on water shortage. In discussing this paper, Robert Dickinson reiterated the importance of examining both increased and decreased streamflow volume scenarios; if climate change brings a warmer atmosphere, increased runoff might occur in the mountainous West.

John Dracup used the Colorado River system to illustrate issues in managing large-scale water resource systems under a changing climate scenario. He emphasized that the historical record is rich in extremes of high and low flow and much can be learned by studying these extremes. During discussion, it was suggested that much knowledge might be gained about system operations for high flow scenarios by using the extremely large 1983 flood volume as a baseline to which alternative operations might be compared. The potential utility of using historical records to explore complex system operations needs further investigation. The Colorado River illustration shows how improved flexibility in the operation of systems offers opportunities to cope with potential long-term change in annual streamflow volume and temporal distribution patterns.

Kenneth Frederick examined the possible economic implications of climate variability on western water supplies. He indicated that the Bureau of Reclamation will need to play a larger role in facilitating water transfers. Demand management and water marketing are both potentially important tools for building system resiliency for managing water resources during droughts under present climatic variability as well as for accommodating possible long-term reductions in "natural" supply. Another significant issue that ties in with the work presented by Leon Hartwell Allen is the biological consequences of water use and climate variability. Economic assessments of biological resources are primitive at best; work is needed to develop appropriate measures for improved economic analyses.

Dan Tarlock discussed western water law in a climate change context. He believes there are now obvious limits to population growth in the West and these may be affected by climate change, especially if the changes reduce river flow volumes and aquifer recharge. In the spirit of Dr. Brown Weiss's keynote address, Mr. Tarlock also addressed the issue of the rights of future generations. He noted that there are no legal obligations to future generations—that aquifers can be mined of their water or soil damaged irreparably without consideration for future needs. Consequently, there is no legal incentive to develop conservation strategies or to make efforts to delay use of water into the future. Mr. Tarlock also said that water transfers from farm use to other uses may prove to be one option to mitigate some of the impacts of a changing climate.

John Schaake discussed the role of streamflow forecasting in managing water resources. Improvements in forecasting are being made using new observation schemes and increased computing power. One approach is the National Weather Service's "Extended Streamflow Prediction" (ESP) activity. Most forecasting assumes statistical stationarity in the scenarios used, but Dr. Schaake noted that a fundamental question of science needs to be addressed: when will it be necessary to abandon the use of statistically stationary precipitation sequences in the ESP technique? Improved forecasting will take advantage of newly developed models and use of denser spatial sampling of precipitation, snow cover, water content, and atmospheric temperature. A crucial concern is for deployment of sufficient measurement stations to take advantage of data handling and modeling capabilities. He also noted that it is important to maintain the existing network of long-term hydrologic measurement stations because, while imperfect, they provide the best available data base for comparison purposes.

MANAGEMENT RESPONSES TO CLIMATE VARIABILITY

Gilbert White chaired the third session, which focused on the types of management responses possible to cope with climate variability. He noted that there are many serious resource management concerns, and that in the long run the maintenance of soil and vegetation worldwide could actually be more important than any increase in atmospheric trace gases. He reminded participants that loss of soil and vegetation is an ongoing and extremely serious problem whether or not there is any climate change. When considering climatic issues and management strategies, soil and vegetation preservation should be a paramount concern.

John Keane discussed water management in Arizona's Salt and Verde rivers. The region has a long history of susceptibility to climate variability and thus has lessons to offer related to climate change. This arid region—once dominated by an Indian culture that disappeared, possibly because of unreliability of water sources—relies on winter generated river flow that is susceptible to warmer atmospheric conditions. The existing water management system of six dams and ground water storage was designed to overcome the region's natural variability.

The Salt-Verde system has been managed to take advantage of the presence of large aquifers by recharging (and pumping) for prolonged periods. This opens up the possibility of joint operation with recharge on the order of years during wet periods and pumping for years during prolonged dry periods. This concept is now known as "cyclic storage" (see, for example, Lettenmaier and Burges, 1982) and may be of considerable importance for making the best use of surface and ground water supplies in arid and semiarid environments where the present climate is naturally variable. Water stored in the ground is a relatively well-known quantity. Water that is to be delivered by precipitation is much less certain. Legal restrictions in the region create difficulties for implementing extensive artificial recharge.

Dale Bucks described present and possible improvements in agricultural water management. He reiterated the fact that agriculture is influenced by both water supply and trace gases. Furthermore, he estimated that agricultural practice contributes approximately 26 percent of the trace gases annually to the atmosphere. He emphasized the need for increased understanding of process hydrology to determine the effects of climatic variability or changes in variability. Increased atmospheric carbon dioxide concentrations can increase plant production provided that there is adequate water. The combined uncertainties presented by the natural variability of western water regimes and the possible consequences of climate change are cause for improved water conservation and management of irrigated agriculture at all levels, from principal supplier to end user.

Daniel Sheer emphasized the need for a renewed examination of how water resource systems are operated. He noted that present natural variability is large, while the rate of long-term climate change is small relative to natural variability. Thus he believes that many factors affect the short-term needs of system operations more than climate change. In considering how water systems are operated, the goal is to have the best possible mix of benefits; unfortunately, there is no clear agreement on what array of benefits to consider. One often forgotten objective concerning urban water supplies is the paramount need for water for fire protection, the basis for many of the water systems initially installed in the United States.

Computer simulations of system behavior under different constraints provides a way to examine how changed operations may be beneficial. Dynamic computer-generated illustrations of water movement throughout a system are an effective way to communicate these possibilities. Such simulations will show in what locations it may be effective to modify existing law to build in resiliency against the effects of uncertain supplies as demand levels increase.

Wayne Marchant and Arnett Dennis discussed weather modification as a possible means of adapting to climate variability. The U.S. Bureau of Reclamation and other organizations have been involved in cloud seeding research for almost 30 years. The greatest potential benefits occur in threshold hydrologic environments where an increase in precipitation goes largely to streamflow production rather than to soil moisture retention. Drs. Marchant and Dennis reported on planned activities for precipitation augmentation in the 13,000 square miles of the upper Colorado basin (largely about 9000 feet elevation), from which most of the river flow is produced. Although doubt remains concerning the benefits of cloud seeding, there is sufficient potential that the opportunity cannot be ignored.

PUBLIC INVOLVEMENT IN WATER RESOURCE DECISIONMAKING

The colloquium concluded with a panel discussion moderated by Helen Ingram. The participants were Jim Carrier, Jim Dyer, and Roger Kasperson. Participants focused on the public's understanding of climate variability and change. Several themes emerged during the panel discussion. One was that the print media have a mission to report news and that they do not have an obligation to build public consensus. Balanced coverage of scientific issues is difficult to achieve because particular individuals may be more quotable and persuasive than others. Other experts with valid information may not convey that information effectively to reporters. Examination of the media's coverage of climate change indicates that most environmental organizations have given increased attention to global climate change since 1986. Although greenhouse warming is of concern to environmental organizations, it was still relatively low on a 1990 list of public concerns. Another issue raised is how climate change might influence life-style. Perhaps the most obvious action will be the need for personal use of water-saving technologies. A personal and community sense of supply-cost relationships will be needed before people will be willing to take actions that reflect avoidable costs.

MANAGING WESTERN WATER UNDER PRESENT VARIABILITY AND FOR AN UNCERTAIN FUTURE

The purpose of the colloquium was to provide a compilation of knowledge for those involved with water management in the West to use to build research and development activities and foster institutional capabilities to handle the multiple facets of climate variability. Many issues merit consideration.

Natural variability in hydrologic processes is all-pervasive. Consequently, resource management allocation decisions are made in an inherently uncertain, and thus risky, setting. Present methods address variability from a statistically stationary perspective. Possible changes in variability and in the level of hydrologic states and fluxes that might accompany climate change must be addressed by nonstationary statistical methods. Unfortunately, there is no basis for doing this in other than a "what-if" manner. The issue is truly "trans-scientific;" that is, there are crucial questions of science involved but the present tools and methods of science are not adequate for answering them (Weinberg, 1972).

Hydrologic variability and its influence on water management was a significant component of many of the papers. In one discussion, Stephen Burges demonstrated examples of different forms of variability from a time series of tree ring data and an annual streamflow volume record from a hydrologic regime sensitive to climatic variations. Based on measures of persistence from the flow record, he demonstrated what that means in terms of supply reliability for a reservoir of fixed capacity, or alternatively how large a reservoir is needed to meet various contracted supply amounts. High reliabilities for large demands can only be attained with impracticably large reservoir capacities. There is little doubt that institutional arrangements for alternatives of supply and demand management must be explored in light of such information.

Research, Development, and Institutional Issues

Despite the difficulties posed by attempting to describe an unknown future, much could be achieved by incorporating what is known about present day variability into water resource management and allocation. Specific observations made by colloquium participants indicate that research, development, and institutional attention should be given to the following issues:

General circulation models (GCMs) should be linked with process level hydrologic representations at the catchment scale. GCMs need to be coupled to models capable of representing hydrologic interactions at scales on the order of a few, to tens, to hundreds and thousands of square kilometers. Modeling present and recent past climate in this context would ensure that the GCM representations are plausible.

Establishment and augmentation of benchmark monitoring networks is an important consideration. Environmental monitoring is conducted throughout the United States and the West but these efforts are not designed to be a sampling network to detect and monitor global climate change. The existing network of climatological, stream gaging, and water quality stations are largely unsuitable for making other than general estimates of what may have happened over a relatively recent period. The limited benchmark stream gaging network operated by the U.S. Geological Survey provides a good model for one component of a broader monitoring program.

Measurements of plant production in carbon dioxide enriched environments should be translated upwards from the laboratory and small field scales. Much work needs to be done for a range of catchment scales to determine the linkages between precipitation, temperature, evapotranspiration, and water yield before alternative water demand and supply scenarios can be considered reliable. Such work needs to be explored in threshold climatic regions where small perturbations could make substantial changes in the local vegetation and temporal water distribution pattern.

Historical streamflow and surrogate records should be examined for their possible use in validating the scenarios created with increasingly refined GCMs. Historical streamflow and surrogate records should be examined for their possible use in validating the scenarios being generated by increasingly refined GCMs. Hydrologic measures have been and continue to be made at the catchment scale, while GCM outputs (estimates of precipitation, evaporation, and runoff) are on a vastly larger scale. Much work needs to be done to determine how to disaggregate GCM outputs to the catchment scale. Also, since hydrologic measurements are of short duration relative to climate change influences, paleo records can act as surrogate hydrological information for longer time periods. There is information in recent hydrologic and climatologic records as well as in paleo records that could be used to calibrate GCMs to make them more compatible with the variable hydrologies of the many catchments typically contained within a GCM cell.

Opportunities exist to improve water resource system operation. To develop improved economic measures, work is needed to include biologic resources in economic benefit evaluations. When considering alternative operation policies for multiple water resource systems, flexibility is a crucial consideration. For example, in the Colorado River system it is unclear if there will be increased or decreased flow and how the flow patterns would be distributed differently in space and time throughout the system, even if the vegetation remained relatively constant. This suggests that many institutional questions regarding mechanisms for using varying amounts of water differently (depending on shortage or excess supply situations) should be examined.

Improved streamflow volume forecasts for water resource system operations are needed. If within-year streamflow variability changes and there is no change in the annual average flow volumes at major locations within a river basin, water managers will need improved methods to forecast near- to intermediate-term streamflow volumes. This need argues strongly for redesigning present river forecast systems to take advantage of present hydrologic and environmental monitoring systems. Such systems must include greatly improved measurements of precipitation (spatial density of measurements) and averaged direct measurements of evapotranspiration to permit closer estimates of the water budget for a given catchment.

Many promising technologies exist that may help reduce uncertainty in water resource supplies. Technologies to increase the resilience of water systems to variability should be evaluated for their utility. Precipitation enhancement, for instance, might prove useful in locations where there would be benefits from maintaining streamflow patterns similar to those that have been experienced during the period of record.

The concept of safe yield engenders a false sense of security. This concept is particularly frail in reduced water availability scenarios; any mechanism that builds in the understanding that shortages will occur is desirable. This may be a key consideration for Bureau of Reclamation staff and members of the multiple publics who will need to confront water allocation policy issues.

Opportunities for the Bureau of Reclamation

Many people consider the Bureau of Reclamation to be the West's primary water resource management agency. As such, the bureau could play a key role in facilitating the region's adaptation to future changes in water supply and demand. Any impediments to reallocating supplies need to be reduced to permit flexible response to changing conditions. The prospect of climate change adds to present uncertainties associated with supply and demand but it does not alter the need to develop flexible and efficient water management. Climate change impacts, if they occur, are uncertain, they are likely to occur slowly and to be relatively small in comparison to natural variability. Thus management flexibility will be critical regardless of the causes of changes in climate. Three institutional issues arose repeatedly during the colloquium that might help provide needed management flexibility:

- Voluntary transfers of federally supplied water might provide temporary response for shorter-term phenomena such as severe drought or permanent responses to supply and demand.
- Water users, principally farmers, can be active partners in efforts to improve water management if provided with incentives to adopt alternative, efficient water use practices.
- Innovative uses of federal facilities, such as making federal storage and distribution facilities available for water transfers even if federal water is not involved, should be evaluated carefully.

To be better prepared to deal with the uncertainties offered by possible changes in climate, water managers should be working now to increase the resilience of existing water systems to normal climatic variability. This flexibility would pay off in the short term when periods of especially wet or dry weather affect water availability. It would also pay dividends over the long term should climate change occur. There appear to be many opportunities to modify how water is managed in the western United States to provide for robust institutional settings that can accommodate hydroclimatic variability and, in the longer term, possible climatic change. Conservation, increased water use efficiency, marginal cost pricing of water, water transfers, alternative uses of existing infrastructure, and explicit consideration of present day variability in determining system supply reliabilities are but a few of the topics discussed in the papers in this proceedings. No single actor can implement a completely effective water management system. Close cooperation among federal, state, and private suppliers and users is essential.

REFERENCES

Howe, C. W. 1990. Technology, institutions, and politics: still out of balance. Water Resources Research 26(10):2249-2250.

Lettenmaier, D. P., and S. J. Burges. 1982. Cyclic storage: a preliminary assessment. Groundwater 20(3):278-288.

National Research Council. 1991. Policy Implications of Greenhouse Warming. Washington, D.C.: National Academy Press.

Weinberg, A. M. 1972. Science and trans-science. Minerva 20:209-222.

2

Sharing Water Resources with Future Generations

Edith Brown Weiss
*Associate General Counsel for International Activities**
U.S. Environmental Protection Agency
Washington, D.C.

The human species has been on this planet only a short time. At least in the western world, we have become accustomed to progress and we expect it. However, because of the grave global environmental problems confronting them, many people today are questioning whether progress will always continue. The subject of this symposium—water resources and climate change—presents this challenge directly. Progress cannot continue if resources essential to our existence are seriously affected by changes in the global climate. This threat raises the issue of what we can do, and what we have an obligation to do, to conserve essential resources, such as fresh water, so that they are available in the quality and quantities needed at a cost that is acceptable.

The threat of climate change raises an important new dimension to the area of water resources: the issue of intergenerational equity. It would be possible to approach this topic by examining the intergenerational implications of either climate change or management of water resources. This presentation focuses on water resources as an intergenerational equity issue, with climate change and the scientific uncertainties regarding it as aspects of the problem.[1]

I will present a theory of intergenerational equity that relates the present generation to past and future generations and relates the human species to the natural system of which we are a part;

---

*On leave from position as professor of law, Georgetown University Law Center. The views expressed in this article are solely those of the author and do not necessarily represent the views of the U.S. Environmental Protection Agency.

with this theory, we can look at issues of intergenerational equity raised by the way we use and care for our planet (Brown Weiss, 1989). I will then apply this theory to questions of water resources, examining future generations' rights to these resources and analyzing possible management strategies in the intergenerational context.

RELATING THE RIGHTS AND OBLIGATIONS OF THE PRESENT GENERATION TO OTHER GENERATIONS AND TO THE NATURAL SYSTEM

The relationship of our generation to other generations in the context of our natural system has two aspects. One is our relationship to other members of the human species—to the human community extended over time. The second is our relationship to the natural system of which we are a part. We should view both relationships in the context of a trust, in which the present generation is simultaneously a beneficiary and a trustee. Each generation is a beneficiary in its use of the planet for its own welfare and well-being and, at the same time, a trustee for conserving the planet for future generations. Similarly, as the most sentient of creatures in the natural system, we are at the same time a part of the system, entitled to use it and benefit from it, and a trustee for its conservation.

There are several models available for analyzing the relationship between our generation and future generations. Two prominent, opposing models are the "preservation" and "opulent consumption" models. The preservation model demands that everything be preserved as it is; the present generation, according to this view, has no right to change anything. This model is reflected in the English "natural flow" theory of water and in preservationist legislation regarding wilderness and other untouched areas. While the model may be appropriate for certain unique natural resources, it is not generally consistent with economic development and improved standards of living. The opulent consumption model encourages immediate consumption of resources, either because there may not be another tomorrow or because the higher level of consumption will make possible greater wealth for this and future generations. This model also has serious limitations; it does not recognize the need to use the environment on a sustainable basis, nor does it consider appropriate environmental costs in its economic calculations.

An alternative approach is to view the human community as a partnership extended over time, in which each generation is a partner. Describing such a theory, Edmund Burke observed that "as the ends of such a partnership cannot be obtained in many generations, it becomes a partnership not only between those who are living but between those who are living, those who are dead, and those who are yet born." The purpose of the partnership is to realize and protect the welfare and well-being of every generation. This requires sustaining the life support systems, ecological processes, environmental conditions, and cultural resources that are necessary for our survival and well-being and for the robustness of the natural system of which we are a part.

The question, then, is how to determine what constitutes "fairness" between generations. As John Rawls (1971) has suggested, we could attempt to imagine what environmental ethic we would espouse as members of a generation if we did not know where along the spectrum of time we are living—whether we are near the first or last generation. In such a position, what would we regard as a fair use of the planet? Presumably, we would want to leave the environment at a minimum no worse off than we found it. That does not mean that we should not use the resources at our disposal to improve our own well-being, or to improve the quality of the environment. It is inevitable that we will change the environment as we use it, but on balance we should leave the planet no worse off than we found it.

This raises the question of the responsibility of the present generation for correcting the environmental abuses of past generations. If a river or lake that may be important to future generations is polluted due to actions taken 50 years ago, can the present generation simply leave it polluted? Surely not. The present generation has some obligation to participate in the removal of that pollution; however, the costs could be distributed over several generations through revenue bonds or other financing mechanisms.

The requirement that each generation must leave the environment no worse off than it found it implies that there is a minimum level of robustness in the natural system that must be passed on to future generations. This minimum floor of robustness is deeply rooted in international law as reflected in the United Nations Charter, the Universal Declaration of Human Rights, and many other documents.

Intergenerational equity, as I have been describing the term, requires an intragenerational dimension. If each community or each country had a completely parochial attitude, caring only about potential impacts on its own citizens and their descendants, it would still be necessary to ensure that other countries adopted policies that take intergenerational rights into account. To care about our own descendants or our own nationals means we have to care about the environment in which they exist for the long term. If we care about that environment in the long term, we have to care about what happens today in the conservation of the environment—including what happens in other communities and countries—because we alone cannot insure a decent environment for our descendants. The actions of other communities and countries may have a profound effect on the future environment of our community and country. That means we have to care about whether other people are willing and able to fulfill their intragenerational obligations.

PRINCIPLES OF INTERGENERATIONAL EQUITY

What are principles of intergenerational equity that we can derive from the intergenerational ethic described above? Four criteria may be helpful in deriving such principles. First, such principles should permit present generations to use resources today, but not at great expense to future generations. Nor should present generations be required to sacrifice greatly for the benefit of future generations. Second, the principles should be reasonably clear in application. Third, the principles should not require that we predict the values of future generations. Fourth, to the extent that such principles apply internationally, they must be consistent with the values and cultural traditions of the different countries of the world. Application of these criteria gives rise to a set of three principles, which I have labeled "conservation of options," "conservation of quality," and "conservation of access."

"Conservation of options" can be defined as the obligation to maintain the diversity of the resource base. This does not necessarily imply maximizing diversity. For example, while a monoculture is vulnerable to climatic change, maximizing diversity can lead to dynamic instability in an ecosystem. The conservation of options principle requires preservation of the components of diversity that provide maximum robustness to the system.

"Conservation of quality" is the obligation to preserve the quality of the environment. This is not a principle against change in the environment, but some balance must be achieved in doing so. Environmental degradation in one area must be balanced by improvements in another. This requires the development of a framework within which that balancing act can occur. Changes in use may make pristine water less clean, but other changes may make polluted water cleaner; as long as there is no unbalanced degradation of quality, overall conservation is met.

Finally, "conservation of access" can be defined as the right of a beneficiary to use a resource. It is essentially a nondiscriminatory, equitable right. Future generations should not be required to pay an extraordinarily high price for an essential resource because the present generation refused to put a price on the resource, deferring the cost of its depletion to future generations.

AN INTERGENERATIONAL APPROACH TO WATER RESOURCE PROBLEMS

Using the three principles of intergenerational equity to look at water resource problems can help us better understand our generation's rights and obligations. In turn, this thinking should suggest appropriate management strategies that could be translated into concrete actions that will benefit future generations.

Water Resource Problems with Intergenerational Impacts

Several problems can exist between our generation and future generations in the use of water resources, some of which may be exacerbated by changing climate and by the scientific uncertainty regarding changing conditions.

Perhaps the most obvious example of an intergenerational water resources problem is toxic contamination of surface and ground waters. Most toxic contamination can be removed, but the flushing times in lakes, such as Lake Superior, may be decades, or even a century, and the costs of removing contamination, as in ground water, may be so high that the contamination is essentially irreversible. With ground water, the high costs of cleaning up or containing toxic contamination are becoming increasingly well understood.

Another intergenerational water problem is saline pollution of fresh water. For example, the pumping of ground water aquifers at rates sufficient to cause saltwater intrusion into the aquifer is hard to reverse at acceptable costs; consequently, it may lead to the abandonment of the ground water aquifer. Rising sea levels, which could result from global climate change, could cause marine water to intrude upon freshwater streams. Some downstream countries may lose that fresh water altogether, because attempting to change the institutional arrangements for allocating those rivers may cause serious conflicts.

Degradation of water resources can have several effects. First, it can damage the ecosystem, which is difficult to reverse. It also can limit the uses of the waters. For example, pollution in the Great Lakes contaminates fisheries and makes the lakes unsuitable for swimming and other forms of recreation. So the resource may exist, but the uses that make the resource desirable to present and future generations may be diminished. Similarly, degradation of the quality of forests may diminish the environmental services they perform in maintaining watersheds.

Depletion of resources, particularly freshwater resources in specific areas, is another intergenerational problem. Depleting resources before their value has been recognized creates higher real prices for future generations. For example, natural gas was burned off of natural helium-bearing deposits before it was recognized as valuable. In terms of freshwater resources, two problems come immediately to mind. One is the problem of pumping ground water faster than the recharge rate, which depletes not only the ground water but also surface water (because of the hydrological links between ground and surface water) and could leave once fertile agricultural areas barren and cause settlements to move. Second is the depletion of nonrechargeable aquifers. This is occurring in the United States, northern Africa, and other parts of the world. In this country, we do not generally put a price on water as a resource; we charge only for transporting it and turning it into a commodity that is usable by communities. This increases the rate at which the resource is depleted. Alternative sources of water are theoretically available, but they may not be readily available and certainly are not available at acceptable costs. Thus, depletion of aquifers raises the real price of water for future generations.

A final intergenerational water problem is restricted access to freshwater resources. For example, changes in global climate may cause areas dependent on the availability of a certain level of fresh water to become warmer and drier. This would raise difficult questions concerning large-scale diversions from water-rich areas such as North America, China, and the Soviet Union, which are expected to have more water. Climate change could also increase pollution concentrations and reduce the amount of water available for fishing and other recreational uses. Further, it could affect downstream communities by creating saltwater intrusion upstream because of insufficient water to hold the saltwater back.

Applying Intergenerational Principles to Water Resource Problems

The water resource problems discussed here raise serious questions of intergenerational equity. The principles defined earlier in this paper provide an approach for analyzing these problems and finding solutions that strike a proper balance between our generation's rights and our obligation to future generations.

Application of the first principle, conservation of options, could mean maintaining a diversity of fresh water supplies—both ground water, surface water, and the technology for turning salt water into useable water. It could mean a moratorium on the mining of aquifers. It could mean tighter restrictions on the depletion of non-rechargeable aquifers. It could also require reexamining our system of subsidizing water for agriculture (which also receives other price supports) in order to make it profitable.

If the second principle, conservation of quality, is taken seriously, it could require further action to prevent irreversible, or effectively irreversible, toxic contamination of water. The principle could require actions to prevent saline intrusion from rising sea levels. It could mean planning the locations of dams by considering projected sea level rises to ensure that downstream waters do not become subject to saline intrusion.

Finally, the third principle, conservation of access, could require the present generation to incorporate the full cost of supplying water, not just the cost of delivery and treatment, so that the real price of the water resources is not significantly higher to future generations than it is to the present generation.

INTERGENERATIONAL RIGHTS AND OBLIGATIONS

The principles of intergenerational equity translate into a set of intergenerational rights and obligations. In this context, obligations and rights are linked together. John Austin (1873), the English legal theorist, said that there may be groups of obligations that exist independently of any correlative rights. Here, however, I believe that rights are correlated with obligations, and the content of the rights is derived from the principles of intergenerational equity set forth above.

Specifically, future generations have rights related to the maintenance of the natural system. (Without such rights, future generations would be subject to Parfit's famous paradox: "Rights are individual rights but until people are born they do not exist and so therefore they cannot have rights.") The existence of these generational rights does not depend upon knowing either the number or kind of individuals who will be present in the future. Rather, we can define rights using objective criteria, flowing from our understanding of the natural system and our responsibility to maintain the robustness of that natural system.

How could such rights be enforced? In part, they could be factored into the decision-making process. It would be possible to have a representative of future generations, similar to a guardian *ad litem*, who could participate in administrative or judicial procedures. But, it is particularly important to consider how one could incorporate the interest of future generations into the marketplace. They are not represented in the marketplace now; the discount rate does not sufficiently represent their interests. Thus, as we think about marketing water rights, we must also consider how the interests of future generations can be represented in this market.

STRATEGIES FOR INCORPORATING INTERGENERATIONAL INTERESTS INTO DECISIONS AFFECTING WATER RESOURCES

There are many strategies for incorporating intergenerational equity into water resource decisions.

First, as noted above, we need to give representation to the interests of future generations: in the marketplace, in political and administrative decisionmaking, and perhaps in judicial bodies. For certain venues, this representation might take the form of an "ombudsman" or "watchdog" for future generations. Such an ombudsman could not only enforce laws relevant to future generations, but could also warn of dangers to the ecosystem, respond to complaints about damage to water resources, and mobilize support for conserving water resources.

Second, we need to have a good monitoring system. Without monitoring, there is no way to judge this or any generation's stewardship of resources. The United States has taken many steps to improve its ability to monitor surface and ground water, but more can still be done. Moreover, monitoring needs to extend to other parts of the world. We also need to be able to monitor social rates of adaptation to changes in water supplies.

Third, we need to think of scientific and technological research as a component of intergenerational equity. Research that would not be supported in the private sector but that is important to maintaining the quality and accessibility of water resources for future generations should be supported. For example, research to promote understanding of the fate and transport of pollutants in ground water, or research to facilitate efficient development and use of alternative water resources (such as research into desalination technologies) should be viewed as part of an intergenerational strategy. If nonrechargeable aquifers are depleted, there may be an obligation to engage in research aimed at making more efficient use of water supplies and making alternative supplies available at equitable prices.

Fourth, we need to focus on assessing the long-term impacts of our water use, particularly in light of the scientific uncertainty regarding changing climate conditions. There should be a process that takes into account the possible impacts of proposed actions on future generations—essentially starting from their interests and working backwards. This process could be incorporated in a "long-term conservation assessment" or as part of an environmental impact assessment.

Fifth, we have to revise how we think about the question of maintenance. Maintenance is now a stepchild to capital investments, but it should be recognized as an intergenerational equity issue. One can understand this issue best by considering capital investments, such as municipal sewage treatment plants, that may be used briefly and not maintained. Failure to maintain such facilities expends resources for our own marginal benefit at the expense of future generations. Recognizing the importance of maintenance to future generations means that the ease and cost of maintenance must be explicit criteria in project design and development. Moreover, it means that facilities that use water resources and infrastructure that transports water should be kept in good repair so that water resources are not unnecessarily wasted. For example, ditches with 70 percent loss rates, like some in the West, may not be a proper way to transport water from an intergenerational perspective.

Sixth, the intergenerational approach suggests that we need emergency assistance arrangements in place in case of hazardous substance spills into water bodies. Our own Clean Water Act, which requires immediate notification of any discharge of oil or hazardous substances, is a useful precedent.[2] Laws in other countries are not always as good; in some cases, it is not the legal provision that is inadequate but its implementation. Even when there is an international agreement with emergency response provisions, responses to a major accident can be too slow. For example, after the Sandos chemical factory spill, it took more than 24 hours for any notice to be given to countries downstream from the plant, even though notification provisions were part of the international agreement covering the Rhine River.[3]

Seventh, the intergenerational approach suggests the need for certain changes to legal arrangements governing water resources. For example, if potential climate change causes people to consider large-scale diversions, it will be essential to develop acceptable criteria for evaluating such diversions—for examining the benefits to the recipient communities, the future needs of the area of origin, the effects on the welfare of present and future generations in both areas, the impacts on the ecosystem, the long-term water loss during transport, alternative sources of water in each area, and the significant effects on weather and climate. The National Water Commission in 1973 devised economic criteria for interbasin transfers that we might update in light of our concern with global climate change and our intergenerational concerns (National Water Commission, 1973).

A number of legal changes suggested by the intergenerational approach arise in the context of managing ground water aquifers. First, we need to protect the recharge areas of ground water aquifers. Within the United States, this means increased attention to land-use planning. Land use thus acquires an important intergenerational dimension. Moreover, in some areas we may need to manage ground water aquifers so that withdrawals do not exceed recharge rates. This requires greater understanding of the recharge rates and the development of monitoring equipment. To prevent the rapid loss of ground water aquifers and ensure that future generations do not have to pay substantially higher prices for ground water, we must develop a system that effectively narrows the existing price differential between ground water and other water supplies. In some cases, it will be necessary to develop guidelines and/or agreements to control the mining of ground water in interstate and/or international aquifers. For this purpose, the work that has been done at the University of New Mexico on an agreement for international ground water management is relevant (Rodgers and Utton, 1985).

We also need to quantify water rights, particularly instream flow rights, from an intergenerational perspective. Potentially, the ecosystem could be further damaged under changing climate conditions if the instream flow rights are not quantified and prioritized. The intergenerational approach also requires moving beyond legal impediments to economically efficient water uses. This may be done, for example, by redefining beneficial use and by increasing the marketing of water rights, while ensuring that the interests of future generations are represented in the marketplace.

Finally, an intergenerational approach may require institutional changes to respond to advances in our scientific understanding of climate change and of the likely socioeconomic effects of such change. Mechanisms for anticipating and responding efficiently to these changes are needed. In international agreements, flexibility to respond to change often takes the form of protocols or annexes to the agreements that can be easily updated at regular meetings of the parties. The Montreal Protocol on Substances that Deplete the Ozone Layer,[4] for example, permits parties to accelerate the reduction or phase-out of a listed substance without formally amending the protocol. Issues involving scientific uncertainty can also be handled through scientific advisory councils and regular scientific assessments, the results of which may be transmitted to the parties for consideration in their periodic meetings. The Great Lakes Water Quality Agreement of 1978 incorporates several of these mechanisms.[5]

Nationally and locally, the important intergenerational issue is how to handle scientific uncertainty with respect to the effects of climate change. For example, in the western United States, possible changes in climate raise important water problems. One problem is how to make water rights flexible in the face of uncertainty over climate change. There are several ways to accomplish this, including leasing or marketing water rights and use of the public trust doctrine. Another problem is how to make water management schemes sufficiently flexible to adapt to climate changes and to protect the interests of future generations. Our management strategy for dams and reservoirs is fairly flexible for short-term management of peak flows and droughts; in the long term, however, these systems may be inadequate, and we must consider seriously how to increase their flexibility.

Finally, to return to the intergenerational perspective, to the extent that current use of water resources imposes disproportionate burdens on future generations, how should we compensate them? Is the greater income that we are generating through our use of water resources sufficient compensation? Should we focus on research and development or provision of facilities to assist in future adaptation to changes in the location, form, and quality of water resources? Or, as I have suggested in a different context, should we consider trust funds for future generations?

Throughout history the human species has been both clever and very lucky in using its natural resources. As resources have become scarce and the prices have risen, uses have become more efficient or substitutes have been discovered or invented that could serve the same function—often better or more economically than the previous resource. Fresh water, however, is different. Our actions now have long-term effects on the quality of fresh water and the access that people will have to fresh water. Technology will help some people in the future, for example, by providing affordable, large-scale desalination plants for marine water, but there are no substitutes for fresh water. If we are not careful today, we can leave our children a huge bill for cleaning up rivers and lakes that we have polluted. We can leave our great-grandchildren a nearly irreversible legacy of eroded watersheds, polluted ground water, and contaminated lake and rivers bottoms. Water is vital. Simple fairness demands that we conserve it for future generations and that we find ways to consider the interests of future generations in the decisions that we make today.

NOTES

1. Approaching the topic from the perspective of climate change would raise a number of additional issues. A definition of intergenerational equity in terms of changing climate might be a rate of climate change sufficiently slow to permit adaptation without excessive costs. This would require knowing the rate at which the physical system and the natural system can adapt. Rates of social adaptation will differ depending on the location, the economic and social conditions, and the country's technological capacity. We understand very little about social rates of adaptation. Even if we understood them better, it would be necessary to select some country's adaptation rate as the norm. Should it be the rate of those least able to adapt or should it be a more moderate rate with assistance provided to those who adapt more slowly?

2. Federal Water Pollution Act, as amended by the Clean Water Act of 1977, 33 U.S.C.A. Sec. 1321(b)(5).

3. Convention on the Protection of the Rhine Against Chemical Pollution. Art. 11, 16 *International Legal Materials* 242 (1977).

4. Protocol on Substances That Deplete the Ozone Layer. 26 International Legal Materials 1550 (1987).

5. Great Lakes Water Quality Agreement, Nov. 22, 1978. U.S.-Canada, 30 U.S.T. 1383, T.I.A.S. 9257 (1978).

REFERENCES

Austin, J. 1873. Austin's Jurisprudence, Lectures on Jurisprudence. London: J. Murray.

Brown Weiss, E. 1989. In Fairness to Future Generations: International Law, Common Patrimony, and Intergenerational Equity. Dobbs Ferry, NY: Transnational Publishers and United Nations University.

Burke, E. 1903-1911. Reflections on the revolution in France. In works of Edmund Burke. London: Bell.

National Water Commission. 1973. New Dimensions in U.S. Water Policy. Washington, D.C.: Government Printing Office.

Rawls, J. 1971. A Theory of Justice. Cambridge, Mass.: Belknap Press.

Rodgers, A., and A. Utton. 1985. The Ixtapa draft agreement relating to the use of transboundary groundwaters. Natural Resources Journal 25:713.

3

The Science of Climate Change and Climate Variability

Roger Revelle
University of California, San Diego

In considering the potential for climate change, we must understand what we mean by climate and what we mean by weather. Weather is the instantaneous condition of the atmosphere at any particular time and place—whether it is raining or the sun is shining, whether the wind is blowing or it is calm, whether it is warm or it is cold. Climate is the average of weather over some time period. We can talk about the climate of a particular season, a particular year, or a particular place. Or, we can talk about the average of weather over several years. Climate has often been defined as the average of weather over 30 years.

We are all familiar with climate change: the biggest change occurs between the summer and the winter. This change occurs in most places in addition to any change due to greenhouse gases or the earth's orbital parameters. Between summer and winter in the United States, we may have a 10 or 15°C difference in average temperature and, of course, great differences in precipitation and atmospheric circulation. When we talk about climate change, we are not really talking about something unfamiliar, because the change that we see between summer and winter in the United States is quite large compared to any expected climate change. However, the possibility of an interannual climate change—a change from year to year as contrasted with seasonal changes—is very important for us from many perspectives, particularly from the perspective of water.

There is good reason to expect that because of the increase of greenhouse gases in the atmosphere there will be a climate warming. How big that warming will be is very difficult to say. The the average temperature will probably increase somewhere between 2 and 5°C at the latitudes of the United States; it will probably change more at higher latitudes and less at lower latitudes.

We can be certain that whatever climate change occurs will have a profound effect on some aspects of water resources. First, we can be sure (there is not much uncertainty about this) that the demand for water will increase. Higher temperatures will cause people, farmers, and industries to use more water. Second, the supply of water will certainly change seasonally. The warming climate will likely bring more rain and less snow; therefore, there will be more winter runoff and less summer runoff and probably other, similar seasonal effects caused by the change in the character of precipitation. Third, it will probably be true that the general circulation of the atmosphere will diminish, because the temperature gradients between high latitudes and low latitudes will be considerably less than they are now.

Beyond these crude predictions, we are faced with a lack of certainty—really a lack of understanding—of what may happen. We know that atmospheric carbon dioxide concentrations will increase and will continue to increase as long as we use fossil fuels and as long as we persist in cutting down our tropical forests. At least in principle, we know what we can do about carbon dioxide: we could cut down on the use of fossil fuels and we could grow trees instead of cutting them down. But there are other gases that contribute to the greenhouse effect and are harder to manage than carbon dioxide. A second greenhouse gas is methane (natural gas), which is not now but may in the future be as important as carbon dioxide. Methane has a variety of sources: it is partly flared off in oil fields (though a good deal of gas comes out of the ground without being burned when oil is produced); it is produced in forest fires; it is produced by the belching of cattle and the burping of termites; it is produced in swamps and in rice paddies. As far as I can determine, none of these sources of methane is controllable, except possibly the methane released in oil fields. So, we expect the atmospheric methane concentration to continue to increase; it is doubling now about every 10 years, and it probably will continue to increase to about 8 parts per million during the next 50 to 75 years.

A third greenhouse gas, nitrous oxide, comes primarily from agricultural use of fertilizers. Its concentration is increasing, though much more slowly than the carbon dioxide concentration, and we can probably do something about it. The fourth greenhouse gas, tropospheric ozone, is really a product of air pollution. If we cut air pollution, in the process we will be cutting tropospheric ozone.

We are quite uncertain about the quantity of carbon dioxide that we will release in the atmosphere in the future. If we keep on at the present rate, carbon dioxide emissions to the air will probably increase from 6 billion tons, which is what they are at present, to 15 billion tons as the less-developed countries develop their economies. We cannot expect, and we probably should not even ask, that these nations not use their fossil fuels. The fossil fuel that both China and India have in abundance is coal, not oil or gas, and coal is a messy, dirty, poisonous substance. I hope we can at least reduce its use to a considerable extent in the United States. But until the economies of India and China develop much more than they are now, we cannot expect much reduction in their use of coal.

Many ways have been suggested for reducing carbon dioxide emissions. The principal ones, from my viewpoint, are the use of nuclear power as a substitute for fossil fuels and the use of hydrogen (produced by the electrolysis of water, with the primary energy coming from nuclear reactors) as a fuel for transportation.

In any case, the concentration of carbon dioxide in the atmosphere will not increase indefinitely, because we will ultimately exhaust the world's reserves of fossil fuels. A quadrupling of atmospheric carbon dioxide compared to mid-nineteenth century levels would be about the highest atmospheric concentration that could be expected in the future and could possibly occur in the twenty-second century.

4

Primer on Climate Change

Robert E. Dickinson
University of Arizona, Institute of Atmospheric Physics
Tucson, Arizona

INTRODUCTION

The climate system is expected to change substantially over the next century as the warming effect of increasing greenhouse gas concentrations is realized. Yet, the general circulation models (GCMs) used to project the future climate generate a description of the future that is rich in physical details but applicable only over areas with spatial dimensions of several hundred kilometers or more. Furthermore, these regional-scale descriptions are poorly understood, vary from model to model, and depend on aspects of the models that have received little or at least inadequate attention. Past studies have shown global warming could have major effects on western water resources. More certain answers require further improvements in the models.

THE BASIS FOR CONCERN ABOUT CLIMATE CHANGE

This paper discusses what the scientific community knows and does not know about how the climate and western water resources will change in the future because of increasing greenhouse gas production from human activities. This is a problem with a decades-to-centuries lead time, similar to the long lead times required for planning for water in the West. Although changes in human behavior could make a very substantial difference in the potential for climate change, they cannot eliminate the problem. Future climate change is inevitable. Because the magnitude and details of this change are still so uncertain, even if we respond to reduce it, we must also still begin to prepare ourselves for the possibility of a different future climate, especially with regard to water resources. At least, we must get ready for a future more uncertain than that suggested by past records.

Most of our present scientific perspective on climate change has been around for more than a decade, and the first scientific discussions of the possible effect on climate of increasing carbon dioxide concentrations date back more than a century. Half a dozen or more assessments of the current scientific consensus have been conducted since 1979, authored by prestigious National Research Council committees and international workshops. It is really striking how little divergence there has been in the conclusions of these studies.

The latest assessment by the scientific community of climate change, just now being released, was carried out by hundreds of scientists and was orchestrated by an organization established just for this task: the Working Group Number One of the Intergovernmental Panel for Climate Change (IPCC) (IPCC, 1990). The IPCC report addresses four topics related to climate change:

1. factors that may affect climate change during the next century, especially those related to human activity;
2. responses of the atmosphere-ocean-land-ice system to climate change;
3. current capabilities for modeling global and regional climate changes; and
4. the past climate record and presently observed climate anomalies.

The IPCC report looks especially at the following questions:

- What factors determine global climate?
- What are the greenhouse gases, and how and why are their concentrations increasing?
- Which greenhouse gases are the most important in climate change?
- How much do we expect the climate to change?
- How much confidence do we have in our predictions?
- Will the climate of the future be very different from today's climate?
- Have human activities already begun to change the global climate?
- How much will the sea level rise if the climate changes?
- How will climate change affect ecosystems?

• What should be done to reduce uncertainties about climate change, and how long will reducing uncertainties take?

This paper examines some of the above questions, focusing on those most relevant to water supplies in the West.

THE CLIMATE SYSTEM AND THE GREENHOUSE EFFECT

The climate system consists of the atmosphere, oceans, land, and ice surfaces and biological processes that interact with these physical elements (Figure 4.1). Many factors determine climate, but foremost among these is solar radiation (Figure 4.2). The amount of solar energy reaching the earth at the top of the atmosphere is almost constant, so variations at the surface are largely controlled by season, latitude, and cloudiness.

Day-to-day changes of weather, as well as climate variations and change, especially the temperatures we feel locally, are all strongly dependent on the disposition of this solar heating. How much is reflected back to space? How is it balanced by terrestrial thermal emissions (that is, by the heat energy that leaks to space to cool our planet to balance the absorbed solar radiation)? How is the internal energy of the atmosphere and oceans sloshed back and forth as these fluids act as heat engines to drive winds and currents?

A less obvious factor, but one very important everywhere, especially for temperatures, is the "greenhouse effect." Thermal radiation cools all terrestrial surfaces: oceans, soils, leaves, rooftops, and so on. Without an atmosphere, this radiation would all escape directly to space. However, over most of the wavelengths in which this energy travels, atmospheric gases and clouds absorb and reradiate it. What is crucial to the greenhouse effect is the fact that atmospheric temperatures are lower at greater heights; that is the reason that mountains are generally colder than lowlands. The colder gases radiate less energy upward than they absorb, and they also radiate downward. The downward radiation directly reduces nighttime radiative cooling at the ground, more so when the sky is moist and especially when it is cloudy.

It is the weakening of the upward radiation to space by high, cold layers that warms the overall climate system. By reducing net atmospheric cooling, the "greenhouse" gases in these cold layers warm the atmosphere, which, in turn, warms the earth's surface. Various greenhouse gases radiate to space mostly over about the

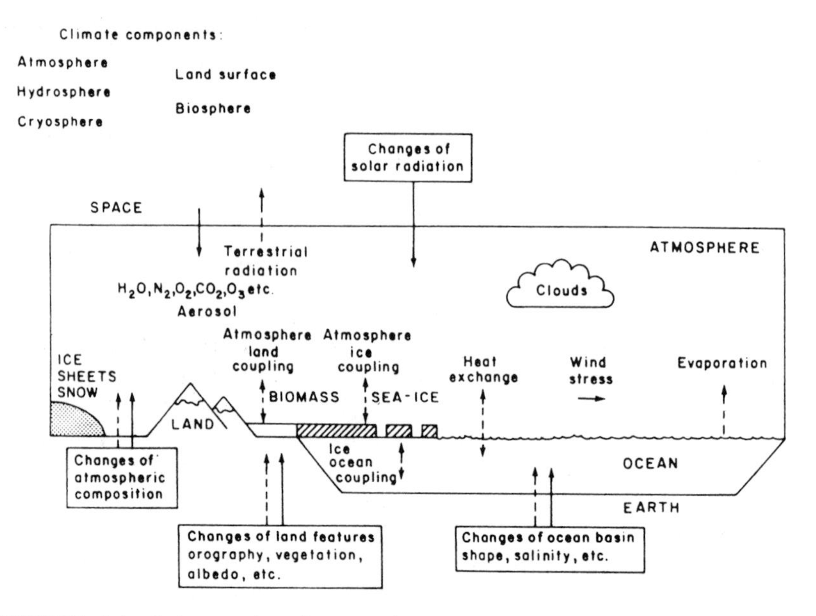

FIGURE 4.1 Schematic of the climate system.
SOURCE: Reprinted, by permission, from IPCC (1990). Copyright © 1990 by Cambridge University Press.

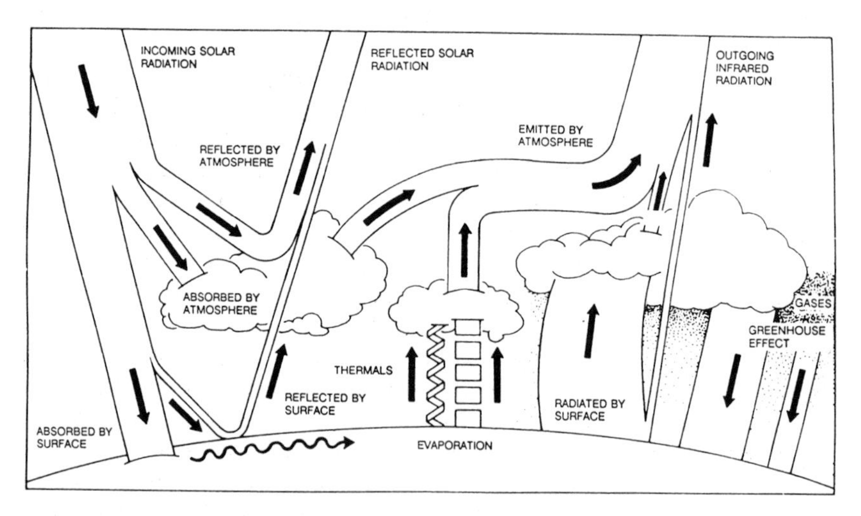

FIGURE 4.2 Schematic of various contributions to the global energy balance.
SOURCE: From Climate Modeling, by S. H. Schneider, 1987. Copyright © 1987 by Scientific America, Inc. All rights reserved.

lowest 15 kilometers of atmosphere, with an average radiating height of about 6 kilometers. They protect us from having a climate that would otherwise be about 33°C colder—a climate much more like that of the Himalayas.

The two most important natural greenhouse gases are water vapor and carbon dioxide, although several other natural constituents and an increasing number of entirely man-made greenhouse gases make significant and generally growing contributions to the greenhouse effect (Figure 4.3). Climatologists have several methods for evaluating the effects of greenhouse gases. Our most accurate information about the greenhouse effect is provided by atmospheric radiation models, derived from fundamental physical theory and measured strengths of the various radiating gases. We also can look at radiation leaving the earth and compare it with the theory. We see the exiting thermal radiation reduced from its surface flux according to characteristic wavelength signatures of the absorbing greenhouse gases and calculable according to greenhouse theory. Another way to appreciate the possible impacts of extreme greenhouse effects is to look at the very hot surface of Venus and the very cold temperature of Mars, also calculable from greenhouse theory.

The two best established factors of the theory of global climate warming are: (1) how much heating results from given concentrations of greenhouse gases (Figure 4.4), and (2) how much concentrations of these gases have increased over the last 100 years, and especially the last 20 to 30 years. We worry especially about methane, nitrous oxide, ozone in the upper troposphere, and the chlorofluorocarbons. The relative number of carbon dioxide molecules has increased from 280 to 355 parts per million (27 percent), methane from 0.7 to 1.7 parts per million, and chlorofluorocarbons from nothing to total concentrations of all species of about a part per billion. The small concentrations of the latter have a surprisingly large effect on greenhouse warming (as well as on stratospheric ozone), contributing about 11 percent of the total increase over the last 100 years (IPCC, 1990). Still, carbon dioxide has contributed somewhat more than half of the total and will continue to dominate increases in greenhouse warming.

Our good records of the international consumption of fossil fuels (coal, oil, and natural gas) indicate that about twice as much carbon each year is put into the atmosphere as that which remains in the atmosphere. What happens to the carbon that does not remain in the atmosphere is still relatively poorly understood. It presumably goes somewhere into the large reservoirs for carbon

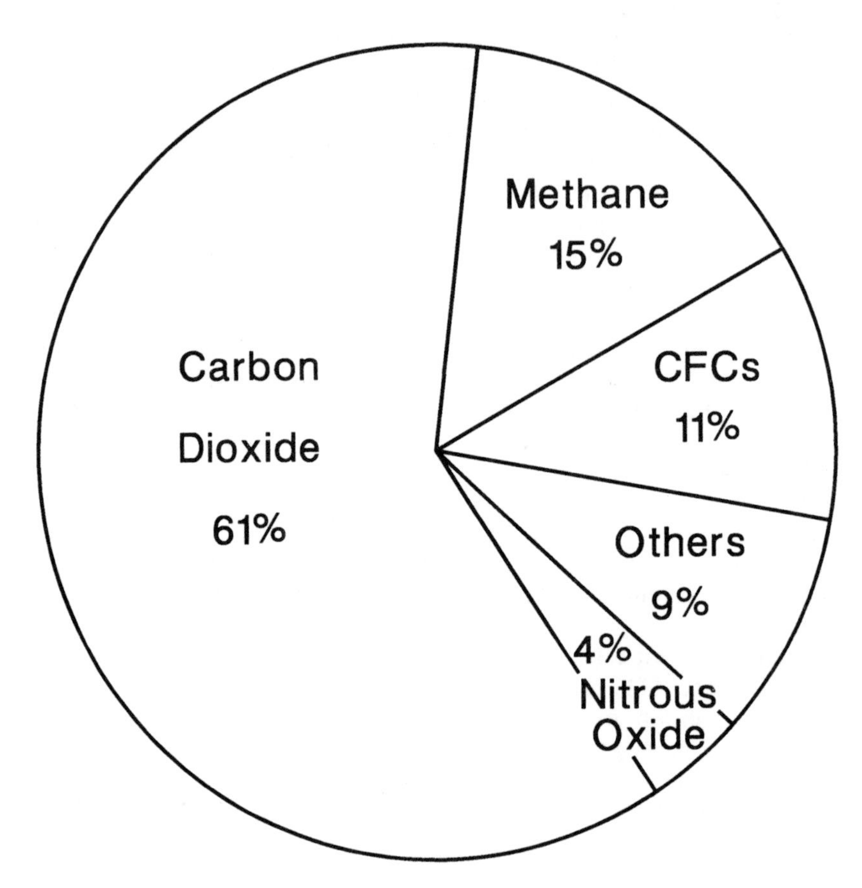

FIGURE 4.3 The relative cumulative effect on climate of greenhouse gases added to the atmosphere in 1990.
SOURCE: IPCC, 1990.

provided by the oceans, land vegetation, and soils (these compartments each hold more carbon than the total in the atmosphere). Difficulty accounting in detail for this lost carbon results from the potentially large additions from burning of tropical forests, the relatively small amounts that can be established as going into the oceans (Tans et al., 1990), and the lack of evidence for carbon going into land reservoirs. Will the fraction of fossil fuel carbon going into the atmosphere increase in the future and so accelerate the atmospheric buildup? We can't say without a better understanding of the carbon cycle.

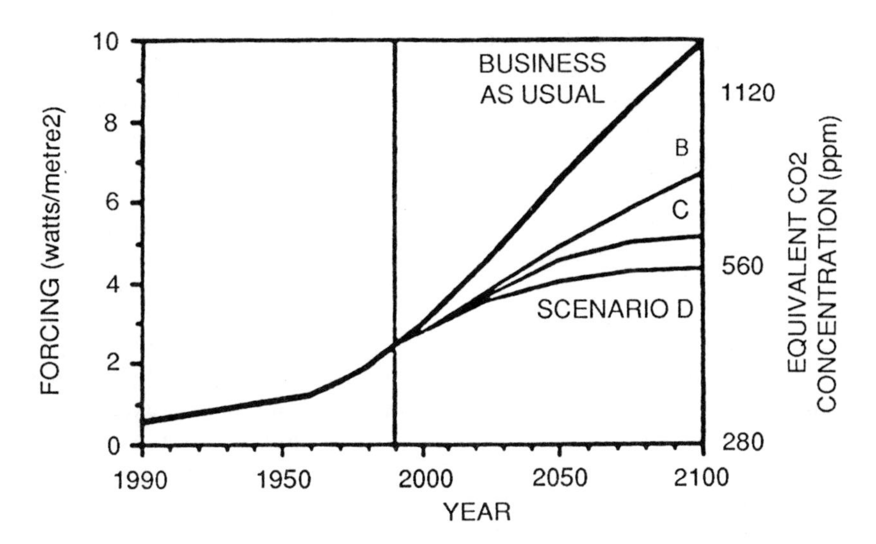

FIGURE 4.4 Radiative forcing by the sum of greenhouse gases, with the future indicated in terms of four scenarios: (A) showing "business as usual" and (B-D) indicating the effects of increasingly stringent regulation.
SOURCE: Reprinted, by permission, from IPCC (1990). Copyright © 1990 by Cambridge University Press.

HOW MUCH WILL OUR HOUSE WARM UP?

Given the well-established increasing concentrations of greenhouse gases and reasonably accurate estimates of the resulting increased heating, what does this tell us about future climate change? Suppose you were to go home and turn on your heating system (but remove the thermostat switch). It is fairly certain that your house would warm up to some new temperature. But what temperature? It would be useful in estimating this new temperature to know how much energy your heating system puts out. This answer would also depend on the temperature outside, how well insulated your walls and roof may be, what ventilation there may be from cracks and open windows (depending on the wind outside), how much additional heat the house may be receiving from the sun, whether or not it is raining, whether the temperature will become warm enough to set off an automatic sprinkler system, and so on. How long it would take to warm up would depend on the thermal mass of your walls.

To estimate what might be the new, warmer temperature of our heated house, we would develop equations for all its energy losses, which depend on the difference between the temperature outdoors and that in the house (Figure 4.5). This model of the house energy exchanges could then be used to infer a new temperature, one for which the energy losses from the house would just balance the heat put out by our furnace. This is what we do or would like to do with climate models. By analogy, we know that we are putting excess heat into our global home (or, equivalently, we are adding to the insulation), and we even know reasonably well how much. It is a fairly obvious conclusion that the planet will warm up, but how much? This depends on how easy it is for this extra heat to leak back out again and on what other climate factors the warmer temperatures will change, either to remove some of the excess heat being supplied or to add yet more. How much will the thermal inertia of the system, especially the oceans, delay the warming?

As with our heated house analogy, the easiest task is to get an overall idea of how much temperature will increase. Determining how this temperature increase is distributed from region to region is more difficult. Again turning to the house calculation, for us to know how temperature might vary from room to room, we would have to understand much about the convection currents carrying the heat between the rooms. Likewise, using a climate model to calculate the change of temperature and water resources in the

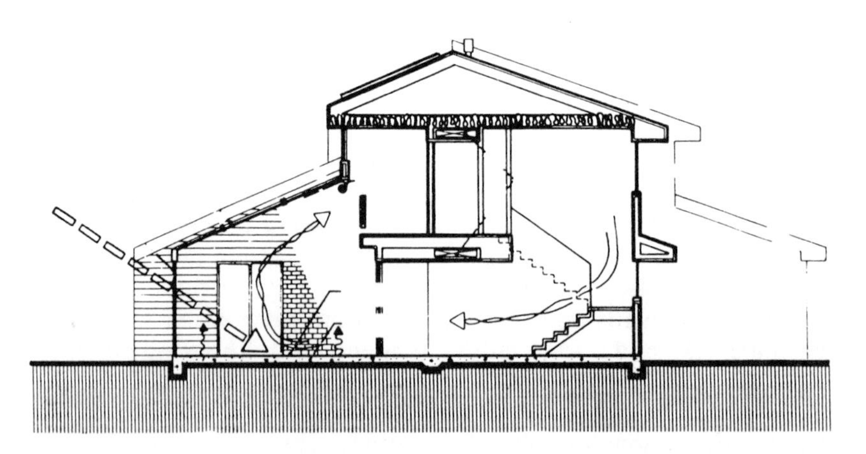

FIGURE 4.5 Metaphor for greenhouse warming: How warm would a heated house become in the absence of a thermostat?

western United States is much more difficult than just determining how much the world as a whole might warm up. This is the challenge the climate modeling community will be attacking over the next decade. The best we can do now is describe a plausible range of possible futures and say what tools must be developed, what research done, to narrow this range of possibilities.

CONSTRAINING POSSIBLE FUTURES THROUGH MODELS

A standard measure of the excess global heat from human activities is the heating that would result from doubling the atmospheric carbon dioxide concentration. This change, expected some time near the end of the next century, adds about the same or a little less heat than would come from a 2 percent increase in solar energy output. This amount is several hundred times the energy added directly from the combustion globally of fossil fuels. If we were to add 2 percent to the sunshine received on a single day, we would see only a slightly warmer temperature, perhaps a few tenths of a degree at most. However, applied over many years, temperatures would rise between 1.5 and 5°C over past normal conditions. We develop our sense of the possible and probable by running the most detailed three-dimensional models of our climate system now available and by exploring how the models' outputs depend on poorly described parts of the system. We are currently especially stuck on what clouds might do to change the reflection of solar energy or to change greenhouse warming as climate changes.

Climate models treat a large number of physical processes and the linkages between these processes. It is only through these models that we are able to combine the knowledge of experts about individual processes into an overall synthesis of how the system behaves and how it might change in the future. The more advanced general circulation climate models give a self-consistent simulation of the atmospheric and surface hydrological cycles (Figure 4.6). These models generate day-to-day weather in half-hour time steps. Atmospheric moisture is carried around by three-dimensional winds generated by numerical solution of the equations for atmospheric hydrodynamics on a sphere. Where atmospheric moisture is determined to be in excess of saturation, this excess is removed from the model atmosphere as rain or snow, depending on the temperature of the lowest model layers.

GCMs use a water budget approach to balance soil moisture change with the difference between precipitation and runoff.

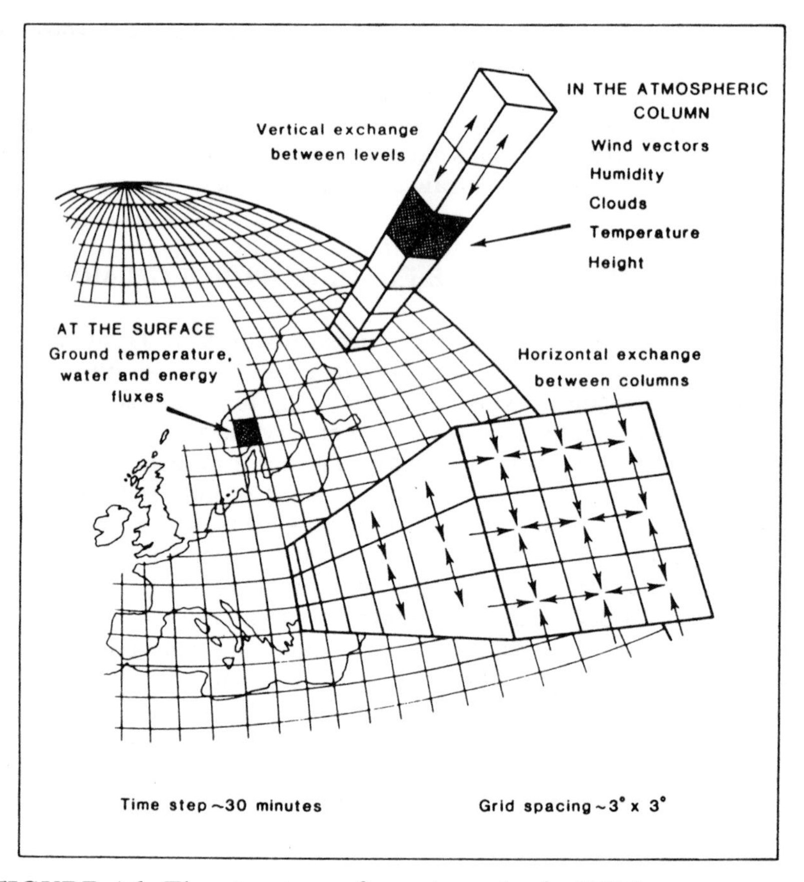

FIGURE 4.6 The structure of an atmospheric GCM.
SOURCE: Reprinted, by permission, from Henderson-Sellers and McGuffie (1987). Copyright © 1987 by John Wiley & Sons.

Thus, the GCMs compute both a runoff and an evapotranspiration. The most developed components of the GCMs are their treatments of large-scale atmospheric hydrodynamic and temperature patterns (treated the same as in weather prediction models) and their modeling of atmospheric radiation. Less well developed are the treatments of atmospheric humidity, precipitation, and clouds (radiative properties and convection), the dynamics of the oceans and sea ice (especially how the atmosphere couples to these systems), variations in atmospheric chemical composition, and the climate aspects of land vegetation and soils, including surface hydrology.

From the viewpoint of western hydrology, we wish to know quantitatively how patterns and amounts of precipitation and evapotranspiration will change. Precipitation will depend on transport of moist air from the Pacific or the Gulf of Mexico, or dry air from elsewhere, and the presence of large-scale upper-air movement and convective destabilization. Evapotranspiration will depend on the amount of solar radiation received by the vegetation, the temperature and relative humidity of overlying air, surface roughness, the net effect of vegetation stomatal resistances, and details of surface runoff processes. All these considerations are already included in current climate models. Large-scale geographical patterns of precipitation as seen on a smooth global map compare quite well with reality. The concern is that some of the treatments are not yet realistic enough and that so far we have not adequately checked their operation against observations.

PRESENT GUIDANCE FROM CLIMATE MODELS AND OTHER ARGUMENTS CONCERNING WESTERN WATER

GCMs generate an extraordinarily detailed description of climate processes. In principle, they convert scenarios for increasing trace-gas concentrations into changes of all the more important surface hydrological parameters: precipitation, surface radiation, surface temperature, evapotranspiration, soil moisture, and runoff. Yet, because the real system is even more detailed and complex than the models, there are still severe limitations on the confidence we can place in our abilities to model future change. Since there is no other plausible method to ascertain future precipitation and surface air temperatures, we must continue to improve and better understand the GCMs.

Two approaches have been used to determine possible future changes of water resources in the West: (1) the use of climate models to project precipitation and temperature, followed by the inclusion of these projections as input into detailed regional hydrological models; and (2) the direct use of climate model calculations of surface hydrological parameters.

Qualitatively, the projections of global average increasing temperature and precipitation are well founded in basic physical principles and are given by all GCMs. However, a wide range of possibilities for regional anomalies has been suggested by simulations for the changes over regions as small as the western United States (Kellogg and Zhao, 1988). Furthermore, surface hydrological processes are much more complex than what is presumed in the GCMs, which, for example, largely or entirely ignore the effects of topography on runoff and infer dynamic effects on precipitation using only a very smooth topography.

The question of how runoff will change given changes in precipitation and temperature provided by GCMs has now been explored by a number of authors. The simplest such approach is to use the mean measured runoff in different basins (Stockton and Boggess, 1979) or the year-to-year variations in a single basin (Revelle and Waggoner, 1983) to infer statistically how runoff might change with changes in precipitation and temperature. This statistical correlation approach does not allow for an examination of changes in the seasonality of runoff or for a distinction between water supplies from snowpack or rainfall. More recent studies of the sensitivity of runoff to precipitation and temperature (Gleick, 1987; Schaake, 1990) have used simple water-balance models. These models resemble the treatments of surface hydrology in GCMs in following a soil moisture budget but differ in their use of monthly (rather than hourly or smaller) time steps and in their use of an empirical relationship between evaporation, soil moisture, and temperature based on the concept of potential evapotranspiration. They employ a physically based formulation for runoff that is comparable in complexity and limitations to that used in the GCMs. However, they have been adjusted to observed runoff records. Gleick (1987) modeled the Sacramento basin, apparently adjusting only a few physical parameters and validating the model against independent data. Schaake (1990) fit data from 52 basins in the southeastern United States by adjusting five parameters. Both these models include snow accumulation and melt.

The water balance studies of Gleick (1987) and Schaake (1990) show that warming of the projected magnitudes would have a large impact on the seasonality of runoff, with heavier winter flows and reduced summer flows (Figure 4.7). The smaller annual changes they find are comparable to those found in the earlier correlative studies.

The examination of surface hydrology modeled explicitly by GCMs to explore the impacts of climate change on water resources has not yet been carried very far. Emphasis has been primarily on the changes in soil moisture (Manabe and Wetherald, 1987; Meehl and Washington, 1988) and on possible implications for mid-continental drought. Kellogg and Zhao (1988) have discussed the wide divergence between past model simulations (Figure 4.8).

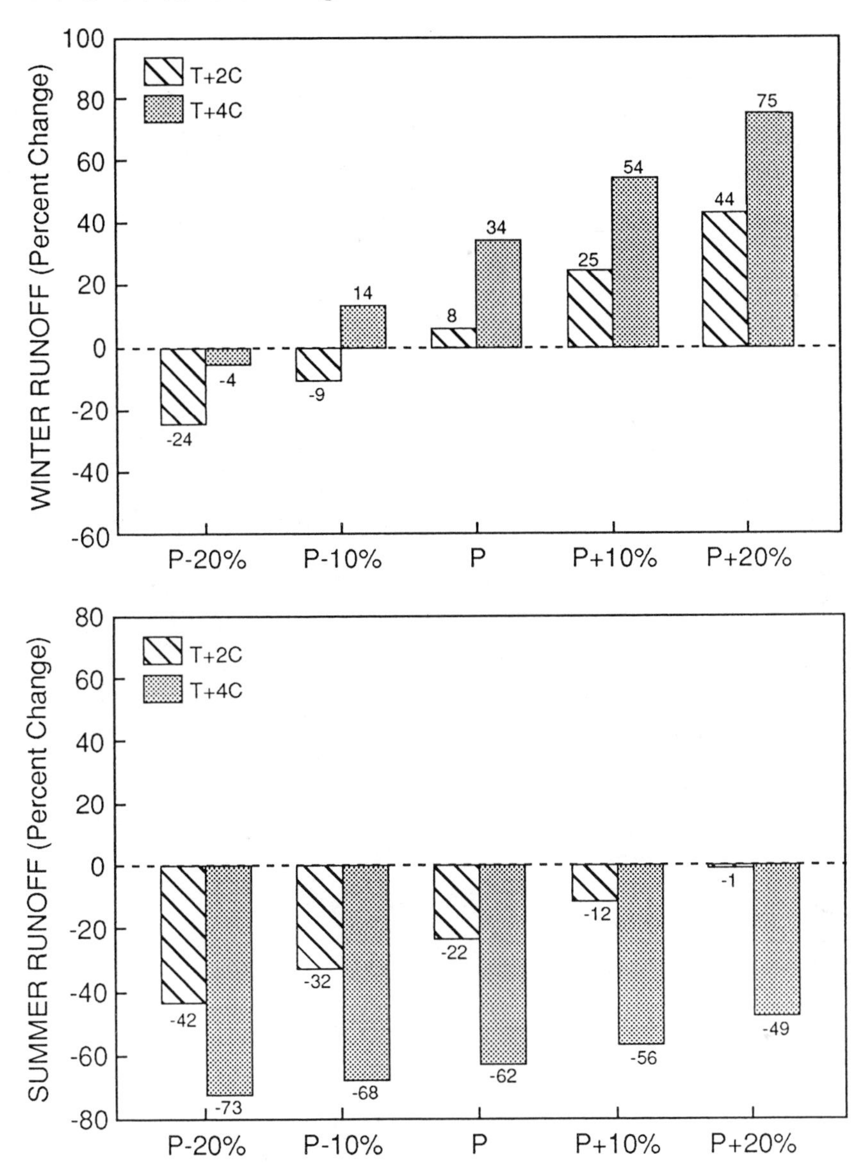

FIGURE 4.7 Changes in winter and summer runoff inferred by Gleick (1987) for the Sacramento Basin under various precipitation changes and warming by 2 and 4°C.

SOURCE: Reprinted from Gleick (1987). Copyright © 1987 by Climatic Change. Reprinted, by permission, of Kluwer Academic Publishers.

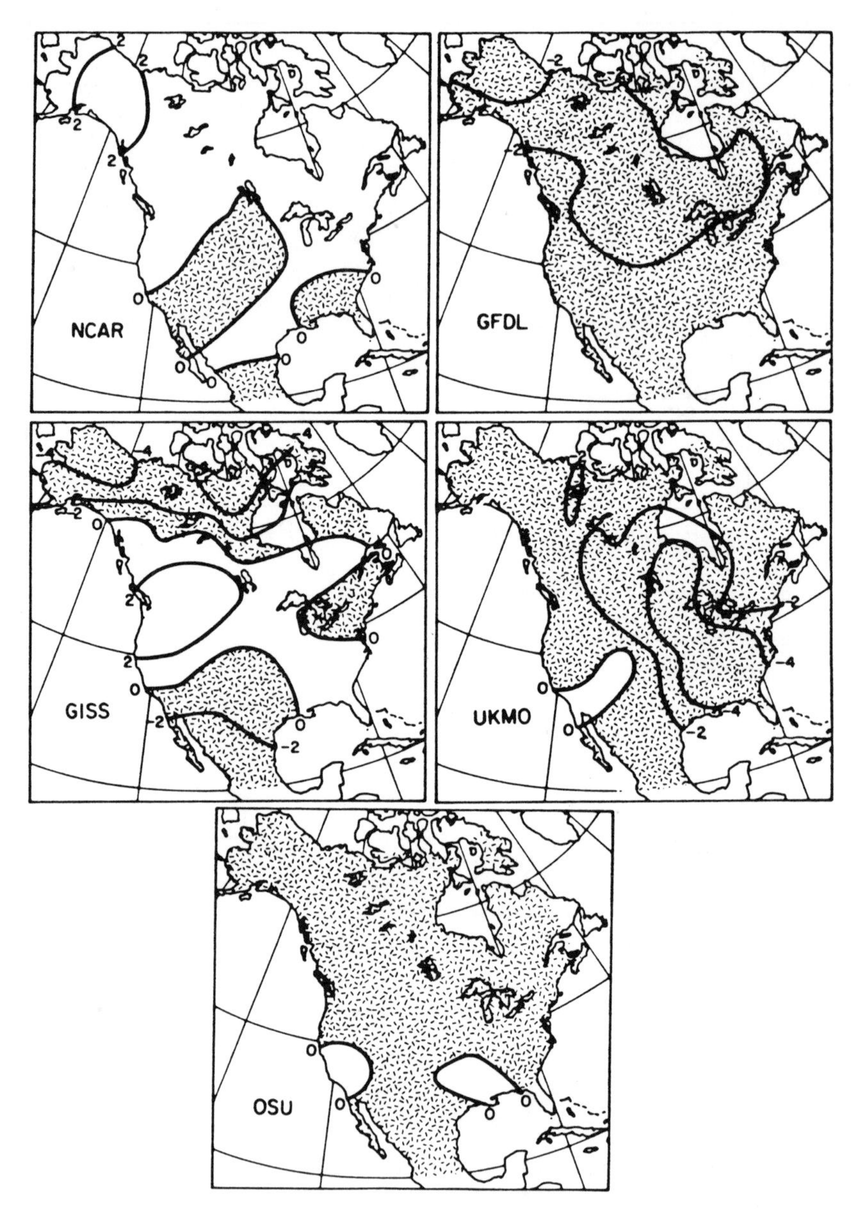

FIGURE 4.8 Changes in summer soil moisture in cm for various GCMs.
SOURCE: Reprinted, by permission, from Gleick Kellogg and Zhao (1988). Copyright © 1988 by Journal of Climate.

CONCLUSION

Climate models are our only tools to project into the future. However, current climate modeling results are significant only in a qualitative sense for surface hydrology as illustrations of physically consistent possible change. They show a wide divergence in their future projections of the atmospheric quantities most important for determining surface hydrology—in particular precipitation and net surface radiation. Furthermore, little attention has yet been given to developing aspects of the models important for surface hydrology. Improvements in these features and means to quantify the relative confidence we can place in these different projections must be a high priority for future work.

REFERENCES

Gleick, P. H. 1987. Regional hydrologic consequences of increases in atmospheric CO_2 and other trace gases. Climatic Change 10:137-161.

Henderson-Sellers, A., and K. McGuffie. 1987. A Climate Modelling Primer. Chichester, England: John Wiley & Sons.

Intergovernmental Panel for Climate Change (IPCC). 1990. Climate Change: The IPCC Assessment. Cambridge, England: Cambridge University Press.

Kellogg, W., and Zong-Ci Zhao. 1988. Sensitivity of soil moisture to doubling of carbon dioxide in climate model experiments, part I: North America. J. Climate 1:348-366.

Manabe, S., and R. T. Wetherald. 1987. Large scale changes in soil wetness induced by an increase in carbon dioxide. J. Atmos. Sci. 44:1211-1235.

Meehl, G. A., and W. M. Washington. 1988. A comparison of soil-moisture sensitivity in two global climate models. J. Atmos. Sci. 45:1476-1492.

Revelle, R. R., and P. E. Waggoner. 1983. Effects of a carbon dioxide-induced climatic change on water supplies in the Western United States. In Changing Climate. Washington, D.C.: National Academy Press.

Schaake, J. C. 1990. From climate to flow. In Climate Change and U.S. Water Resources. New York: John Wiley & Sons.

Schneider, S. H. 1987. Climate modeling. Scientific American 256:72-80.

Stockton, C. W., and W. R. Boggess. 1979. Geohydrological Implications of Climate Change on Water Resources Development. Ft. Belvoir, Virginia: U.S. Army Coastal Engineering Research Center.

Tans, P. P., I. Y. Fung, and T. Takahashi. 1990. Observational constraints on the global atmospheric carbon dioxide budget. Science 247:1431-1438.

5

Climate Change and Climate Variability: The Climate Record

Kevin E. Trenberth
National Center for Atmospheric Research
Boulder, Colorado

INTRODUCTION

Global warming is a prominent part of the climate change expected in association with the observed increases in greenhouse gas levels brought about by man's activities. Yet natural variations, such as those associated with El Niños, make it difficult to detect anticipated climate changes. In addition, detection in the observational record of the expected global warming is confounded by flawed and patchy observations and because observed climate change is not geographically uniform. This paper discusses the difference between weather and climate and analyzes examples of recent climate anomalies, including the 1988 North American drought. It also summarizes the climate changes expected from increased greenhouse gas concentrations in light of the observational record.

The observed patterns of temperature change over the globe are not yet well accounted for by climate models. Nevertheless, climate models are the best tool we have for quantifying the complex processes involved in climate change and for making future projections. Large natural variability will continue to be a major factor to be planned for, and any changes in the incidence or kind of El Niños could have profound impacts in the western United States.

THE POTENTIAL FOR GLOBAL WARMING

There are many uncertainties in future projections of climate change: uncertainties about what form the climate change will take, how quickly it will develop, and how extreme it will be.

With the recognition that climate changes are inevitable as a result of the observed increase in greenhouse gas concentrations in the atmosphere, there is a great need for better observations and analyses to document the past and current climate and how it has changed with time. The need is for both regional and global evidence, to the extent that it exists, for many climatological variables.

While there are many uncertainties in any future projections, there are also some certainties. The levels of several greenhouse gases are increasing and will continue to do so. Carbon dioxide is the best known greenhouse gas. For example, observations at Mauna Loa show that since 1958 the carbon dioxide concentration has increased from 315 to approximately 350 ppm (Figure 5.1), and bubbles of air in ice cores reveal that the preindustrial carbon dioxide levels of the last century were around 280 ppm. So concentrations have already increased by 25 percent, largely because of man's burning of fossil fuels and clearing of forests. The other greenhouse gases of note are methane, nitrous oxide, and the chlorofluorocarbons (CFCs); the concentrations of all these gases

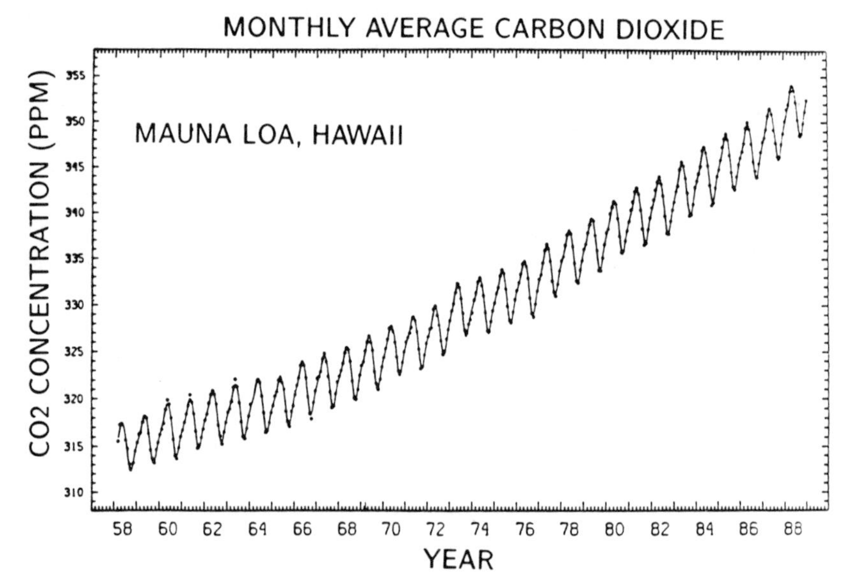

FIGURE 5.1 Monthly average carbon dioxide from Mauna Loa, Hawaii in parts per million by volume.
SOURCE: Reprinted, by permission, from Keeling et al. (1989). Copyright © 1989 by American Geophysical Union.

are observed to be increasing due to man's activities and will continue to increase. The pressures of population growth, more than any other factor, will guarantee this. It is not my purpose here to delve into this aspect any further, though it is a subject of interest in itself.

Because these gases are greenhouse gases—that is, they interfere with the outgoing radiation (Figure 5.2)—they will cause climate change. In the midst of all the uncertainties, we must not lose sight of this. The greenhouse gases are well mixed and globally distributed and the increase will undoubtedly cause climate change. The main prospect of this gas accumulation is indeed global warming.

THE 1988 DROUGHT

During the summer drought of 1988, there was a proliferation of news media speculations about a possible link between the drought and the greenhouse gas effect. In fact, the best assessment we can make indicates that the drought was, more likely, essentially a natural phenomenon brought about primarily by changes in the tropical sea surface temperatures (SSTs) in the Pacific Ocean. The theory of the drought (Figure 5.3), which has been tested with models (Trenberth et al., 1988), goes as follows:

- In the tropical Pacific, atmospheric winds drive the ocean currents and thus determine the SST patterns; simultaneously, the SSTs determine where the main convergence in winds occurs in the atmosphere and thus where the tropical storms and thunderstorms are most apt to form in organized patterns. The names given to the different extremes of these changes are El Niño, for the times when the tropical Pacific is warmer than normal, and La Niña, for the times when it is colder than normal.
- In April 1988, a strong La Niña developed, creating very cold water along the equator. Warmer water, a residual from the previous El Niño that occurred in 1986 and 1987, was pushed up to a region southeast of Hawaii and elsewhere. This pattern was favorable to a northward displacement of the Inter Tropical Convergence Zone, a region of organized thunderstorms, and resulted in a change of heating patterns in the atmosphere through the latent heat released in rainfall.

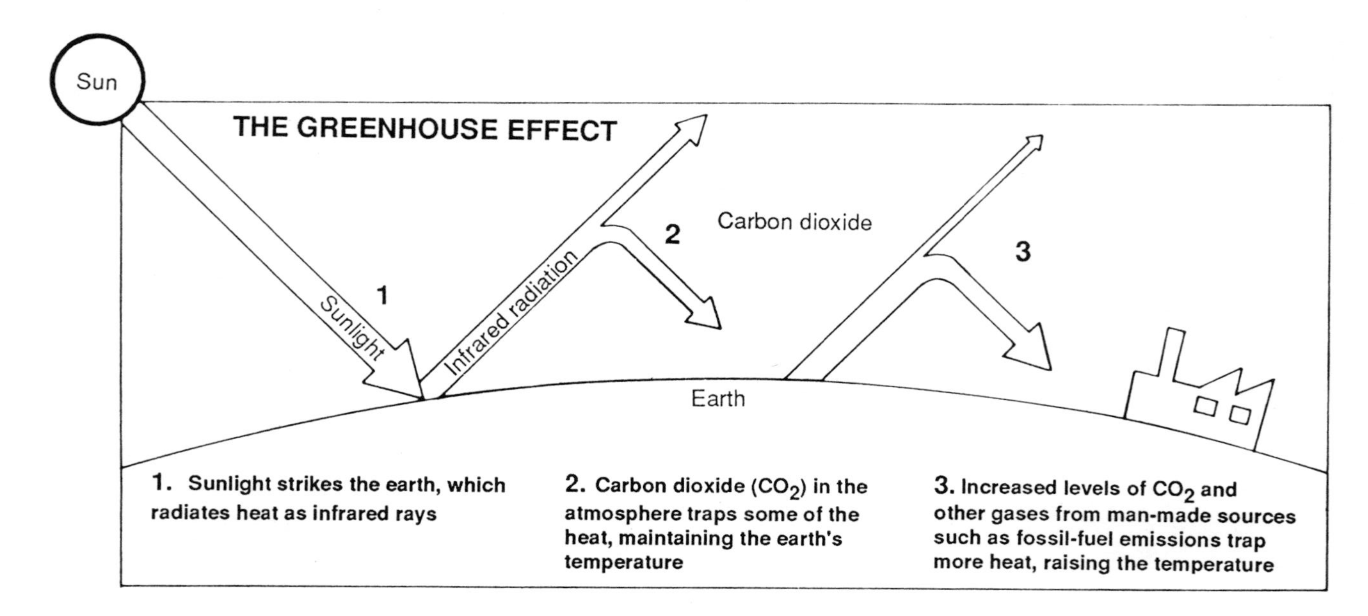

FIGURE 5.2 Schematic illustration of the greenhouse effect.

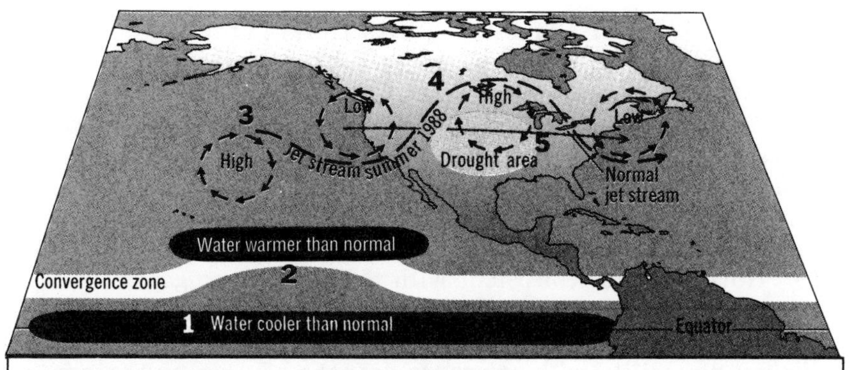

A NEW THEORY OF THE DROUGHT

Last summer the cold water of La Niña (1) in conjunction with a patch of warmer water caused the atmospheric convergence zone (a stormy region that forms over the warmest water) to shift north from its usual position near the equator (2). This displaced zone, like a rock in a stream, set up waves of high and low pressure in the atmosphere (3). These unusual highs and lows caused the jet stream to shift northward over central North America (4), robbing the Midwest of moisture and causing the drought (5).

Source: Kevin Trenberth, National Center for Atmospheric Research — TIME Map by Joe Lertola

FIGURE 5.3 Schematic outline of the theory for the 1988 drought. **SOURCE**: Reprinted, by permission, from TIME Magazine (October 31, 1988). Copyright © by TIME.

- In turn, the change in heating disrupted the atmospheric flow by setting up anomalous waves in the atmosphere, manifested as a change in the jet stream and associated storm tracks in mid-latitudes. In the spring of 1988, this led to anticyclonic conditions over North America—conditions favorable for the development of the drought.
- The dry conditions led to heat waves, because the incoming solar radiation in summer goes mostly into heating the atmosphere rather than to evaporating moisture, and this, along with the moisture shortage, led to a perpetuation of the drought.

The greenhouse effect may well have made the heat waves hotter and the drought somewhat more severe than it would have been otherwise, but these effects were probably relatively small. The drought would almost certainly have occurred without any enhanced greenhouse effect; droughts are essentially a natural phenomenon and have occurred throughout history.

WEATHER VERSUS CLIMATE

In 1989, there were other weather extremes to talk about. While it was listed as the sixth warmest year on record for the globe, the warmest year ever in Great Britain (where the instrumental records, which begin in 1659, are the longest anywhere), and also the sunniest year, there were some contraindications, including record cold outbreaks in the United States in February 1989 and again in December 1989. How are these consistent with the greenhouse gas effect?

While one possible explanation of these signs relates to whether or not there might be increases in extremes in a greenhouse gas world—a distinct possibility with respect to droughts, floods, and things like hurricanes—another likely prospect is that these extreme events were caused simply by "weather" or natural variability. There is a lot of confusion in the mind of the public on this issue. Weather arises from internal instabilities in the atmosphere that produce anticyclones and cyclones, and cold and warm fronts, as the equator-to-pole temperature gradient is continually being eroded by cold polar outbreaks and warm southerlies in these weather systems, even as the incoming solar and outgoing earth radiation help maintain those temperature gradients. Weather, then, is an atmospheric phenomenon, and it has an infinite range of variability.

Climate, on the other hand, is more persistent, and we think of it as essentially forced, or caused, by things external to the atmosphere. The ocean is the primary link that produces climate anomalies; the largest interannual changes in world-wide weather patterns arise from the El Niño events in the tropical Pacific. Changes in greenhouse gas concentrations is another phenomenon external to the weather system. Other possibilities are changes in the sun, volcanic eruptions that blot out the sun, and interactions with snow, sea ice, and glaciers. Climate then, involves not only the atmosphere but also the oceans and interactions with other parts of the climate system.

CLIMATE OBSERVATIONS

There are three areas of concern related to the use of climate observations to detect climate change:

1. the quality of the observational data base—its spatial and temporal distribution, biases, and random errors;
2. the validity of spatial averages—of the analysis methods used to compile regional and global numbers; and

3. the best ways to interpret results—to account for natural variability while detecting trends in climate signals.

Surface Temperatures

The problems of homogenizing the original data are severe because over time there have been changes in the following:

- instrumentation, instrument exposure, and measurement techniques (e.g., standard thermometers versus electronic thermistors);
- station location;
- observing times and methods of computing daily and monthly means;
- station environment, such as urbanization; and
- analysis methods for computing area averages.

For land temperatures, station locations have moved, such as from the central city to the airports in the 1950s, and observation times have changed. Because daily averages sometimes use averages of hourly or six-hourly readings and sometimes simply average the daily maximum and minimum temperatures, changes in observation time can introduce a bias. These effects can be corrected if the changes in instrumentation, station site, time, and method of measurement are known, but this is not the case for many places outside of the United States.

Some studies use only land-based stations, but such an approach is hardly global. Climate data in Antarctica began only in 1957. Observations at many southern ocean islands were initiated in the 1940s. For SSTs there are huge gaps in coverage over the ocean as we go back in time (Figure 5.4). Global coverage combining SST and land data is estimated as shown in Table 5.1.

Other problems with SSTs are the changes in the kinds of buckets used to bring up water for temperature measurements and the switch, especially around World War II, to use of engine room readings of temperatures of water brought in to cool the engines. Because of heat in the engine room, the latter tend to be 0.3 to 0.7°C warmer than bucket temperatures, but the difference varies greatly depending on the draft of the ship and the engine room configuration. More recently, satellite data have introduced other kinds of errors because of contamination by water vapor and aerosols in the atmosphere and because the satellite sees only the surface—the skin temperature.

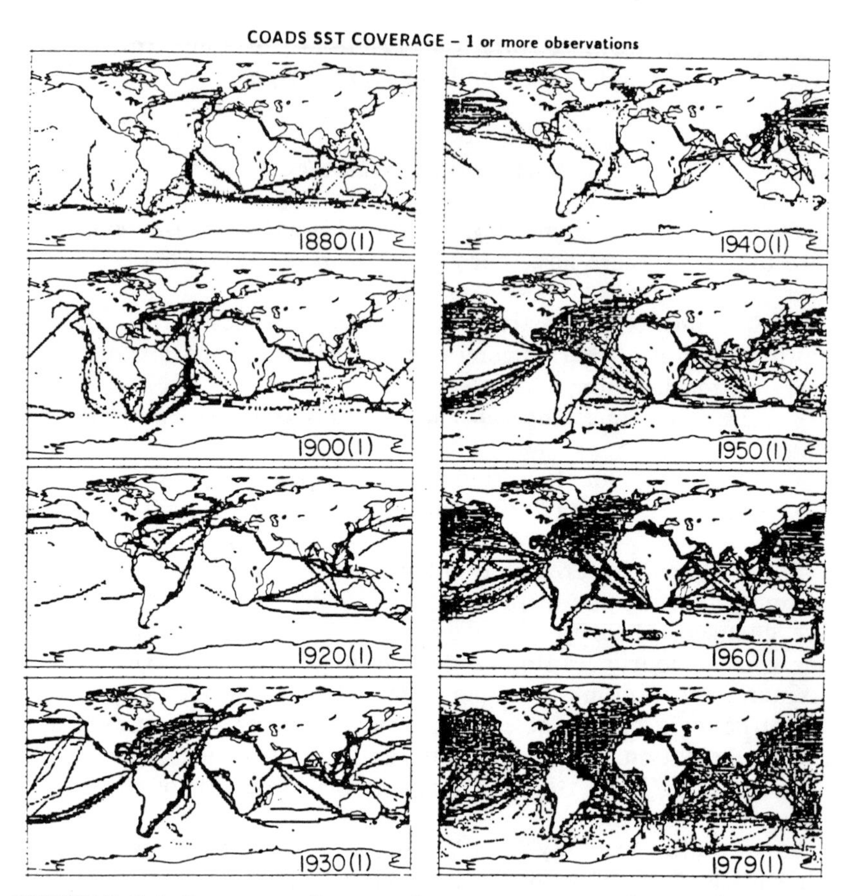

FIGURE 5.4 Coverage of sea surface temperature from ship observations during January of the years shown. To be plotted, one or more observations were required in a 2° latitude by 2° longitude square with the small *x* plotted at the center of gravity of the observations. **SOURCE**: Adapted from Comprehensive Ocean-Atmosphere Data Set (COADS).

Figure 5.5 shows uncorrected and corrected SSTs averaged over the Northern Hemisphere as a function of time. Of the two corrected versions here, one shows almost no trend, while the other shows an upward trend and is more similar to what land data show. The point here is that the corrections that need to be made to the data are uncertain, at least to 0.2°C, and the magnitudes of the corrections are larger than the signal we are looking for.

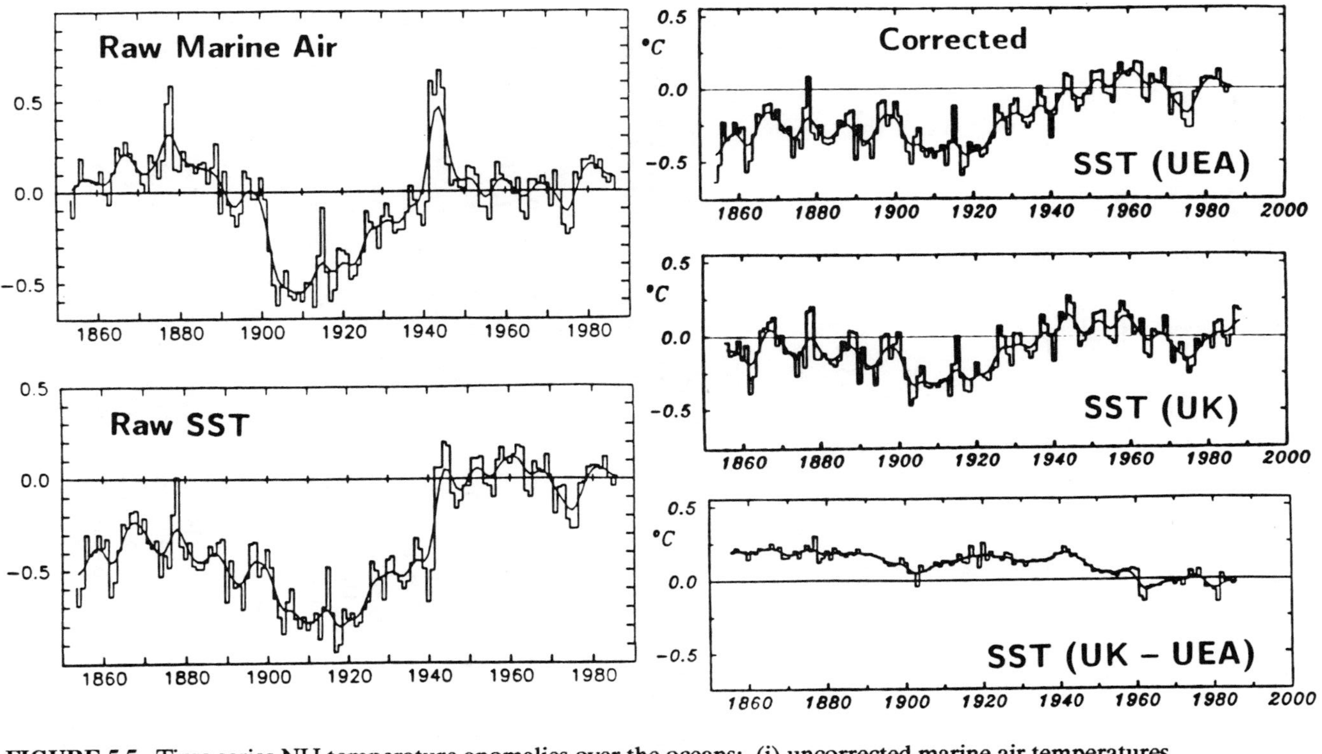

FIGURE 5.5 Time series NH temperature anomalies over the oceans: (i) uncorrected marine air temperatures, (ii) uncorrected sea surface temperature (SST), (iii) SST as corrected by the University of East Anglia (UEA), (iv) SST as corrected by the U.K. Meteorological office (UK), and (v) differences in SST (UK-UEA).
SOURCE: Adapted from Farmer et al., 1989.

TABLE 5.1 Portion of the Globe for which Surface Temperature Data are Available for Various Decades (in percent).

	Percentage Covered by Temperature Data		
Decade	Globe	Northern Hemisphere	Southern Hemisphere
1960s	70	85	55
1930s	50	65	35
1910s	33	45	20

The best estimates we have of global and hemispheric temperatures (Figure 5.6), keeping in mind the uncertainties discussed above, do reveal a modest warming of approximately 0.5°C, with the warmest years in the 1980s. Note that in the Southern Hemisphere, the temperature trend is what we might expect—the largest temperature increases occurred after World War II when greatest increases in greenhouse gases occurred. In contrast, the biggest jump in the Northern Hemisphere occurred in the 1920s. The details of why the record is like this are not yet well understood. A big question is whether the overall warming will continue, as we might expect because of the greenhouse effect, or whether temperatures will decline as they did in the 1960s. Of this overall warming, it is possible that 0.1°C of the increase is from the urban heat island effect.

However, regionally, there are large departures from this pattern, and there has not been much of a trend for the United States (Figure 5.7). For the United States, the highest temperatures are found in the 1930s, the dust bowl era in which droughts and heat waves prevailed. For comparison, many years have been warmer than 1988, the second-to-last point on Figure 5.7 and the year of our most recent major drought.

When we look regionally at the decade from 1977 to 1986 (Figure 5.8), we see that the largest warming—more than 1.75°C—relative to the mean for the period 1951 to 1980 occurred over

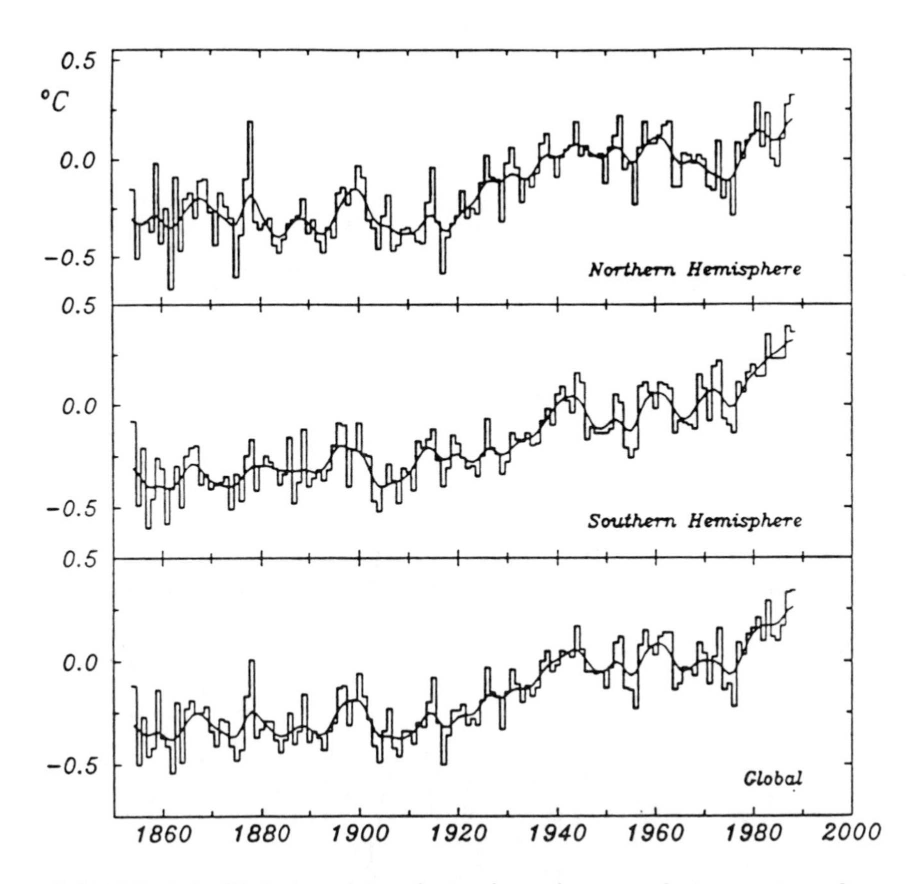

FIGURE 5.6 Global and hemispheric estimates of the total surface temperature variations from 1884 to 1988.
SOURCE: Reprinted, by permission, from Farmer et al. (1989). Copyright © 1989 by University of East Anglia Climate Research Unit.

found in the 1930s, the dust bowl era in which droughts and heat waves prevailed. For comparison, many years have been warmer than 1988, the second-to-last point on Figure 5.7 and the year of our most recent major drought.

When we look regionally at the decade from 1977 to 1986 (Figure 5.8), we see that the largest warming—more than 1.75°C—relative to the mean for the period 1951 to 1980 occurred over Alaska. The

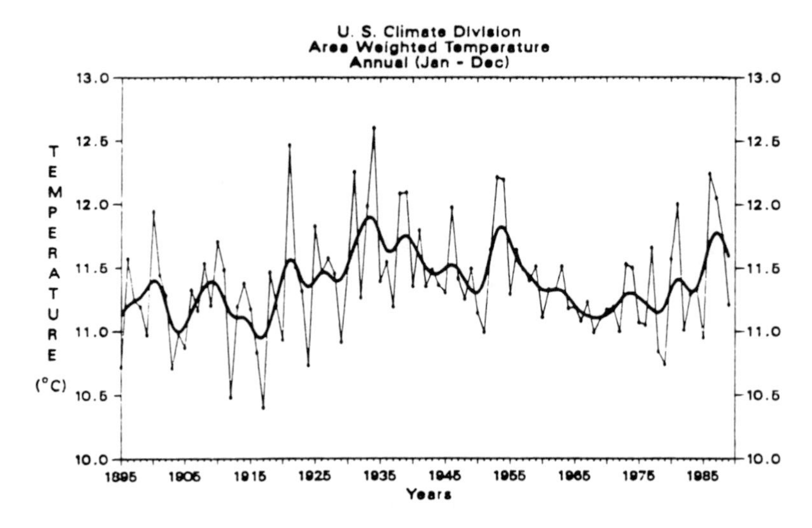

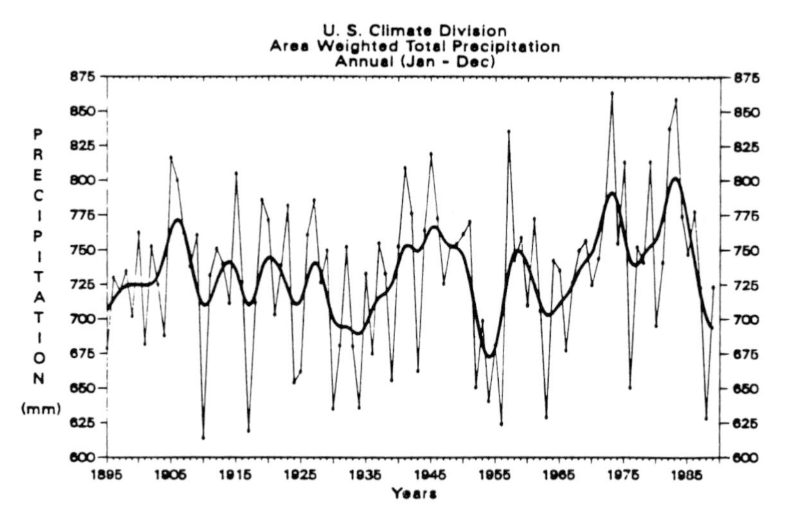

FIGURE 5.7 Time series of annual mean temperature and precipitation for the United States from 1895 to 1989.
SOURCE: Updated from Karl, 1988.

Alaska. The western states were warmer but the eastern states colder, providing little change for the United States as a whole. Note also the cooling in the North Atlantic and North Pacific, which would not be well captured using only land data. It turns out that this pattern

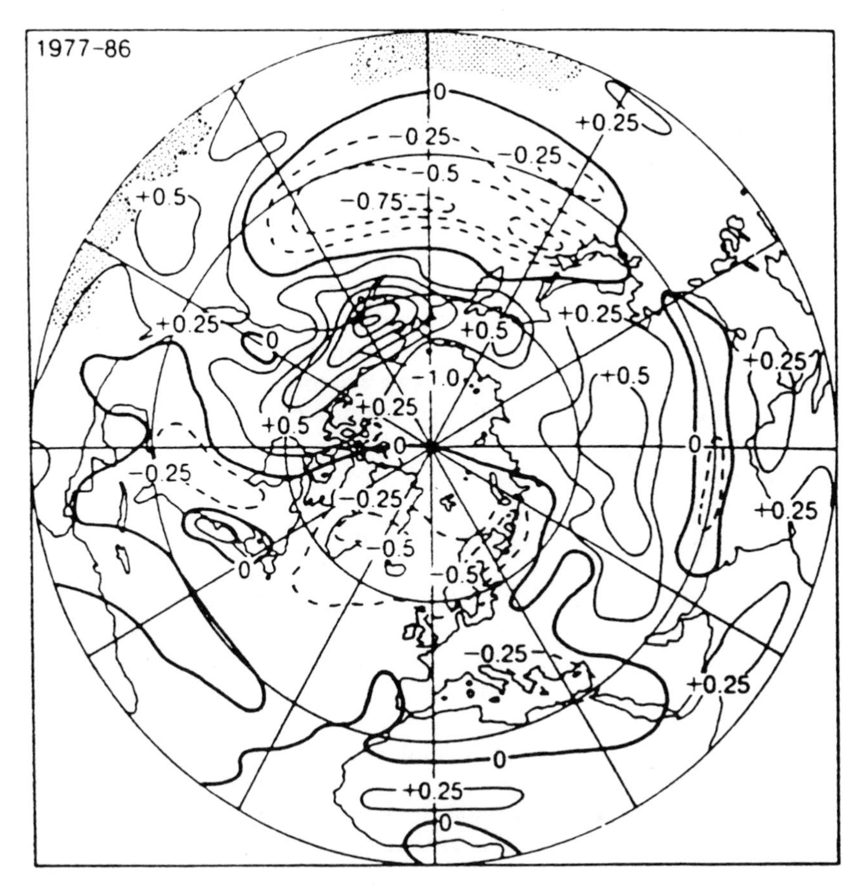

FIGURE 5.8 Decadal mean surface temperature anomalies for 1977 to 1986 relative to a 30-year mean (1951-1980). Contours are every 0.25°C and stippled areas indicate inadequate data.
SOURCE: Trenberth, 1990.

occurred for a reason related to changes in atmospheric circulation (Trenberth, 1990). Over the North Pacific in winter (November through March), the Aleutian Low pressure system was more intense than normal from 1977 to 1988, bringing more southerly winds over the West Coast and warmer, moister air into Alaska, while on the back side of the Aleutian Low colder, drier air caused cooling over the North Pacific. This change in circulation takes the form of a wave pattern in the jet stream, and the North Pacific changes are also linked to the cooling over the East Coast.

Why did this occur? A big part of it may be related to changes in the tropics and what meteorologists refer to as "teleconnections." Teleconnections come about because the atmosphere responds to forcing by setting up waves that may propagate over great distances. In this case, the corresponding changes in the tropics are manifest as the El Niño-Southern Oscillation (ENSO) phenomenon (Figure 5.9).

A major question, at present unresolved, is whether there may be a change in intensity or frequency of ENSO events with global warming. From 1977 to 1988 there were three major El Niños (warm events) in a row but no corresponding La Niñas (cold events) in the tropical Pacific. This is unprecedented in the past 108 years. Consequently, the question of why it was warm in Alaska becomes one of why there were three ENSO events in a row. Is this a signal of the greenhouse effect?

Precipitation

Reliable precipitation climatologies are not available over the oceans because even where rare measurements from ships do exist, they are unreliable. Over land, the climatologies are undoubtedly useful, but still subject to inhomogeneities, and it is difficult to obtain good estimates of changes with time, except regionally. The records that do exist exhibit strong variability on all time scales. Interannual variations, often associated with ENSO, are strong, and there are also large decadal scale variations that can be verified by changes in lake levels. In the United States, record high levels in Lake Michigan and the Great Salt Lake in 1985 provide evidence of the persistently above normal rainfalls from about 1970 to 1985 (Figure 5.7). Some lower frequency variations and trends also appear in several records, but their reality is often less certain. One example of a strong and persistent reduction in rainfall is in the Sahel region of Africa, which has caused Lake Chad in the central Sahel to fall substantially in recent decades.

For precipitation data, the large spatial variability and inherent small scales make coverage, analysis, and averaging methods important considerations. Small changes in station location and changes in exposure, such as tree growth, commonly corrupt rainfall records, but these changes are generally not detectable unless detailed station history records are kept. Because of the small-scale spatial variability, adjustment of rainfall records for inhomogeneities using an

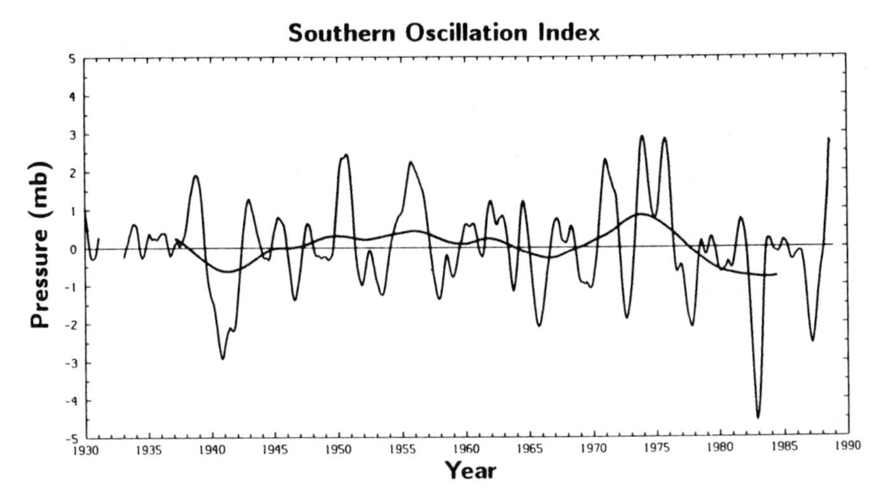

FIGURE 5.9 Time series of the Southern Oscillation index.

adjacent station comparison requires a much denser network of stations than adjustments of temperature records. The latter also means that many more stations are needed to adequately monitor rainfall than are needed to monitor temperature, especially in complex terrain areas.

In addition, changes in instrumentation through alterations in rain gauge design to address factors such as wind, wetting losses, and evaporation that affect the collection efficiency and measurement of rain, and especially snow, introduce biases with time that must be corrected. National practices for converting frozen precipitation into liquid equivalent vary. In the past, discontinuities in isohyets* across the borders of adjacent countries have often been noted. Correcting of the pervasive inhomogeneities in rainfall records with complete station records is essential if confidence is to be gained in how rainfall is changing with time.

For the United States, regional reconstructions of precipitation variations with time have been made by the National Climatic Data Center (NCDC) at the National Oceanic and Atmospheric Administration (NOAA). Data for years prior to 1931, however, are much more uncertain and have a different basis. Prior to 1931, data from all

* Lines of constant liquid precipitation amount.

reporting stations were averaged with equal weighing, taking no account of the distribution of stations. After 1931, statewide estimates were made by areally averaging the values from each climatic division (divisions represent climatically homogeneous subregions). Within each division, each station is weighted equally, although the number of stations varies with time. In 1982, 5000 stations were used across the United States. Values prior to 1931 have been adjusted in an ad hoc fashion to be compatible with post-1931 methods (Karl et al., 1983), but this process may alter long-term trends.

The states are amalgamated such that the nation is divided into nine broad regions (Figure 5.10). Time series of precipitation for the three western regions are shown in Figure 5.11. In the Northwest, low-frequency fluctuations lasting several decades are in evidence, with the 1920s appearing as particularly dry. Precipitation in the Northwest was fairly consistently above the long-term mean from about 1945 until 1984 but has been very low since 1985. Year-to-year fluctuations of 10 to 20 percent of the mean are quite common.

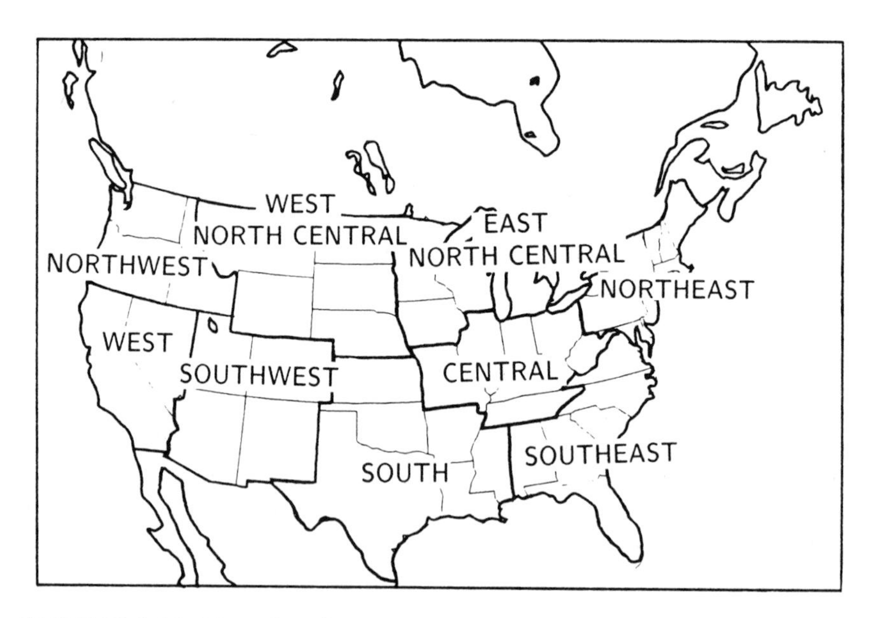

FIGURE 5.10 Map showing the major climatic regions of the United States.

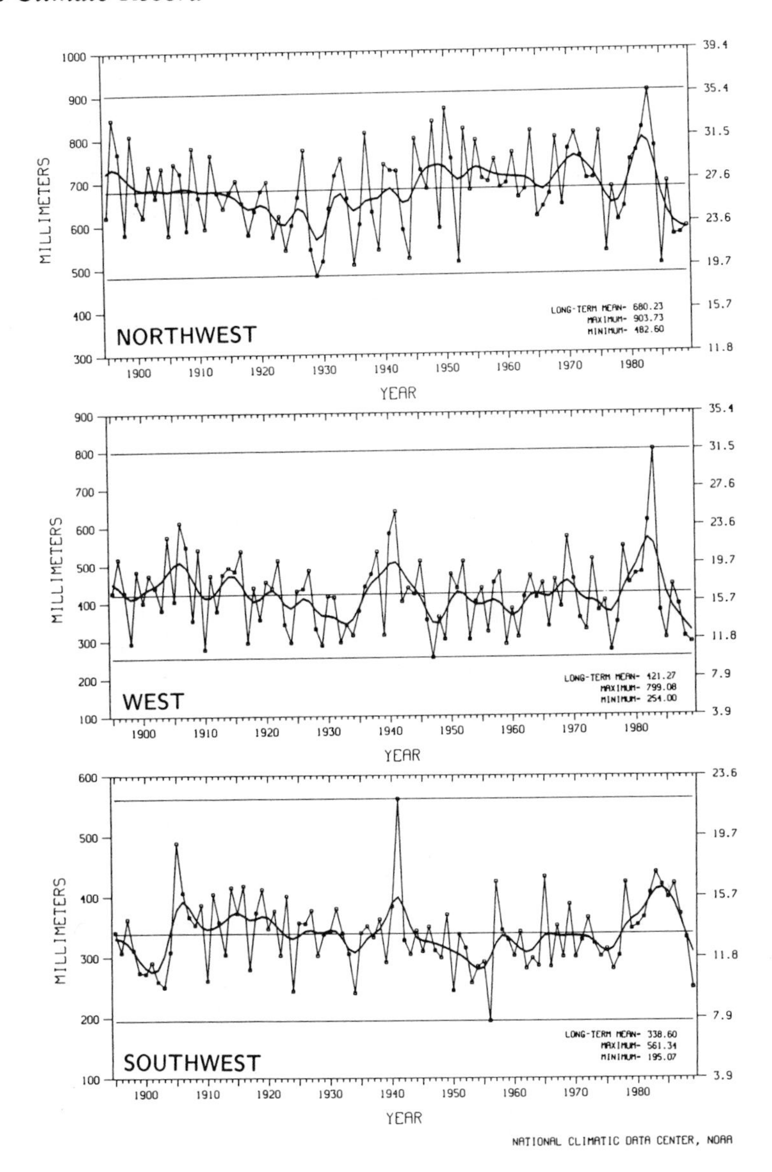

FIGURE 5.11 Time series of annual mean precipitation amounts in the Northwest, West and Southwest regions in mm (left) and inches (right). The long-term mean and extremes are given at the lower right and lines are plotted for each. A low pass smoother shows interdecadal fluctuations.
SOURCE: Courtesy of T. Karl.

In the West and Southwest, the annual mean precipitation is much lower but fluctuations of 10 to 20 percent are still common. Low-frequency (greater than decadal) fluctuations are less in evidence, but occasional extremely high rainfalls are recorded, in particular for the years 1941 and 1983 in both the Southwest and the West. Both of these years were times of major El Niño events (Figure 5.9). As a result, the rainfall distribution tends to be skewed, and more years tend to be below "normal" (i.e., the median is less than the average).

The net amount of water available for agriculture or hydrology, however, depends also on evapotranspiration. It is important to take this factor into account in assessing drought. Nationally, the Palmer Hydrological Drought Index is one measure of the cumulative effects of evapotranspiration and precipitation anomalies, and, for the United States, the 1930s and 1950s are seen as the main periods of drought (Figure 5.12).

Longer time series of droughts or precipitation have been reconstructed from tree ring analysis, such as that for Iowa (Figure 5.13), from Duvick and Blasing (1981). Although the method of reconstruction may remove very long-term trends, the reconstruction indicates that five decades in the past 300 years have been comparable to or drier than the 1930s. It appears that planning should take account of the expectation that such extreme droughts may occur about twice per century.

MAJOR FINDINGS FROM THE OBSERVATIONAL RECORD

Below is a summary of the major findings at this point from the observational record:

- The best estimates of global mean surface temperatures reveal increases of about 0.5°C over the past century.
- The main increase in Northern Hemisphere temperatures occurred between 1920 and 1940, with a temporary reversal after 1960 and a recovery to the highest levels in the 1980s.
- In the Southern Hemisphere, the main upward trend has been more steady, beginning after about 1930.
- El Niño events influence global mean temperatures by 0.1°C.
- The best estimates of global temperature change are not very good. Coverage is not global, especially prior to World War II.
- Large spatial variations exist in regional temperature trends associated with changes in atmospheric circulation.

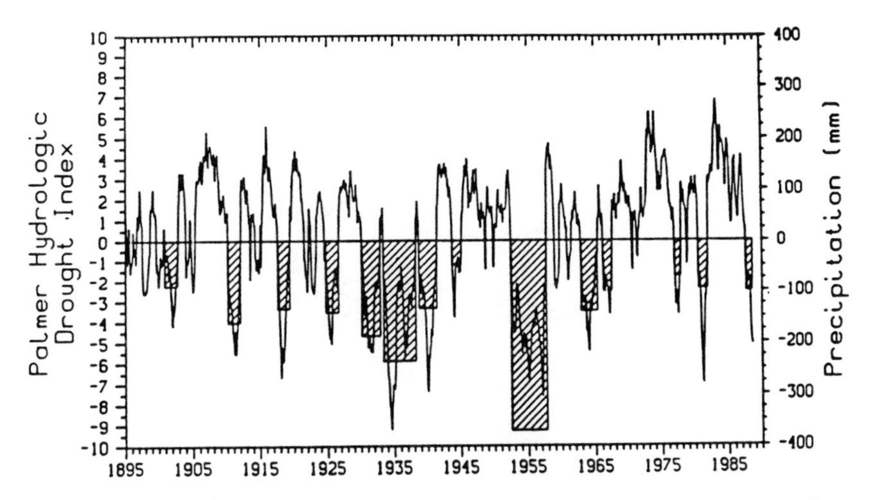

FIGURE 5.12 Palmer Hydrological Drought Index for the United States from 1895 to 1988. Periods of severe drought are cross hatched.
SOURCE: National Oceanic and Atmospheric Administration (NOAA), and National Climatic Data Center (NCDC), 1988.

- Substantial uncertainties exist in the overall trends, from sampling error and changes in instrumentation and measurement methods. Overall global trends are uncertain by about ±0.20°C per 100 years.

COMPARISON WITH MODEL RESULTS FOR GREENHOUSE GAS INCREASES

Climate models attempt to represent all the complex processes involved in and interactions between the atmosphere, ocean, and land (Figure 5.14) as mathematical expressions—representing the physical laws—that can be solved in a supercomputer. In surveying the modeling results for simulations with increased greenhouse gas concentrations, looking for common features, and assessing the reliability of different models, I have drawn the following conclusions concerning the model results and the observational record:

- The models indicate warming of 2 to 5°C for an equilibrium climate with doubled carbon dioxide. Effective doubling of carbon

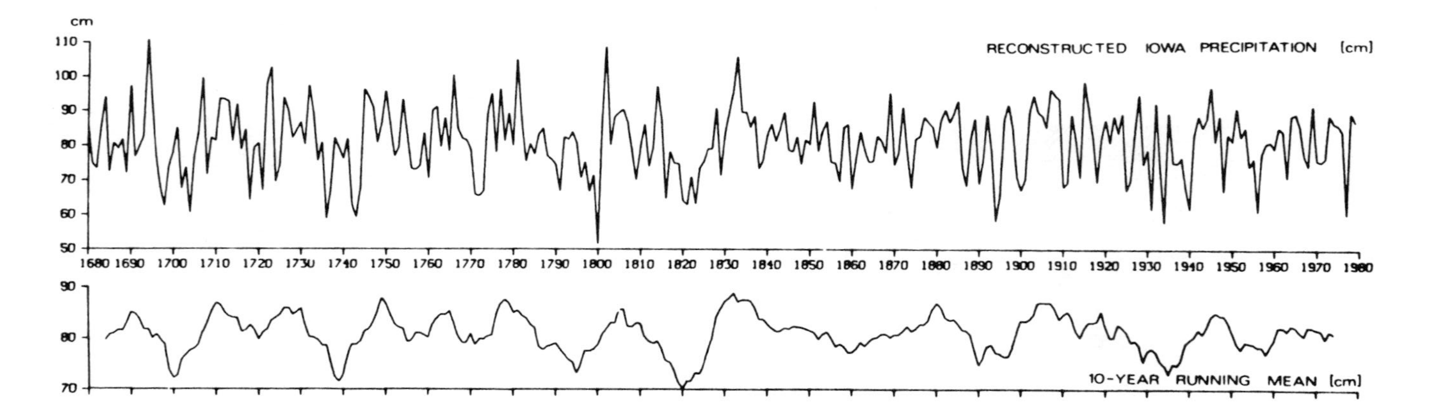

FIGURE 5.13 Reconstructed Iowa precipitation from tree rings from white oak.
SOURCE: Reprinted, by permission, from Duvick and Blasing (1981). Copyright ©1981 by the American Geophysical Union.

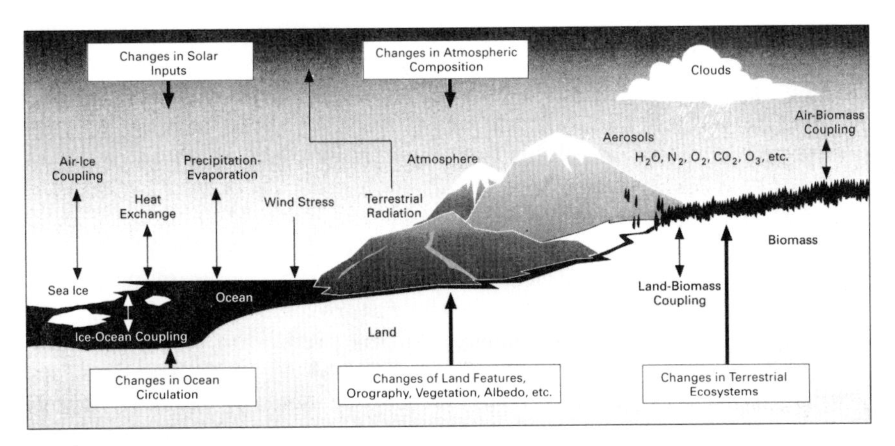

FIGURE 5.14 A schematic representation of the components of the climate system and the interactions between them which should be represented in climate models.

dioxide, taking into account the other greenhouse gases, will likely occur around the years 2030 to 2050, with corresponding manifestations of Northern Hemisphere climate change occurring 20 to 30 years later. Because of uncertainties in the models, particularly uncertainties about the role of clouds, I think the best estimate is at the low end, perhaps a 2°C warming for these conditions. The larger values do not appear to be tenable given the observational record unless the oceans are playing tricks on us. Moreover, a big factor in producing the larger values is a positive feedback in clouds, which is not proven. Nevertheless, the error bars on this number are significant. Also, substantial discrepancies exist between the observational temperature record and that expected from models—discrepancies in the timing of Northern Hemisphere warming in the 1920s and in spatial patterns between the Northern and Southern hemispheres and between the land and ocean.

• Increases in water vapor (a greenhouse gas) and decreases in snow cover and sea ice (lower albedo) provide positive feedbacks that should enhance the effects as time goes on. One result of this is that largest warming should occur in the Arctic. Evidence for an enhanced response in the Arctic can be found in the temperature rises from 1900 to 1940 but not in recent years. Significant trends in snow and ice do not emerge from the climate record, but the record is short.

• The hydrologic cycle is likely to speed up by about 10 percent with carbon dioxide doubling, causing increased evaporation and increased rainfall in general. Because more precipitation is apt to fall as rain in winter and because snow melt is likely to be faster in spring, soil moisture availability going into summer over mid-latitude continents is likely to be less. Combined with increased evapotranspiration in summer, this implies that any natural tendency for a drought to occur is likely to be enhanced. Increases in rainfall in winter with drier summer conditions imply that water management will become a prominent issue in the future. However, there is not good agreement among the models on this prediction. Precipitation is well known only over land. Real regional decadal variations appear to exist, but overall trends are quite uncertain. Meaningful comparisons between model results and the observational record are difficult to come by.

• Because warming results in the expansion of the ocean and the melting of snow and glacial ice, one real threat is a rise in sea level. However, there may be some compensation through increased snowfall on top of the major ice caps (in Greenland and Antarctica) so that they could increase in height even as they melt around the edges. Sea level has risen by 1 to 1.5 mm per year or 2.4±0.9 mm per year when corrected for isostatic adjustment—values that are reasonably consistent with alpine glacier melting and thermal expansion of the upper ocean. The rate of sea level rise should increase, but the main impacts will likely be felt much later next century.

• Stratospheric cooling is another likely effect of increased greenhouse gas concentrations, and this has important implications for ozone, because the kind of heterogeneous chemistry responsible for the Antarctic ozone hole is more effective at lower temperatures. The chlorofluorocarbons that are responsible for the ozone hole are also a greenhouse gas. Large interannual variability confounds temperature trend detection in the lower stratosphere. The record is fairly short (30 years) and inhomogeneous. Detection of greenhouse effects is complicated by ozone-induced changes.

• Another possibility, because of increased SSTs, is increased frequency and/or increased intensity of tropical storms and hurricanes. This is because SSTs above about 27°C are needed before hurricanes can sustain themselves by feeding on the extra water vapor and latent heat available in such regions.

• Greatest confidence exists on the global scale; regional climate changes are much more uncertain.

A central task in climate research over the next few years will be to reduce these uncertainties and thereby provide more reliable projections into the future. At the same time, we must recognize that climate has changed throughout history on its own through natural variability and that there is a large "noise" component due to weather that can mask, for a considerable time, any man-made effects. At present, any planning must recognize these factors and uncertainties.

ACKNOWLEDGMENTS

Tom Karl kindly provided material for Figures 7, 11, and 12.

REFERENCES

Duvick, D., and T. J. Blasing. 1981. A dendroclimatic reconstruction of annual precipitation amounts in Iowa since 1680. Water Resources Research 17:1183-1189.

Farmer, G., T. M. L. Wigley, P. D. Jones, and M. Salmon. 1989. Documenting and Explaining Recent Global-Mean Temperature Changes: Final Report to NERC. University of East Anglia, U.K.: Climate Research Unit.

Karl, T. R. 1988. Multiyear fluctuations of temperature and precipitation: The gray area of climate change. Climatic Change 12:179-197.

Karl, T. R., L. K. Metcalf, M. L. Nicodemis, and R. G. Quayle. 1983. Statewide Average Climatic History. Historic Climatology Series 6.1. Asheville, N.C.: National Climatic Data Center, National Oceanic and Atmospheric Administration.

Keeling, C. D., R. B. Bacastow, A. F. Carter, S.C. Piper, T. P. Whorf, M. Heimann, W. G. Mook, and H. Roeloffzen. 1989. A three dimensional model of atmospheric $C0_2$ transport based on observed winds, part I: analysis of observed data. Pp. 165-236 in D. H. Peterson, ed., Aspects of Climate Variability in the Pacific and the Western Americas. Washington, D.C.: American Geophysical Union.

National Oceanic and Atmospheric Administration (NOAA) and National Climatic Data Center (NCDC). 1988. Drought of 1988 and Beyond. Proceedings of a Strategic Planning Seminar, October 18. Washington, D.C.: National Academy of Sciences and U.S. Department of Commerce.

6

Climate Change and Climate Variability: The Paleo Record

David Meko, Malcolm Hughes, and Charles Stockton
University of Arizona, Laboratory of Tree-Ring Research
Tucson, Arizona

INTRODUCTION

The written and instrumental record is far too short to give an adequate account of the range of possible behavior of the earth's climate. Instrumental climate records in many regions extend back only a few decades, and direct observations of ecological or geomorphological processes are usually even shorter. These records are all limited to the period since atmospheric carbon dioxide started its upward climb in the early nineteenth century. A window on the preindustrial environment is needed. The techniques of paleoecology and paleoclimatology offer that window and have yielded unique insights to the behavior of the earth's systems.

The geological record reveals that the earth's climate has differed radically from today's during the earth's history. The study of these past conditions provides insights into the possible range of behavior. It has also provided important tests for our understanding of the causes of climate change. It has been possible to use models of climate, such as general circulation models (GCMs), to calculate the expected conditions on our planet at times when its geography was quite different from today's—when, for example, there was only one huge continent, or when the atmosphere had a different composition. Comparison of these expected conditions with those actually revealed by the geological record has helped advance knowledge of the mechanisms driving climate change.

Climate variations on time scales of decades to centuries are particularly important to consider for water resources management. The paleo record has contributed greatly to our understanding of long-term climate variations and enabled quantitative reconstruction of hydrologic variables with annual resolution. In this paper, we first review highlights of climate history as gleaned from the paleo record. We then provide an example of application of tree-ring data to the study of hydrologic variability in the western United States.

WHAT HAS BEEN FOUND SO FAR FROM THE STUDY OF PAST CLIMATE?

The greatest changes have occurred over the longest time scales: millions to hundreds of millions of years. For much of its four-billion-year history, the earth has been relatively warm and free of ice sheets such as those now found in Greenland and Antarctica. Yet, there have been major changes that happened relatively rapidly. For example, there is evidence of major cooling about 37 million years ago, when a long ice-free period ended, and 2.4 million years ago. This second change marked the start of the generally cooler epoch (the Quaternary) that has continued to the present. The best documented processes leading to such major long-term shifts have to do with changes in the distribution of continents and oceans, the rise and decline of major mountain ranges, and variations in sea level. Changes in the relative fluxes between the biological and geological components of the carbon cycle have also played a part by modifying the carbon dioxide concentration of the atmosphere. Such changes in the processes regulating global climate may be thought of as changes in the boundary conditions of the climate system. Large changes in these boundary conditions have taken place on time scales of hundreds of thousands, millions, and tens of millions of years. It is possible that some have occurred more rapidly, but many of the dating techniques used do not possess the resolution necessary to reveal rapid change in the distant past.

In the relatively cool world of the most recent 2.4 million years, the distribution of continents and relief has been much as at present. Since about 875,000 years ago, the climate has undergone repeated major excursions between long ice ages (about 110,000 years each) and much shorter periods of warmer climate (5,000 to 15,000 years). Such a warmer period commenced about 15,000 years ago.

Ocean Sediments

The main evidence for these periodic climate fluctuations

comes from the floor of the deep ocean, which holds sediments made up of the remains of various microscopic animals. The shells of one group, the foraminifera, are made up of calcium carbonate containing oxygen derived from the seawater. Cores have been collected from these sediments at many locations in the oceans, the deepest sediments in each core being the oldest. Hence, we have access to oxygen that was part of the ocean's water at some known time in the distant past. The proportion of the heavy isotope of oxygen (^{18}O) to the more common isotope (^{16}O) in the world's oceans has a strong link to the total volume of water locked up as ice in glaciers and the polar caps. This is because the heavy oxygen is left behind disproportionately when water evaporates from the ocean surface. When this water falls onto the glacier ice as snow, it is correspondingly short of heavy oxygen. As the global volume of ice grows, so the small proportion of heavy oxygen in the ocean water increases. Hence, we can calculate global ice volume from the record of the ratio of the oxygen isotopes ($^{18}O/^{16}O$) in foraminiferan shells from dated ocean sediments.

This record of ice volume changes (Figure 6.1) has been confirmed many times (Emiliani, 1955; Imbrie et al., 1984), not only from analysis of ocean sediments but from studies of the sea level recorded by coral reefs and land-based records such as those from loess. Loess is the deposit formed by fine, wind-blown dust derived from cold desert areas such as the Gobi Desert. Such cold, dry areas are found around the major ice sheets of the ice ages, and so there are regions (e.g., part of Czechoslovakia) where loess is deposited during ice ages but not during the warmer periods between them (Kukla, 1977). The number and timing of such loess deposits was found to coincide with the periods of greatest ice volume calculated from the ocean $^{18}O/^{16}O$ record, providing powerful evidence for the validity of those calculations.

Milankovitch's Four Frequencies

Just as it is possible to separate out the contributions of notes of different pitches or frequencies to a musical sound, so it is possible to break down the global ice volume record into its frequency components (Imbrie et al., 1984). It turns out that there are four major frequencies in the record: one every 100,000 years; one every 41,000 years; one every 23,000 years; and one every 19,000 years. These are the very periodicities predicted for the earth's climate by Milankovitch (1941) as recalculated by Vernekar (1972)

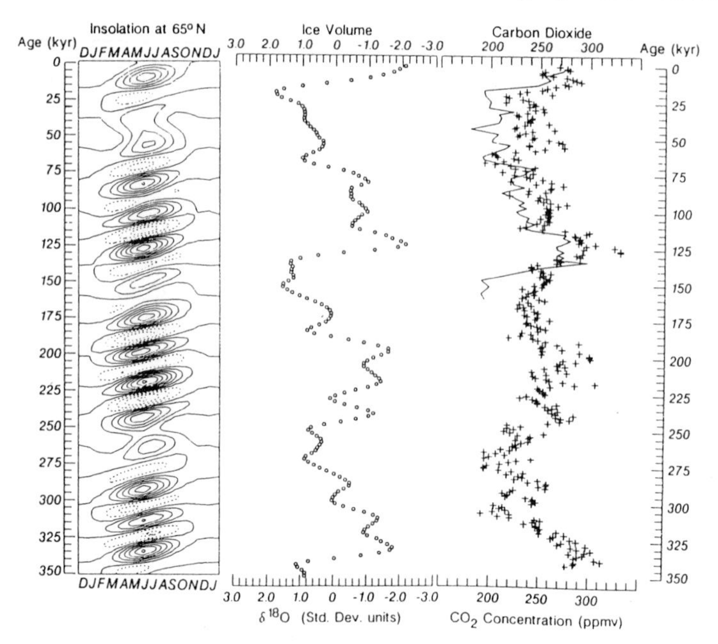

FIGURE 6.1 Central panel: Global ice volume estimated from oxygen-18 deficit in foraminiferan shells over the last 350,000 years. Left panel: Calculated insolation at 65 degrees north (contour interval 10 Wm^{-2}, dashed lines negative). Right panel: Estimated atmospheric carbon dioxide concentration (+, from isotopic ratios in marine sediments; continuous line, as measured in Vostok ice core).
SOURCE: Reprinted, by permission, from Bartlein and Prentice (1989). Copyright © 1989 by Elsevier Science Publishers, Ltd.

and Berger (1979). Milankovitch showed that the amount of solar energy reaching the top of earth's atmosphere varied slightly on these time scales as a result of changes in the way the earth revolves around the sun. The 100,000-year periodicity in solar receipts is driven by changes in the eccentricity ("stretch") in the earth's orbit, the 41,000-year periodicity by changes in the obliquity ("tilt") of the earth's axis, and the 19,000- and 23,000-year changes by the precession ("wobble") of the axis.

The fit between the global ice volume record and these variations in solar receipts calculated from celestial mechanics is not perfect. In particular, the earth's climate as recorded by the foraminifera shows a much stronger periodicity at 100,000 years than the astronomical calculations would predict. Further, the actual changes in the amount of energy received from the sun are small—unlikely that these changes in the earth's orbit alone are big enough to explain the onset and end of the repeated ice ages of the Quaternary; the greater-than-expected observed variation at the 100,000-year periodicity is also a puzzle. On the other hand, the fit between the timing of the orbitally induced changes in solar receipt at high northern latitudes and the major features of the global ice volume record suggests at least a pacemaker role for the orbital variations. It is important to remember that the orbital or "Milankovitch" changes produce marked variations in the seasonal pattern of solar receipts at different latitudes. This will be referred to below in a discussion of the last 20,000 years. The recurrent glaciations of the Quaternary may well have been associated with major extensions equatorward of the mid-latitude arid regions (Dickinson and Virji, 1987; Lézine, 1989). As land surface conditions in arid regions change, so does their contribution to suspended dust in the atmosphere. An extensive effort to analyze ocean cores for dust originating on the continents now underway (Rea and Leinen, 1988) indicates major changes in the atmospheric transport of dust associated with changes between glacial and interglacial periods. Some possible explanations for the importance of the 100,000-year periodicity have to do with the growth, inertia, and effects of the major continental ice sheets associated with the ice ages.

Polar Ice

Remaining polar ice provides the most remarkable record of past climate, particularly for the last glacial cycle, which started more than 110,000 years ago. The snow falling on the central parts of the Greenland or Antarctic ice sheets is trapped for hundreds of thousands of years because the temperature at the base of the ice is too low to allow melting. Hence, by taking core samples from the top of the ice downward, it is possible to sample snow (now ice) that fell in the distant past. A number of techniques are used to calculate the age of ice at a particular depth, including mathematical models of the mechanical processes taking place within the ice. As discussed above, the ratio of the oxygen isotopes ($^{18}O/^{16}O$) in the falling snow and hence in the ice is related to general climate conditions. Both this ratio and that between heavy hydrogen, or deuterium, and ordinary hydrogen ($^{2}H/^{1}H$ or D/H) correlate well with temperatures above the ice at the time the snow fell. The D/H record has been used to construct a record of temperature at the Soviet Vostok station in Antarctica that extends back about 160,000 years, to the end of the ice age before last (Lorius et al., 1990). A range of temperatures of 5 to 6°C over Antarctica is reconstructed. This record corresponds to a remarkable degree to the record of global ice volume derived from foraminiferan $^{18}O/^{16}O$ in ocean cores (Figure 6.2). Not only does the

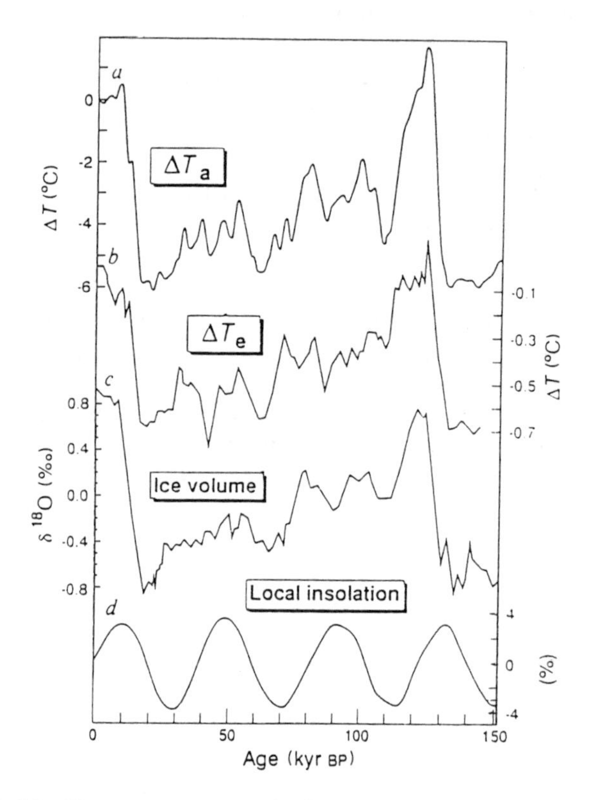

FIGURE 6.2 Temperature deviation, ice volume, and local insolation at Vostok station in Antarctica over the last 150,000 years. Top curve: Temperature (shown as the deviation from modern temperature) calculated from ice D/H ratios. Second curve: Temperature deviation calculated from the greenhouse effect of carbon dioxide and methane. Third curve: Ice volume, derived from the oxygen isotope ration in formaniferan shells of ocean sediments. Lower line: local insolation at 78 degrees south.
SOURCE: Lorius et al. (1990). Reprinted, by permission, from NATURE Vol. 347 pp. 139-145. Copyright © 1990 Macmillan Magazines Ltd.

polar ice contain ancient water, but trapped within it is the air that was circulating over the ice sheet at the time the snow fell and became transformed to ice. This air can be extracted from the ice and analyzed for gases such as carbon dioxide and methane. Figure 6.3 indicates that the atmospheric concentrations of these

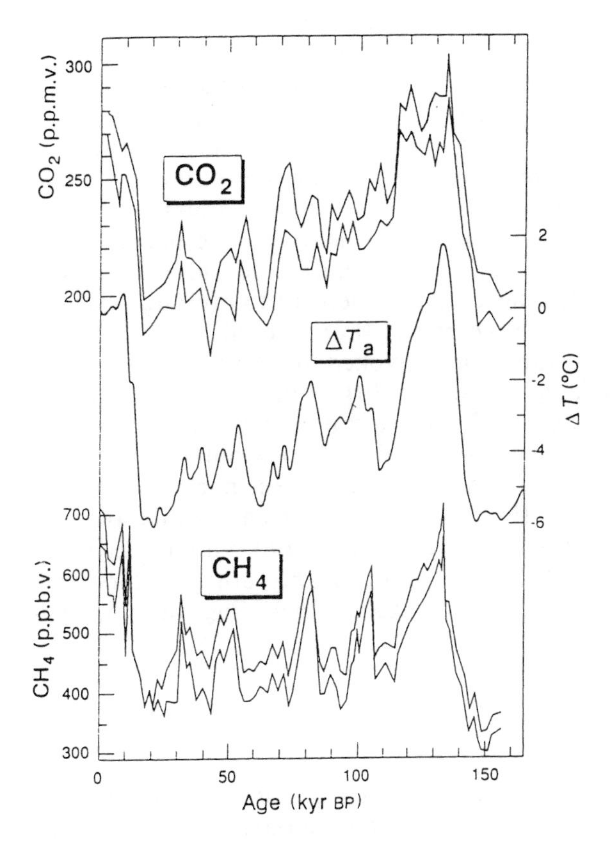

FIGURE 6.3 Temperature deviation, carbon dioxide concentration, and methane concentration at Vostok station, Antarctica, over the last 160,000 years. Middle curve: Temperature deviation from modern temperature calculated from ice D/H ratios. Top curve: Reconstructed atmospheric carbon dioxide concentration. Lower curve: Reconstructed atmospheric methane concentration.
SOURCE: Lorius et al. (1990). Reprinted, by permission, from NATURE Vol. 347 pp. 139-145. Copyright © 1990 Macmillan Magazines Ltd.

gases have shown variations remarkably similar to those of temperature over that last 160,000 years. The range of concentrations of atmospheric carbon dioxide—from close to 300 parts per million by volume (ppmv) in the warm periods between ice ages to as low as 180 pp mv in the depths of an ice age—is particularly striking. A calculation of the direct greenhouse effect (with no climate feedbacks) produced by the observed concentrations of carbon dioxide and methane shows that this can account for between 40 and 65 percent of the observed Antarctic temperature change (Lorius et al., 1990) (Figure 6.2).

The most recent ice age ended suddenly, having reached its most severe depths around 18,000 to 20,000 years ago. There were extensive continental ice sheets in both the Northern and Southern hemispheres that started to shrink approximately 14,000 years ago (Broecker and Denton, 1990). The great North American ice sheet centered on the Canadian Shield had disappeared by about 7,500 years ago. Atmospheric carbon dioxide and methane concentrations had increased more than 25 percent to preindustrial levels between 11,000 and 9,000 years ago (Lorius et al., 1990). At the same time, the orbital variations invoked by Milankovitch were producing marked changes in seasonality at high latitudes (more or less equal and opposite in the Northern and Southern hemispheres). At around the time of the glacial maximum (approximately 20,000 years ago), solar receipts at 60 degrees north in summer and winter approximated those of the present day, while around 9,000 years ago the contrast between summer and winter was markedly stronger (Figure 6.4). These and probably other factors produced major changes in regional climates, including that of western North America.

The Western Climate

The middens of pack rats provide invaluable information on past environments in arid and semi-arid regions such as most of the West. These small mammals collect plant materials from a very limited area (about 30 meters in radius). The plant fragments (e.g., twigs, leaves, needles, fruits, and seeds) are often preserved in crystallized urine called "amberat." If identifiable remains of a plant are found in a midden, they constitute firm evidence that the species was present locally. This has made it possible to reconstruct the vegetation of the West over tens of thousands of years, with the help of radiocarbon dating.

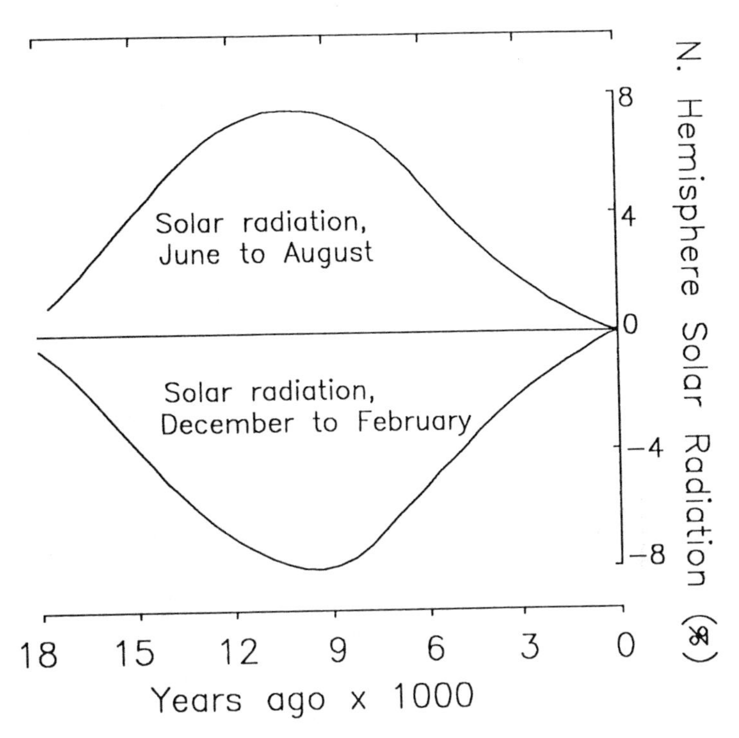

FIGURE 6.4 Insolation at 60 degrees north since the last glacial maximum (after COHMAP, 1988).

Betancourt (1990) has brought these records together in a review of the vegetation history of the Colorado Plateau (Figure 6.5). He demonstrates major changes in plant species distributions between the Late Glacial (15,000 to 11,000 years ago) and Late Holocene (4,000 to 800 years ago) periods. During the Late Glacial period, the upper tree line was several hundred meters lower than at present and boreal forest (spruce and true fir) was found some 900 meters lower. The vegetation patterns were not, however, simply shifted downhill in response to a colder climate. Ponderosa pine (*Pinus ponderosa*) and Colorado pinyon (*Pinus edulis*) are major features of the present vegetation of the Colorado Plateau, but both were absent in the Late Glacial period. Ponderosa was replaced by limber pine (*Pinus flexilis*) in many cases. The absence of ponderosa pine may have been the result of cooler summers and orbitally determined lower insolation in the latter part of the growing

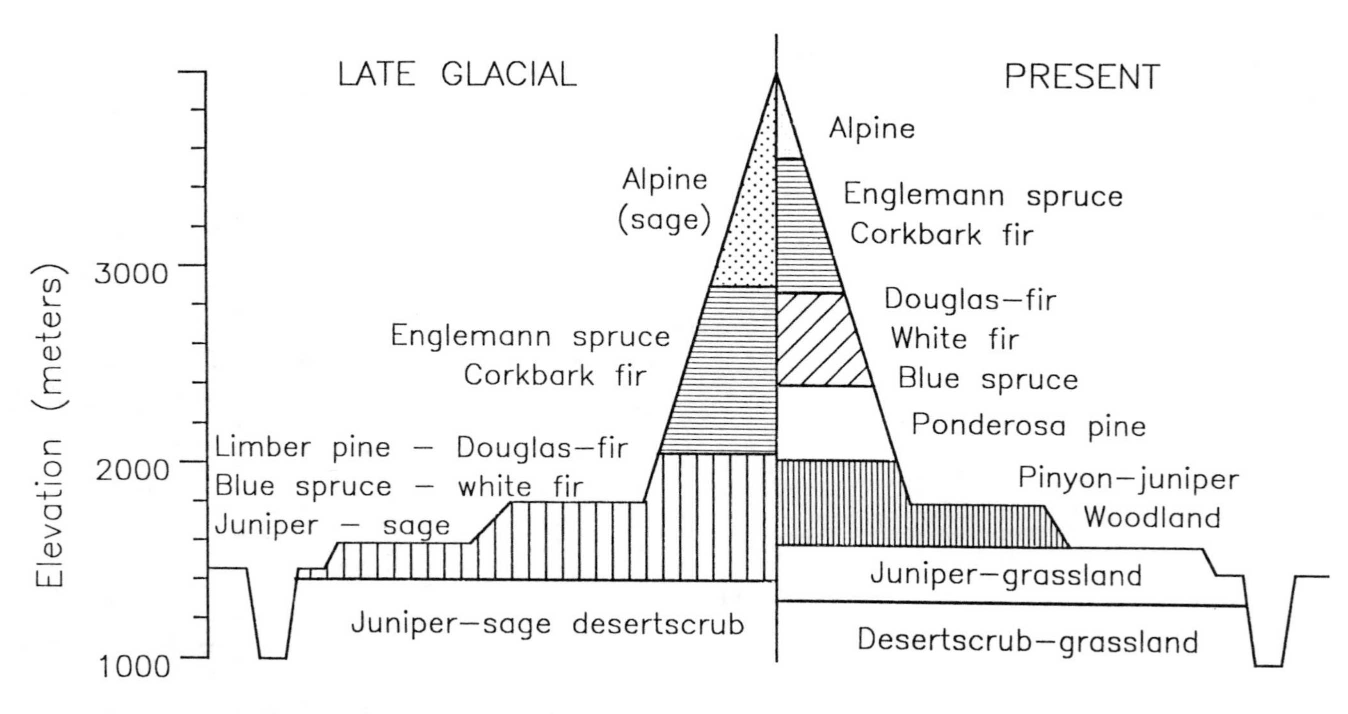

FIGURE 6.5 Generalized late-glacial and present plant zonation on the Colorado Plateau (after Betancourt, 1990).

season. Utah juniper (*Juniperus osteosperma*) was present, probably because of its greater cold tolerance than Colorado pinyon. Betancourt inferred from these and other plant species distributions that the Late Glacial summers on the Colorado Plateau were more than 6°C cooler than at present (Betancourt, 1990)—about the same temperature difference as noted from ice D/H ratios at Vostok in Antarctica. He indicated that the relative proportion of annual precipitation falling in summer (less than 10 percent) was even smaller than in recent times (20 to 30 percent). Note that different climates do not simply have different means, but also different variabilities and seasonalities.

Other evidence of past climate in the West may be extracted from the chemical composition of plant remains found in dated pack rat middens. Just as the D/H ratio in polar ice reflects certain environmental temperatures, so the same ratio in plant cellulose is related to that in the plant's source water and hence, in part, to growing-season temperatures. Long et al., (1990) measured plant cellulose D/H ratios in materials from pack rat middens in the Great Basin and on the Colorado Plateau and from them inferred July temperatures over more than 30,000 years. The general form of the record is very similar to that calculated from a GCM, although the GCM calculated that the difference between present and ice-age temperatures (about negative 2.5°C) is lower than other evidence would suggest.

It has been possible to reconstruct conditions at several times in the last 20,000 years over much of the world by combining information on the composition of marine plankton from ocean cores, pollen from lakes and bogs, plant remains from pack rat middens, and natural records of the water levels in lakes (COHMAP, 1988). Twelve thousand years ago (during the Late Glacial period), with extensive ice sheets still present, such a reconstruction shows that the whole of the American West was wetter than it is now. Between 9,000 and 6,000 years ago, the Pacific Northwest was drier than at present, whereas the Southwest was wetter. Summer temperatures were between 2 and 4°C higher in the West, and winters were probably cooler. The Arizona monsoon started to appear about 9,000 years ago. GCM results using the known boundary conditions (orbital position, carbon dioxide content of air, global ice volume, and sea surface temperatures) at these times produce similar patterns of regional climate. This gives increased confidence in the physical understanding used to build GCMs and hence in their usefulness when considering the effects of changed boundary conditions on the climate system.

Rapid Climate Shifts

Superimposed on the relatively smooth global shift into (gradually) and out of (rapidly) repeated ice ages or glaciations, there have been shorter lurches of widespread, if not global, extent. The most studied of these is an event known as "the Younger Dryas." During the rapid emergence from the last glaciation about 11,000 years ago, northern Europe and northeastern North America plunged back from interglacial conditions broadly similar to today's to the chill of an ice age (a temperature decline of about 6°C). This may have taken only 100 years or less. The period is named for *Dryas octopetala*, a plant of the high Arctic that appeared with other Arctic plants in parts of Europe where forest formerly stood. Perhaps no more than 1,000 years later, this massive "cold snap" ended, maybe even more rapidly than it started. There is tantalizing, but inconclusive, evidence that other major climate changes occurred at the same time elsewhere, for example on the Loess Plateau of China. Even if the Younger Dryas event was strictly a North Atlantic affair, it is of enormous significance because of its scale and speed. Our present state of knowledge would not enable us to anticipate such a change even if it were to start next year. One persuasively argued explanation links the Younger Dryas to massive changes in ocean circulation induced by a change in the point at which most of the meltwater from the retreating North American ice sheet emptied into the sea (Broecker and Denton, 1990). This explanation requires the Atlantic circulation to shift rapidly between alternate modes, resulting in sudden changes in the northward transfer of heat by the ocean that currently gives the North Atlantic periphery climates that would otherwise be found ten degrees further south. It is possible that other such sudden and drastic changes in the last 15,000 years will be found elsewhere as other regions are studied as intensively as Europe and eastern North America.

Other drastic changes, albeit more short lived (lasting 1 to 10 years) than the Younger Dryas, may be associated with very large explosive volcanic eruptions or groups of eruptions. There is very strong evidence for much larger and climatically effective eruptions at various times in the last 10,000 years than have occurred during times of historical record (Hammer et al., 1980; Baillie and Munro, 1988). Even if we understood and could predict the effects of such an eruption on climate, we are unlikely to be able to forecast the eruption itself.

Less drastic changes in climate, lasting one to a few centuries, have been discussed by many authors. The Little Ice Age, a cooler period starting some time between A.D. 1450 and A.D. 1650 and ending perhaps as late as A.D. 1890 (Grove, 1988) has been discussed as if it were a global cooling of about 1.5°C. In fact, the quality of the evidence for this is very mixed and is much more dense in some parts of western Europe and eastern North America than elsewhere. Support comes from traces of glacier advance and retreat in some regions (Denton and Karlen, 1973), oxygen isotope records from annually layered ice caps in Peru and Tibet (Thompson and Mosley-Thompson, 1990), temperature-sensitive tree rings of high elevation trees in California (LaMarche, 1974) and Washington (Graumlich and Brubaker, 1986), and from boreal forest trees in Canada (Jacoby and d'Arrigo, 1989) and the polar Urals (Graybill and Shiyatov, 1989) but not, for example, from tree rings in northern Scandinavia (Briffa et al., 1990). Little evidence is available for lower latitudes or the oceans. Another much-discussed global change is the "Medieval Warm Epoch." There is certainly evidence for warmer conditions in the circum-North Atlantic region at times between A.D. 800 and A.D. 1350 and some indications of the same effect in polar and high-elevation ice cores elsewhere. As in the case of the Little Ice Age, the evidence is far from complete. It is not yet possible to say with confidence whether these much-discussed cold and warm periods were really global or whether the suggestion of a name and a date, usually from European experience, has attracted evidence that really records unconnected events.

In order to resolve these problems it is necessary to use records of environmental conditions that are reliably dated to the calendar year. The most extensive such natural record is provided by the annual rings of trees in the temperate and boreal forests. In dry regions, the width or thickness of the annual ring is often controlled by the availability of moisture, whereas in cool, moist regions ring width or maximum wood density records summer temperatures. An extensive literature documents the techniques used to extract climate information from tree rings (Fritts, 1976; Hughes et al., 1982; Cook and Kairiukstis, 1990). The direct application to hydrologic problems of records of tree growth derived from tree rings is of particular relevance to the topic of this meeting. An example of such a study follows.

TREE RINGS AS HYDROLOGIC INDICATORS

There are better natural records of climate of the last few hundred years in the western United States than anywhere else on earth. These include an extraordinary wealth of climatically sensitive tree-ring records that can be used to place the instrumental period in a wider perspective.

The physical basis for using tree rings as hydrologic indicators in semiarid regions is well understood. The most frequently used tree-ring variable in drought and hydrologic studies has been the ring-width index, a measure of departures from normal of annual diameter growth of the tree. Both the growth increment of a tree and the annual or seasonal flow of a river are closely related to the water balance of the soil integrated over days, weeks, or months. Statistical studies have repeatedly shown that hydrologic variables and annual growth indices from properly selected trees are highly correlated (Schulman, 1956; Stockton, 1975; Smith and Stockton, 1981; Cleaveland and Stahle, 1989).

Spatial Patterns of Drought From Tree-Ring Networks

Where the spatial coverage is sufficiently dense and time coverage sufficiently long, networks of tree-ring data can convey important hydrologic information on the joint space-time variation of moisture anomalies. Without tying the tree-ring patterns to a specific hydrologic variable, we can infer the year-by-year development of drought patterns by mapping tree-ring indices. Spatial and temporal coverage by tree-ring data is especially favorable for such an analysis in the southwestern United States, where some 121 tree-ring sites provide continuous coverage for the years A.D. 1600 to A.D. 1962. The cut-off years for this period are dictated largely by the age distribution of suitable tree-ring sites and the history of field collections in the Southwest. Species of dubious quality for drought reconstruction (e.g., bristlecone pine from high elevations) can be excluded from the analysis.

To summarize drought patterns using tree-ring data, we divided the southwestern United States into a grid of 35 cells, each with a dimension of 3 degrees latitude by 2 degrees longitude (Figure 6.6), grouped tree-ring sites by their enclosing cell, and averaged individual series in each cell together. We then analyzed the resulting 28 "cell-average" tree-ring series (seven cells contained no tree-ring sites) to produce annual maps of relative growth anomalies in two of the more severe multiyear droughts of the A.D. 1600 to 1962 period: A.D 1667 to 1670 and A.D. 1843 to 1848. These droughts are particularly relevant for their widespread coverage of runoff-producing regions. The first drought, A.D. 1667 to 1670, has been reported as particularly extreme in the watersheds of the upper Colorado River and

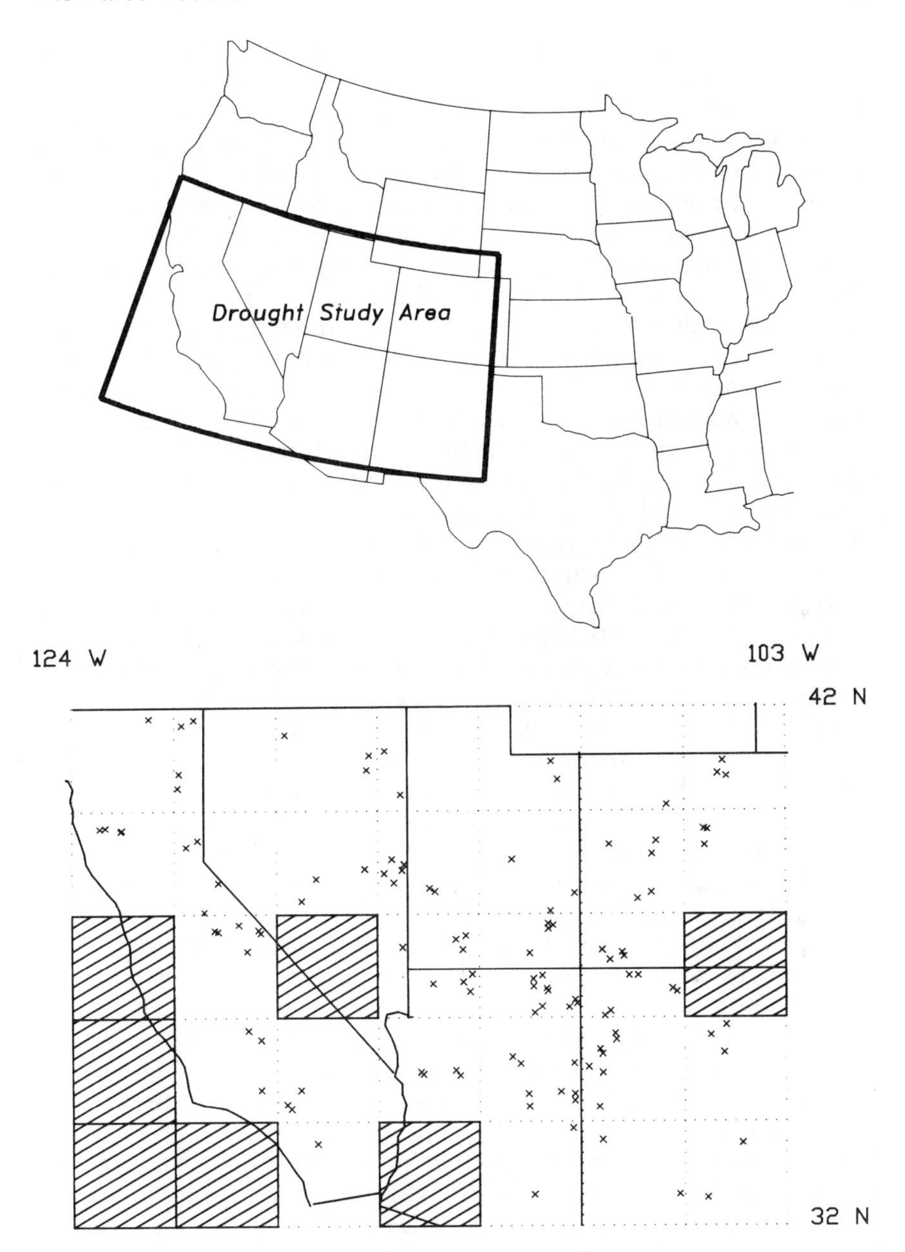

FIGURE 6.6 The grid network we used to group tree-ring sites (designated with an *x*) and tree ring sites. Cells containing no suitable tree ring chronologies are hatched.

the Salt and Verde rivers, which drain the central and eastern highlands in Arizona (Stockton, 1975; Smith and Stockton, 1981). The second drought, A.D. 1843 to 1848, has been noted for its severity in reconstruction of drought and streamflow in the Four Rivers area of the Sierra Nevada of northern California (Earle and Fritts, 1986). This area is a major source of water supply to southern California.

The annual cell-average tree-ring series for the two droughts have been coded on maps by circles of varying size within each cell (Figures 6.7 and 6.8). The radius of a circle is proportional to the exceedance probability, p, of growth index for the year if the index is less than the long-term median, or to 1-p if the index is greater than the long-term median. Circles have been scaled such that the largest or smallest growth value in the 1600 to 1962 period yields a circle with a diameter equal to the width of the cell. In terms of inferred moisture conditions, therefore, a dotted circle filling the cell would indicate the driest (lowest growth) year on record, while a hatched circle would indicate the wettest year on record. No circle (zero radius) implies median moisture conditions.

A practical hydrologic consideration in attempting to quantify drought regionally is the spatial distribution of droughts relative to major runoff-producing areas. The maps in Figures 6.7 and 6.8 clearly show that the spatial scale of drought in individual years of the 1660s and 1840s droughts was generally smaller than the entire Southwest. In most years, therefore, severe deficits in runoff would not be expected over all major runoff-producing areas simultaneously. For example, the Sierra Nevada of northern California appear to have been normal or wetter than normal in 1669 and 1670, when severe drought appears to have occurred in the Colorado Rockies and the central Arizona highlands.

In the 1840s drought, however, the year 1847 stands out as an exception to this generalization and points to the possibility of synchronous severe drought over all major watersheds of the Southwest. The unusually extensive drought of 1847 appears to be imbedded in a generalized 6-year drought that would best be characterized as a "Far West" drought. A temporal pattern to the drought of the 1840s is hinted at: anchoring in the Far West in 1843 and 1844, shifting inland in 1845 so that the extreme northwestern part of the study area was out of the drought pattern, returning to the Far West mode in 1846, expanding dramatically to cover the entire Southwest in 1847, and again shifting to the Far West in 1848. A steep northwest-southeast gradient to wetter than normal conditions is inferred toward the far northern part of the West region. A persistent ridge, probably narrow in longitudinal extent, cells for

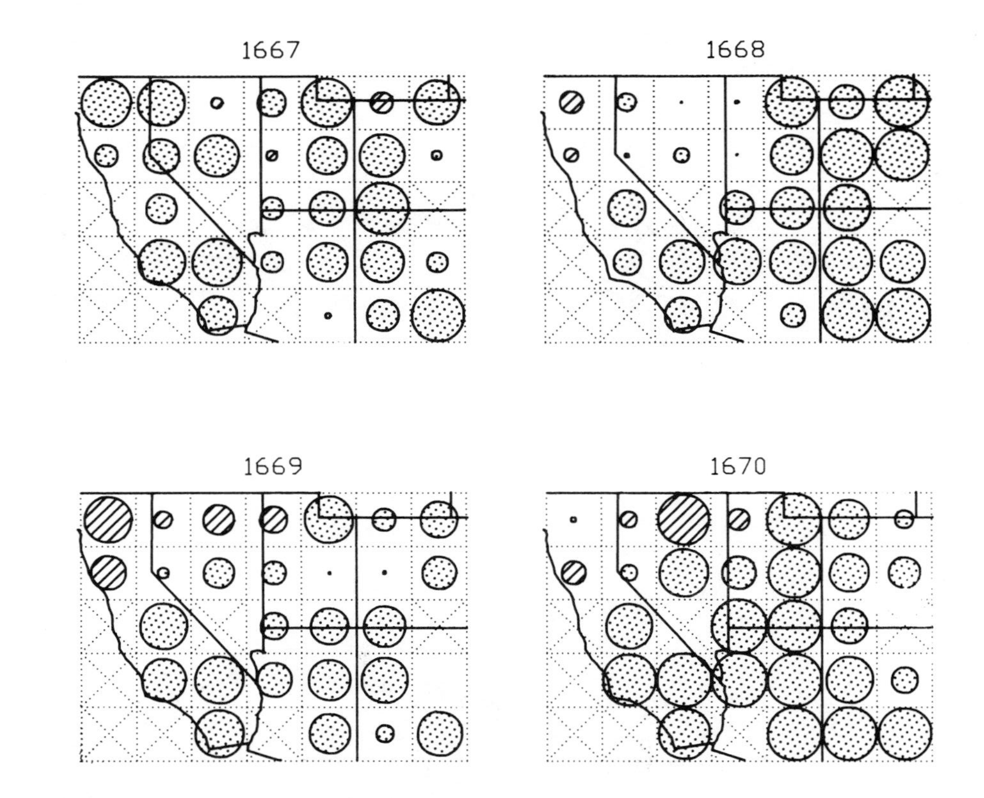

FIGURE 6.7 Maps showing relative departures of growth by grid individual years of the 1660s drought. Low-growth anomalies are dotted; high-growth anomalies are hatched. Scaling of circles is discussed in the text.

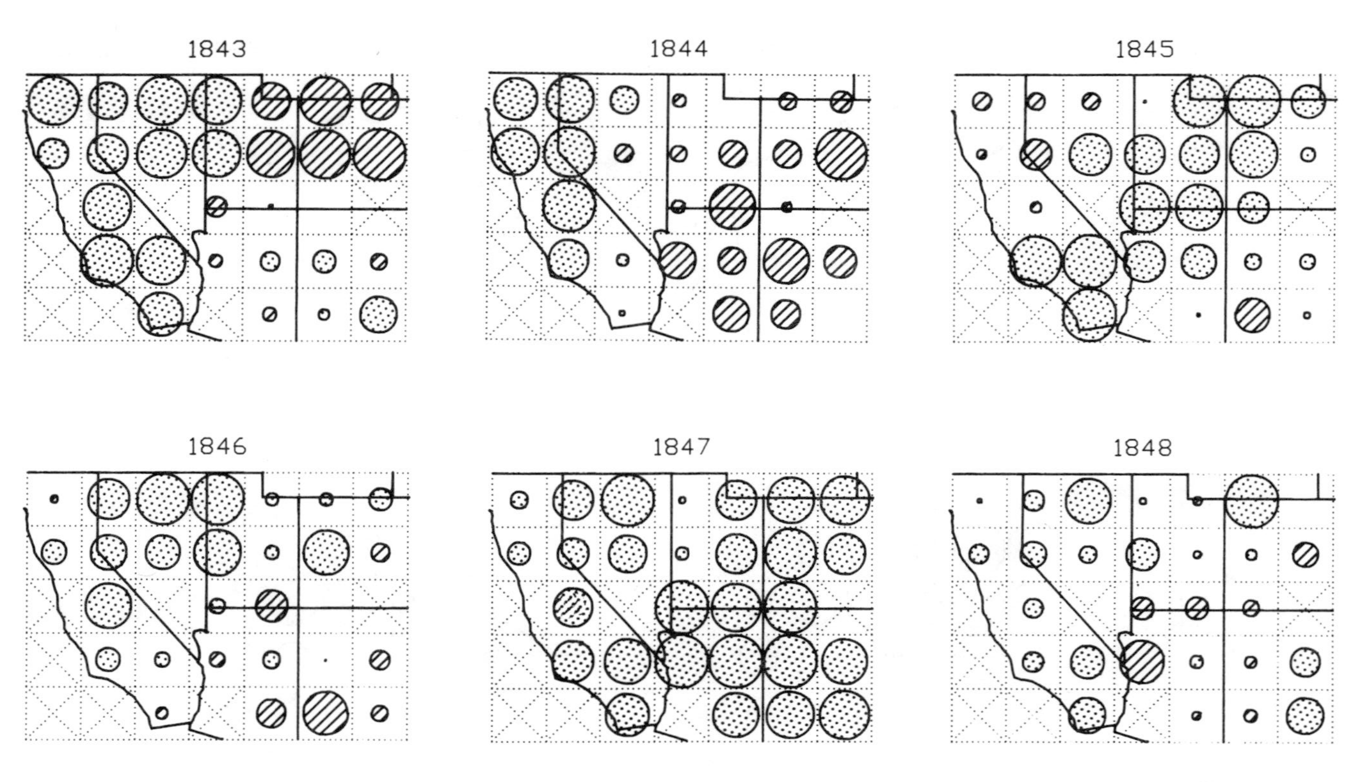

FIGURE 6.8 Map showing relative departures of growth by grid cell for individual years of the 1840s drought. Low-growth anomalies are dotted; high-growth anomalies are hatched. The text explains the scaling of circles.

along the West Coast is a plausible meteorological scenario for most years of this drought. Another possibility is a ridge in the eastern North Pacific whose effects on suppression and diversion of storms did not extend far enough inland, except in 1847, to affect states east of Nevada and California. The very wet conditions in the Colorado Rockies during some years of the Far West drought could possibly have resulted from the movement of storms southeastward from Canada and the Pacific Northwest with intensification over the Rockies. The dry tier of cells along the southern boundary of the Southwest region suggests that movement of storms and moisture from the Southwest under the ridge was probably not the source of wetness in Colorado.

Basin-Specific Streamflow Reconstructions

Where the water resources of a particular watershed or river system are in question, it is often useful to go beyond the mapping of tree-ring variations, as described above, to the quantitative reconstruction of specific hydrologic time series. Tree-ring reconstructions of streamflow have been conducted by the Laboratory of Tree-Ring Research at the University of Arizona for three major river-basin systems in the western United States: the upper Colorado River basin, with parts in Colorado, Wyoming, Utah, and New Mexico; the Salt and Verde rivers in central and eastern Arizona; and the Four Rivers group (Yuba, Sacramento, American, and Feather) of northern California (Figure 6.9).

The four phases in a typical tree-ring reconstruction of streamflow include: (1) planning and collection of hydrologic and climatic data, (2) field sampling and physical preparation of tree-ring samples, (3) selection and calibration of a reconstruction model, and (4) generation of reconstructed streamflow along with cross-validation or verification of the reconstruction (Figure 6.10). Descriptions of available methodology can be found elsewhere (Stockton et al., 1985; Cook and Kairiukstis, 1990). The planning and field sampling steps are especially critical in a hydrologic reconstruction to ensure that the tree-ring data provide an optimum signal for the climatic input governing streamflow. In the West, the strategy includes concentrating sampling as much as possible in the major runoff-producing areas of the watershed.

Stockton's (1975) Colorado River reconstruction serves as a good example of a hydrologic tree-ring study whose results have important implications for water resources. The importance of the

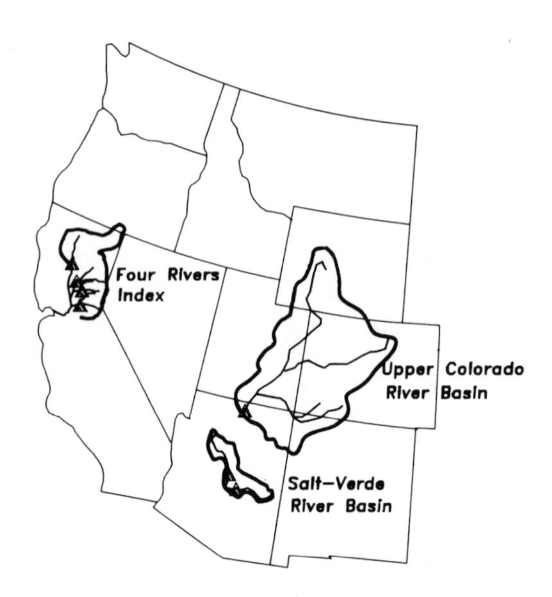

FIGURE 6.9 Map showing locations of three areas for which regional or basin-specific tree-ring reconstructions of streamflow have been conducted by the Laboratory of Tree-Ring Research.

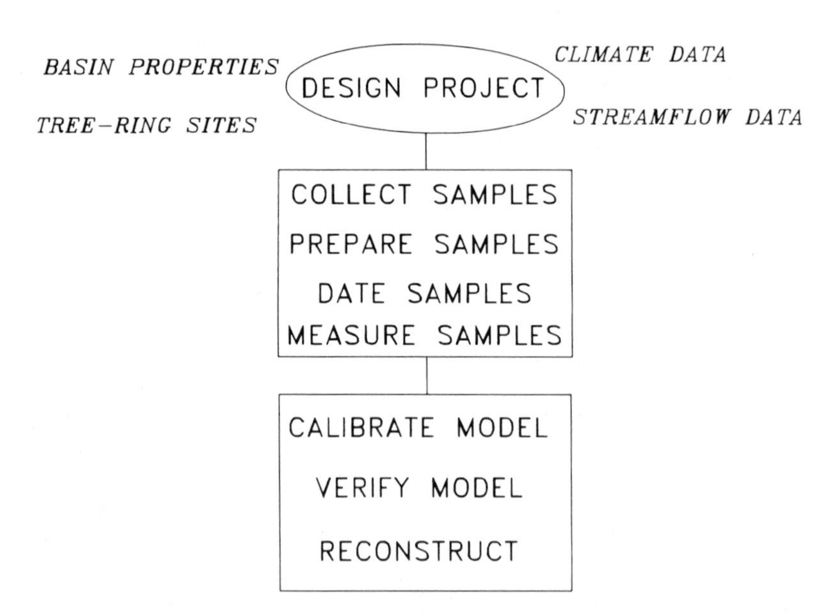

FIGURE 6.10 The major steps in a tree-ring reconstruction of streamflow.

Colorado River as a source of water for agriculture, hydroelectric power generation, and municipal and industrial uses in the southwestern United States cannot be overstated. This 1440-mile-long river flows through some of the most arid lands in the country, and its 244,000-square-mile drainage area includes parts of seven states and a small portion of Sonora and Baja California in Mexico. The Colorado has an average annual flow of just under 14 million acre-feet (maf), a small amount when compared to such rivers as the Columbia and Mississippi. In spite of this relatively low flow, more water is diverted from the basin than from any other river basin in the United States. The river is an important source of supply for southern California and, with the nearly completed Central Arizona Project, for the metropolitan areas of Phoenix and Tucson in Arizona.

Most of the flow for the Colorado originates in the river's upper basin (the area north of Lee's Ferry, Arizona), which includes some 109,300 square miles. About 85 percent comes from only 15 percent of the area—the high mountains of Colorado, Wyoming, and Utah (Stockton and Jacoby, 1976).

The tree-ring data for the study comprised 30 different sites from the major runoff-producing regions. Stockton (1975) calibrated statistical models to reconstruct annual and seasonal flow series at several gage locations in the upper basin from these tree ring data. The reconstruction for the outflow point of the upper basin—Lee's Ferry, Arizona—extends back to A.D. 1520. The period from 1906 to 1930 had the highest sustained flows in the entire reconstruction. The average annual flow, 16.2 maf, for that period was used as a basis for the Colorado River Compact. If the tree-ring reconstruction is accepted as accurate, the design period was simply not representative of the long-term flow of the river. Thus, the division of water between states of the upper and lower Colorado basins, as well as Mexico, is based on an anomalously high value and is apt to result in shortages when all of the entities involved demand their allocated share of the available water.

The terms of the Colorado River Compact specify that no less than 75 maf will be delivered at Lee's Ferry in any consecutive 10-year period (Holburt, 1982). For this reason, 10-year moving averages of the Lee's Ferry reconstruction are of particular interest. Nonoverlapping 10-year means of the reconstruction beginning in 1521, 1531, and so forth until 1951 are graphed in Figure 6.11 along with similar averages for the actual natural flow series as provided by the U.S. Bureau of Reclamation. The droughts of the 1580s through 1590s and the 1660s are again prominent—the earlier especially so for its combined intensity and duration.

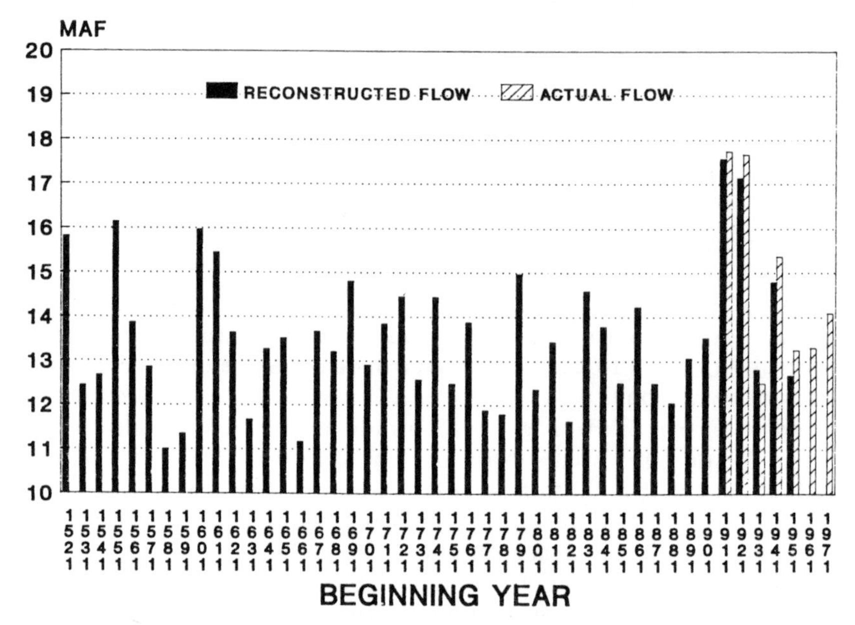

FIGURE 6.11 Bar plot of nonoverlapping 10-year means of reconstructed and actual flow at Lee's Ferry. Actual data are virgin flow obtained from the U.S. Bureau of Reclamation.

The designation of the lowest reconstructed 10-year mean depends on the yearly grouping. The second year of each decade (e.g., 1951) was used to begin each 10-year period in Figure 6.11 because the minimum flow in the actual natural-flow series happened to occur during the period 1931 to 1940. The minimum nonoverlapping 10-year-average reconstructed flow by this grouping is 11.0 maf for 1581 to 1590. In terms of running 10-year means, however, the lowest value is 9.71 maf for 1584 to 1593. Whether the implied 10-year total natural flow of 97.1 maf would lead to noncompliance with the contract terms specifying delivery of 75 maf depends, of course, on the 10-year total depletions for the period. These depletions would reduce the 97.1 maf by some unknown amount depending on water use in the upper basin.

We emphasize that data such as those plotted in Figure 6-11, while in the proper units for streamflow, are merely estimates. Such estimates represent our best possible information on what streamflow conditions may have been, but the inherent uncertainty should always be kept in mind. The magnitude of the uncertainty for reconstructed values in individual years is appreciable: the standard error of the estimate for the Lee's Ferry reconstruction is about 2 maf. It is reasonable, however, to expect somewhat less uncertainty in values averaged over several years. For example, a simple regression of 10-year running means of observed Lee's Ferry flow against the reconstructed flow yields a standard error of 0.46 maf—only about 25 percent of the standard error of the annual reconstructed values.

Uncertainties in Regional Water Supplies

It is not unusual for metropolitan water utilities in the West to import water from regions separated by hundreds of miles. The impact of drought on water supply then depends on the synchrony of climate anomaly patterns over the various source regions. An example of an area depending on several widely separate runoff sources is metropolitan southern California.

The Los Angeles area draws much of its supply from two source regions: the upper Colorado River and northern California. We have examined jointly two hydrologic tree-ring reconstructions for information on extended low-flow periods in two source runoff areas for the Los Angeles water supply: the upper Colorado River and the Sierra Nevada of California. The corresponding tree-ring studies are by Stockton (1975), for the Colorado River at Lee's Ferry, and Earle and Fritts (1986), for the Four Rivers (Sacramento, Feather, Yuba and American) index of northern California (Figure 6.9). Both reconstructions were conducted specifically to maximize the signal for annual runoff, and both extend over several centuries: to A.D. 1520 for the Colorado River and to A.D. 1560 for the Four Rivers index.

Storage acts as a buffer against the effects of meteorological drought, and the considerable capacity of surface and underground storage in this case suggests that droughts must last several years to significantly impact water supply. We have adopted the 20-year moving average of reconstructed annual flow as the data unit to be analyzed for drought. Shorter droughts, especially those of great intensity, can, of course, cause hardship to water users not tied in to the larger storage network. A sustained deficiency of precipitation over a period of 20 years or longer, however, would likely affect even those major water utilities that have elaborate contingency plans to counter drought.

Twenty-year moving averages for both reconstructions are shown in Figure 6.12. The 1660s and 1840s droughts, referred to previously in the discussion of spatial patterns of tree-ring growth departures, are prominent in these plots. The years before A.D. 1600, however, hold by far the lowest reconstructed 20-year flows on the Colorado River. The most severe sustained droughts inferred from lowest 20-year moving average reconstructed flows were as follows for the two series:

- For the Colorado River at Lee's Ferry, flow dropped to 10.95 maf for the years 1579 through 1598.
- For the Four Rivers index, flow dropped to 13.55 maf for the years 1918 through 1937.

Figure 6.12 clearly shows that the 1579 to 1598 period was drier than any other on the Colorado. Considering that the long-term mean of the Lee's Ferry reconstruction is 13.5 maf, the most severe sustained drought for the Colorado River represents a cumulative deficiency of 51 maf over 20 years. The designation of most severe sustained droughts is more uncertain for the Four Rivers index. The Four Rivers area apparently experienced a drought in the 1840s only slightly less severe than that of the 1918 to 1937 period. Moreover, the standard error of the estimate for the annual values of the Four Rivers reconstruction is 5.5 maf, compared with 2.0 maf for the Colorado River reconstruction.

A comparison of the two curves in Figure 6.12 suggests a lack of consistent synchrony between 20-year average flow departures in the upper Colorado River basin and northern California. Annual (unsmoothed) flow series in the two basins are positively correlated, though the coefficients are small:

$r = 0.40$, actual series, 1906-1985
$r = 0.23$, reconstructed series, 1560-1961
$r = 0.37$, actual series, 1906-1961
$r = 0.25$, reconstructed series, 1906-1961

All except the 0.25 value are significant at the 95 percent confidence level. The last two coefficients listed suggest that the reconstruction may underestimate interbasin correlation. Although like-sign departures occur from time to time in the 20-year moving average curves of Figure 6.12, periods of contrast are frequent. For example, the 20-year period of highest average flow for the Four Rivers index, beginning in about 1800, was a time of below-

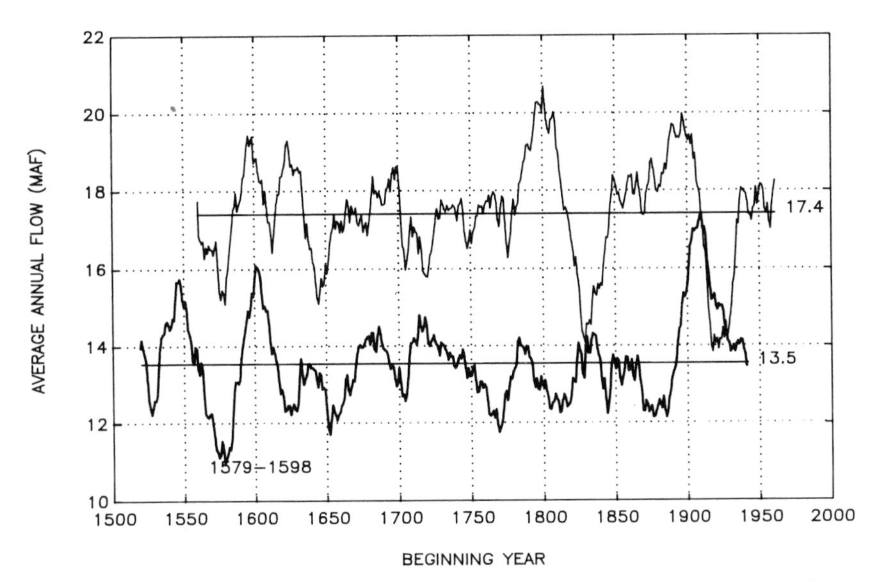

FIGURE 6.12 Time series plots of 20-year running means of reconstructed flows for the Colorado River at Lee's Ferry (lower line) and for the Four Rivers index, northern California (upper line).

average reconstructed flow for the Colorado River. On the other hand, the aforementioned 1579 to 1598 drought on the Colorado River coincided with the third lowest nonoverlapping 20-year mean flow on the Four Rivers index.

The synchrony in droughts and wet periods between the two regions can perhaps be judged more clearly from a scatter plot of the 20-year moving averages (Figure 6.13). A consistent relationship between 20-year departures in the two areas is clearly lacking. In fact, the correlation coefficient for the scatter plot is essentially zero. The lack of correlation does not mean, however, that severe droughts or wet periods have not occasionally been synchronous over the two regions. For example, the period from 1579 to 1598 was notably dry in both regions, and the period from 1904 to 1923 was notably wet in both regions. On the other hand, the years between 1918 and 1937 were a time of contrasting anomalies in terms of 20-year averages—dry in the Four Rivers index but wet in the Colorado River series. The line connecting the

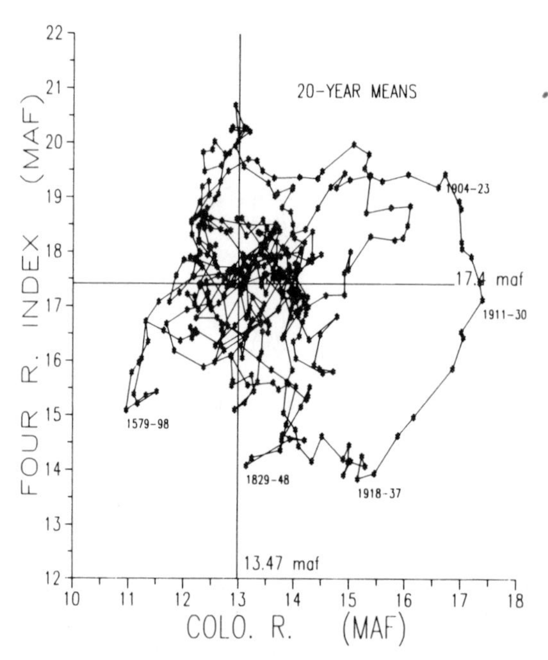

FIGURE 6.13 Scatterplot of running, 20-year means of reconstructed flow of Four Rivers against reconstructed flow of Colorado River at Lee's Ferry. Symbols are connected by lines to show development of anomalies in time. Times of selected key anomalies are annotated.

points on the scattergram indicates a transition from the 1904 to 1923 period to the 1918 to 1937 period, in which the Four Rivers index was becoming increasingly dry relative to the Colorado River series.

The tree-ring record as represented by data used in this study represents about a 450-year time window. On this time scale, moisture anomalies in the two regions are apparently neither consistently synchronous nor compensating. Climate change, whether due to increasing levels of atmospheric carbon dioxide or other influences, could conceivably produce monotonic trends in decadal rainfall totals over larger regions and on longer time scales than discernible from tree-ring data or gaged streamflow records. Indeed, the conversion of ring widths to tree-ring indices involves removal of any trend on the order of one-half the length of the tree-ring series itself, which places a practical limit on the climatic wavelengths that can be inferred. Research is currently ongoing to make use of tree-ring specimens covering thousands of years to extend our record of climatic and hydrologic variations.

CONCLUSIONS

Major changes in global climate have occurred on geologic time scales. So far as their causes are understood, they arose from changes in external boundary conditions such as solar receipts, the gross composition of the atmosphere, and the configuration of the oceans and continents. Although these changes are usually described as occurring rather slowly—that is, over millions of years—it should be remembered that they may have occurred more rapidly in some cases but appear slow because of the coarse temporal resolution of geological records.

Within the overall cool period of the last 2.4 million years, there have been a number of short intervals (up to 15,000 years long) in which the climate has broadly resembled that of the present. The mechanisms controlling the switch between full glacial and such interglacial conditions have been subject to intense scientific interest in recent years. Although they are not fully understood, it is reasonable to state that changes in boundary conditions have played an important part in driving the recurrent pattern of glacials and interglacials. The combination of boundary conditions projected for the next few decades has not occurred before. Consequently, there is great uncertainty as to whether the climate system will continue to behave much as it has in the last 9,000 years.

The Younger Dryas episode demonstrates that major climate change (almost as big as the difference between an ice age and modern climate and covering a large region, such as the North Atlantic basin), can occur in a few decades. Very rapid but less persistent changes to conditions outside the range experienced in the last few hundred years have also taken place since the last retreat of the ice. Such changes may result entirely from the internal mechanisms of the atmosphere and oceans, or they may be caused by events such as very large explosive volcanic eruptions.

Other than the El Niño-Southern Oscillation (see Trenberth, Chapter 5), understanding of decade- to century-scale variations in climate is limited. Reconstructions of streamflow from tree-ring data indicate that such variations are of a magnitude that cannot be ignored in planning for management of water resources in the West. Reconstructions for the upper Colorado River basin and the northern Sierra Nevada of California both emphasize that hyrologic variations of the current century have been unusual in a 400-year context. The highest-flow period on the Colorado and the lowest-flow period in the Sierra Nevada are found in the current century. The modern gaged streamflow record may therefore be an unrepresentative sample for estimating water availability. The large range of departures of reconstructed flows averaged over 20-year periods also suggested that hydrologic models for annual flow simulation incorporate nonstationarity in the mean.

ACKNOWLEDGMENTS

This work was supported in part by National Science Foundation Grant ATM-88-14675.

REFERENCES

Baillie, M. G. L., and M. A. R. Munro. 1988. Irish tree rings, Santorini and volcanic dust veils. Nature 332:344-346.

Bartlein, P. J., and I. C. Prentice. 1989. Orbital variations, climate and paleoecology. Trends in Ecology and Evolution 4:195-199.

Berger, A. 1979. Insolation signatures of Quaternary climatic changes. Nuovo Cim. 2(c):63-87.

Betancourt, J. L. 1990. Late Quaternary biogeography of the Colorado Plateau. Pp. 259-292 in J. L. Betancourt, T. R. van Devender, and P. S. Martin, eds., Fossil Packrat Middens: The Last 40,000 Years of Biotic Cange in the Arid West. Tucson: University of Arizona Press.

Briffa, K. R., T. S. Bartholin, D. Eckstein, P. D. Jones, W. Karlen, F. H. Schweingruber, and P. Zetterberg. 1990. A 1,400-year tree-ring record of summer temperatures in Fennoscandia. Nature 346:434-439.

Broecker, W. S., and G. H. Denton. 1990. What drives glacial cycles? Scientific American 262:49-56.

Cleaveland, M. K., and D. W. Stahle. 1989. Tree ring analysis of surplus and deficit runoff in the White River, Arkansas. Water Resources Research 25(6):1391-1401.

Cook, E. R., and L. A. Kairiukstis. 1990. Methods of Dendrochronology: Applications in the Environmental Sciences. Boston: Kluwer Academic Publishers.

Cooperative Holocence Mapping Project (COHMAP). 1988. Climatic changes of the last 18,000 years: observations and model simulations. Science 241:1043-1052.

Denton, G., and W. Karlen. 1973. Holocene climatic variations—their pattern and possible cause. Quat. Res. 3:155-205.

Dickinson, R. E., and H. Virji. 1987. Climate change in the humid tropics, especially Amazonia, over the last twenty thousand years. Pp. 91-105 in R. E. Dickinson, ed., The Geophysiology of Amazonia: Vegetation and Climate Interactions. New York: John Wiley & Sons.

Earle, C. J., and H. C. Fritts. 1986. Reconstructing Riverflow in the Sacramento Basin Since 1560. Report prepared for the California Department of Water Resources under agreement number DWR B-55395.

Emiliani, C. 1955. Pleistocene temperatures. J. Geol. 63:536-578.

Fritts, H. C. 1976. Tree Rings and Climate. London: Academic Press.

Graumlich, L. J., and L. B. Brubaker. 1986. Reconstruction of annual temperature (1590-1979) for Longmire, Washington, derived from tree rings. Quat. Res. 25:223-234.

Graybill, D. A., and S. G. Shiyatov. 1989. A 1009-year tree-ring reconstruction of mean June-July temperature deviations in the Polar Urals. Pp. 37-42 in R. D. Noble, J. L. Martin, and K. F. Jensen, eds., Symposium on Air Pollution Effects of Vegetation. Broomall, PA: U.S. Department of Agriculture, Northwestern Forest Experiment Station.

Grove, J. M. 1988. The Little Ice Age. London: Methuen.

Hammer, C. U., H. B. Clausen, and W. Dansgaard. 1980. Greenland ice sheet evidence of post-glacial volcanism and its climatic impact. Nature 288:230-235.

Holburt, M. B. 1982. Colorado River water allocation. Water Supply and Management 6(1-2):63-73.

Hughes, M. K., P. M. Kelly, J. R. Pilcher, and V. C. LaMarche, Jr., eds. 1982. Climate from Tree Rings. Cambridge: Cambridge University Press.

Imbrie, J., J. D. Hays, D. C. Martinson, A. MacIntrye, A. C. Mix, J. J. Morley, N. G. Pisias, W. L. Prell, and N. J. Shackleton. 1984. The orbital theory of Pleistocene climate: Support from a revised chronology of the marine $\delta^{18}O$ record. Pp. 269-305 in A. Berger et al., eds., Milankovitch and Climate. Boston: D. Reidel Publishing Company.

Jacoby, G. C., and R. D'Arrigo. 1989. Reconstructed northern hemisphere annual temperature since 1671 based on high-latitude tree-ring data from North America. Climatic Change 14:39-59.

Kukla, G. J. 1977. Pleistocene land-sea correlations, Part I: Europe. Earth Sci. Rev. 13:307-374.

LaMarche, V. C., Jr. 1974. Paleoclimatic inferences from long tree-ring records. Science 183:1043-1048.

Lézine, A. M. 1989. Late Quaternary vegetation and climate of the Sahel. Quaternary Research 32:317-334.

Long, A., L. Warnecke, J. L. Betancourt, and R. S. Thompson. 1990. Deuterium variations in plant cellulose from fossil packrat middens. Pp. 259-292 in J. L. Betancourt, T. R. van Devender, and P. S. Martin, eds., Fossil Packrat Middens: The Last 40,000 Years of Biotic Change in the Arid West. Tucson: University of Arizona Press.

Lorius, C., J. Jouzel, D. Raynaud, J. Hansen, and H. Le Treut. 1990. The ice-core record: climate sensitivity and future greenhouse warming. Nature 347:139-145.

Milankovitch, M. M. 1941. Canon of insolation and the ice-age problem. Königlich Serbische Akademie. Translated by the Israel Program for Scientific Translations, 1969. Jerusalem, Israel: Program for Scientific Translations.

Rea, D. K., and M. Leinen. 1988. Asian aridity and the Zonal Westerlies: Late Pleistocene and Holocene record of eolian deposition in the northwest Pacific Ocean. Paleogeog., Paleoclim., Paleoecol. 66:1-8.

Schulman, E. 1956. Dendroclimatic Changes in Semi-arid America. Tucson: University of Arizona Press.

Smith, L. P., and C. W. Stockton. 1981. Reconstructed streamflow for the Salt and Verde rivers from tree-ring data. Water Resources Bull. 16(6):939-947.

Stockton, C. W. 1975. Long-term Streamflow Records Reconstructed from Tree Rings. Tucson: University of Arizona Press.

Stockton, C. W., W. R. Boggess, and D. M. Meko. 1985. Climate and tree rings. Chapter 3, A. D. Hecht, ed., in Paleoclimate Analysis and Modeling. New York: John Wiley & Sons.

Stockton, C. W., and G. C. Jacoby. 1976. Long-Term Surface-Water Supply and Streamflow Trends in the Upper Colorado River Basin. Lake Powell Research Project Bulletin No. 18. Washington, D.C.: National Science Foundation.

Thompson, L. G., and E. Mosley-Thompson. 1990. One-half millennia of tropical climate variability as recorded in the stratigraphy of the Quelccaya ice cap in Peru. Pp. 15-31 in D. H. Petersen, ed., Aspects of Climate Variability in the Pacific and the Western Americas, Geophysical Monograph 55. Washington, D.C.: American Geophysical Union.

Vernekar, A. D. 1972. Long-period Global Variations of Incoming Solar Radiation. Boston: American Meteorological Society.

7

Effects of Increasing Carbon Dioxide Levels and Climate Change on Plant Growth, Evapotranspiration, and Water Resources

Leon Hartwell Allen, Jr.
U.S. Department of Agriculture
Gainesville, Florida

The atmospheric carbon dioxide concentration has risen from about 270 parts per million (ppm) before 1700 to about 355 ppm today. Climate changes, including a mean global surface temperature rise of between 2.8 and 5.2°C, have been predicted by five independent general circulation models (GCMs) for a doubling of the carbon dioxide concentration. The objectives of this paper are to examine plant responses to rising carbon dioxide levels and climatic changes and to interpret the consequences of these changes on crop water use and water resources for the United States.

BACKGROUND: PLANT RESPONSES TO ENVIRONMENTAL FACTORS

The main purpose of irrigation is to supply plants with adequate water for transpiration and for incorporating the element hydrogen in plant tissues through photosynthesis and subsequent biosynthesis of various tissues and organs. Transpirational flux requires several hundred times more water than photosynthesis.

In a series of U.S. Department of Agriculture studies beginning in 1910 in Akron, Colorado, Briggs and Shantz (1913a,b; 1914) showed that the water requirement of plants is linearly related to the biomass production of plants. They established this linear relationship by growing plants in metal containers filled with soil. Throughout the period of growth, they monitored water use carefully by weighing and adding measured amounts of water to maintain a desirable soil water content as water lost by plant transpiration was replenished.

The findings of Briggs and Shantz have been confirmed repeatedly (Allison et al., 1958; Arkley, 1963; Chang, 1968; Hanks et al., 1969; Stanhill, 1960). Figure 7.1 shows the linear relationship between biomass produced and rainfall plus irrigation water used by Sart sorghum and Starr millet in Alabama, as adapted from data of Bennett et al. (1964). De Wit (1958) examined the relationships among climatic factors, yield, and water use by crops. He found the following general linear relationship to be true, especially in semiarid climates:

$$Y/T = m/T_{max} \quad (1)$$

where

Y = yield component (e.g., total above-ground biomass or seed production)

T = cumulative actual transpiration

T_{max} = maximum possible cumulative transpiration

m = constant dependent on yield component and species, especially on differences among photosynthetic mechanisms

Pan evaporation was used to represent T_{max}, which is proportional to climatic factors, especially air vapor pressure deficit (VPD):

$$T_{max} \alpha \ VPD = (e_s - e_a) \quad (2)$$

where

e_s = the saturation vapor pressure at a given air temperature

e_a = the actual vapor pressure that exists in the air.

Combining these relationships, we see that yield is proportional to cumulative transpiration divided by vapor pressure deficit:

$$Y/T = k/(e_s - e_a) \quad (3)$$

where k is a constant with units millibars • g (dry matter) • g^{-1} (water). Like m, k depends on yield component, species, and photosynthetic mechanisms.

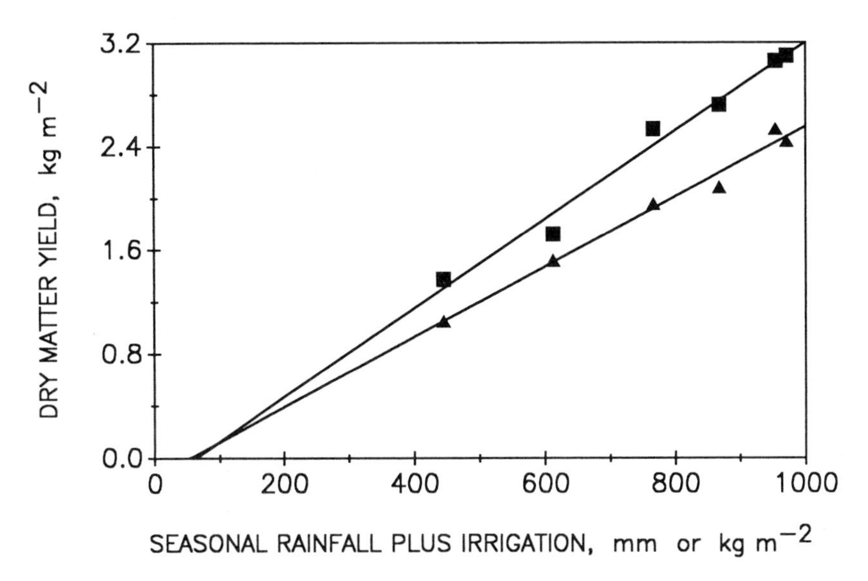

FIGURE 7.1 Linear relationship between biomass production and water use for two forage crops in 1956 and 1957 at Thorsby, Alabama. Squares: Sart sorghum. Triangles: Starr millet.
SOURCE: Adapted from Bennett et al., 1964.

Thus, we can see that theory predicts that yield will be proportional to cumulative transpirational water use, divided by vapor pressure deficit. There are several ways of calculating the VPD; it can be computed by aggregating seasonal daytime average VPD, or by using approximation methods based on daily maximum and minimum temperatures (Jensen, 1974). As pointed out by Tanner and Sinclair (1983), the maximum e_s can be computed from the daily maximum temperature, and e_a can be estimated from the daily minimum temperature. Tanner and Sinclair estimated that the effective daytime e_s falls at a point two-thirds to three-quarters of the distance between the e_s computed at the daily maximum temperature and the e_a computed at the daily minimum temperature. The effective daytime VPD values then must be averaged over the growing season of the crop. Regardless of the method used to compute a representative VPD, yield versus cumulative transpiration linear relationships vary with the aridity of the climate—specifically with the temperature and vapor pressure regime under which the crop is grown. Figure 7.2 (modified from Stanhill, 1960) shows water used versus dry matter yield of pastures from the latitude of Denmark (which has a cool, humid atmosphere) to the latitude of Trinidad (which has a hot, dry atmosphere). Based on comparisons among existing climates, we can expect that transpirational water requirements of plants will increase if climates get warmer.

Atmospheric carbon dioxide is known to affect plant yield. Kimball (1983) reviewed 430 observations of carbon dioxide enrichment studies conducted prior to 1982 and reported an average yield increase of 33 percent, plus or minus 6 percent, for a doubling of the carbon dioxide concentration. This value has been generally confirmed by many other studies since that time. The yield increases seem to apply for both biomass accumulation and grain yield. Thus, plants may grow larger and, considering Figure 7.1, they may use more water as the global carbon dioxide concentration increases.

Transpirational water use is clearly related to ground cover (Jensen, 1974; Doorenbos and Pruitt, 1977). Daily water use soon after crops are planted on bare soil is typically only 10 to 20 percent of water use after effective ground cover is reached. Water use rises sharply as the crop's leaf area increases. Similarly, water use drops 60 to 70 percent when hay crops such as alfalfa are cut. As leaf regrowth occurs, transpiration rates recover rapidly as the ground cover of leaves is restored. Ground cover can be quantified with a leaf area index (LAI): the ratio of leaf area per unit ground area. Therefore, any carbon dioxide-induced stimulation of early growth of leaf area or increase of total leaf area growth may increase transpiration.

Increased carbon dioxide concentrations are known to cause smaller stomatal apertures and hence to decrease the leaf conductance for water vapor (Morison, 1987). This is a second mechanism whereby increased carbon dioxide concentrations may affect plant transpiration.

Another effect of rising carbon dioxide concentrations is the change in water-use efficiency (WUE). Water-use efficiency has a range of definitions. For whole-season processes, it is best defined as the ratio of dry matter (or seed yield) produced to the amount of water used by crops. For shorter-term whole canopy processes, it is best defined as the ratio of the photosynthetic carbon dioxide uptake rate per unit land area to the transpiration rate per unit land area. Figure 7.2 demonstrates the effect of climate on WUE.

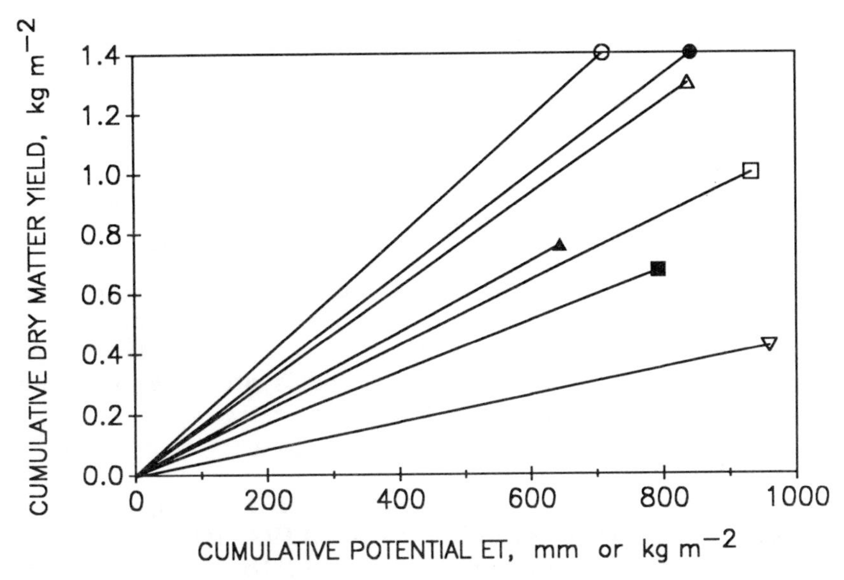

FIGURE 7.2 Cumulative dry matter yield versus cumulative potential evapotranspiration (ET) of pastures under a range of climatic regimes. Open circle: Denmark. Filled circle: The Netherlands. Open triangle: England. Filled triangle: New Jersey. Open Square: Toronto, Canada. Filled square: Gilat, Israel. Open inverted triangle: Trinidad, West Indies.
SOURCE: Adapted from Stanhill, 1960.

Equation 3 quantifies the relationship between WUE and vapor pressure deficit.

In summary, the following relationships have been established by research:

1. Transpiration is linearly related to biomass accumulation and yield.

2. Transpiration is also linearly related to the aridity of the climate—in other words, to the vapor pressure deficit. Thus, rising global temperatures would increase transpiration by increasing the atmospheric vapor pressure deficit.

3. Transpiration is affected by the degree of ground cover.
4. Rising carbon dioxide concentrations will increase plant growth. More rapid leaf area development and more total leaf area could translate into more transpiration.
5. Rising carbon dioxide concentrations will decrease leaf stomatal conductance to water vapor. This effect could reduce transpiration.
6. Rising carbon dioxide concentrations and rising global temperatures could change WUE.

The following sections of this chapter will examine more closely the effects of rising carbon dioxide concentrations and climate change on vegetation, providing qualitative and quantitative assessments of how these changes will affect photosynthesis, growth, and transpiration water requirements of crops.

DIRECT EFFECTS OF CARBON DIOXIDE ON PHOTOSYNTHESIS, TRANSPIRATION, AND GROWTH OF PLANTS

Atmospheric Carbon Dioxide

The carbon dioxide concentration of the earth's atmosphere has varied throughout geologic time. Ice core data from Antarctica and Greenland have been obtained and, from entrapped air bubbles, used to show carbon dioxide and methane concentrations of the atmosphere throughout the past 160,000 years (Barnola et al., 1987; Lorius et al., 1990). Changes in the deuterium content within ice crystals have been used to establish temperature changes over this same time period (Jouzel et al., 1987). In general, carbon dioxide concentrations were as low as 180 to 200 parts per million (ppm) 13,000 to 30,000 years ago and 140,000 to 160,000 years ago during the coldest parts of the last two ice ages (Barnola et al., 1987). Carbon dioxide concentrations rose to about 270 ppm during the last interglacial period (116,000 to 140,000 years ago) and during the current interglacial period (beginning about 13,000 years ago). Ice core data since about 1700 A.D. and direct atmospheric sampling data since 1958 show that the carbon dioxide concentration increased to 315 ppm by 1958 and to about 355 ppm by 1990 (Keeling et al., 1989). The rate of increase of atmospheric carbon dioxide is about 0.5 percent per year, which means that the change is accelerating.

These changes in atmospheric carbon dioxide have important implications for plants and the global carbon cycle as well as for climate. Atmospheric carbon dioxide is the raw material for terrestrial green plant photosynthesis, and thus it represents the first molecular link in the food chain of the whole earth. In later sections, we will examine the importance of carbon dioxide for photosynthesis and plant growth, as well as the importance of potential climate change on water resources for the future.

Plant Photosynthetic Mechanisms

Three types of photosynthetic mechanisms of terrestrial green plants have been identified: C3, C4, and CAM. Responses of these three photosynthetic mechanisms to carbon dioxide have been reviewed by Tolbert and Zelitch (1983). The biochemical pathway of photosynthetic carbon dioxide uptake was first determined for C3 plant photosynthesis. This pathway involves the use and subsequent regeneration of ribulose 1,5-biophosphate in a cyclic series of reactions, and it is frequently called the Calvin cycle. The first product of photoassimilation of carbon dioxide is 3-phosphoglyceric acid, a three-carbon sugar—hence the term C3 pathway of photosynthesis.

The C4 plants begin their carbon dioxide uptake in a different process sometimes called the Hatch-Slack pathway. In mesophyll cells of leaves, these plants form a four-carbon molecule, oxalacetate, in the first step of incorporation of carbon dioxide. This four-carbon compound is changed into aspartic acid or malic acid and then transported immediately to bundle sheath cells. Here, the carbon dioxide is released and utilized in the C3 biochemical pathway. Thus, the C4 plant mechanism first traps carbon dioxide in the mesophyll cells, and then transports and concentrates the carbon dioxide in the bundle sheath cells, where it is utilized in C3 plant metabolism (Tolbert and Zelitch, 1983).

Crassulacean acid metabolism, or CAM, is a mechanism whereby plants typically take up and store carbon dioxide during the night and use it in photosynthetic carbon dioxide fixation during the day, when sunlight is available. Pineapple and "air plants," such as Spanish moss and orchids, have this photosynthetic mechanism. Since few agricultural crops are CAM plants, they are not important in the process of managing water resources under conditions of climate uncertainty.

Since C4 plants have a mechanism for concentrating carbon dioxide in bundle sheath cells of leaves, their photosynthetic rates will not respond to rising carbon dioxide levels to the same extent as C3 plants. Irrigated crop or turf plants that fit into the C4 category include maize (corn), sorghum, millet, sugar cane, and bermuda grass. Plants that fit into the C3 category include: wheat, rice, potato, soybean, sugar beet, alfalfa, cotton, tree and vine crops, and most vegetable crops and cool-season grasses.

Plant Growth Responses to Carbon Dioxide

Increasing atmospheric carbon dioxide levels have caused increasing photosynthetic rates, biomass growth, and seed yield for all of the globally important C3 food and feed crops (Acock and Allen, 1985; Enoch and Kimball, 1986; Warrick et al., 1986; Allen, 1990). Some plants, such as cucumber, cabbage, and perhaps tomato, have shown a tendency to first increase leaf photosynthetic rates in response to elevated carbon dioxide concentrations, and then to decrease photosynthetic rates after several days. This behavior is called "end-product inhibition of photosynthesis," and it is caused by the failure of translocation of photoassimilates to keep up with photosynthetic rates (Guinn and Mauney, 1980).

A few experiments have been conducted with carbon dioxide concentration maintained across a range of 160 to 990 ppm. Figure 7.3 shows the results of one study with soybean canopy photosynthetic rates across the 90 to 900 ppm carbon dioxide concentration range. A nonlinear hyperbolic model was used to fit soybean photosynthetic rate data to carbon dioxide concentration (Allen et al., 1987). Photosynthetic rates at the various carbon dioxide concentrations were divided by the photosynthetic rate at a carbon dioxide concentration of 330 ppm to normalize the data to a common condition. Data sets of biomass yield and seed yield from four locations over three years were also fit to the model (Allen et al., 1987). Relative yields with respect to yields at 330 to 340 ppm were used.

The form of the model fit to the experimental data was:

$$R = R_{max} * C/(C + K_c) + R_{int} \quad (4)$$

where

R = relative response of photosynthetic rate, biomass yield, or seed yield

R_{max} = asymptotic upper limit for R from baseline R_{int}

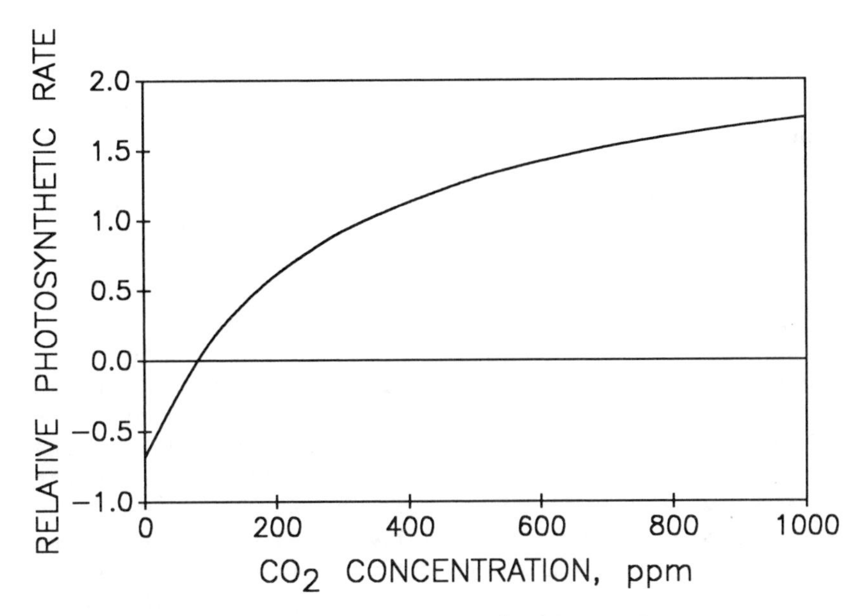

FIGURE 7-3 Photosynthetic carbon dioxide uptake rate responses of a soybean crop canopy exposed to carbon dioxide concentrations ranging from 110 to 990 ppm. All data points are relative to the response obtained at 330 ppm.
SOURCE: Adapted from Allen et al., 1987.

C = carbon dioxide concentration (ppm)

K_c = Apparent Michaelis constant (ppm)

R_{int} = Y-axis intercept for zero C

From the parameters of this equation, photosynthetic rate, biomass accumulation, and seed yield changes of soybean due to carbon dioxide concentration changes can be estimated (Allen et al.,

1987). Table 7.1 shows the changes predicted across three time periods: from the last ice age (when the carbon dioxide concentration was at a minimum) to the preindustrial revolution era (about 1700), from 1700 to 1973, and from 1973 to about a century into the future. The modeled data show that there should have been large increases in productivity between the ice-age (when carbon dioxide concentration was about 200 ppm) and the beginning of the industrial revolution (when the carbon dioxide concentration was about 270 ppm). Likewise, there should have been a 12 percent increase in grain-yield productivity between 1700 and 1973, when the carbon dioxide concentration increased from about 270 to 330 ppm.

Most of the recent concerns about rising atmospheric carbon dioxide concentrations have been quantified by predicting changes for a doubling of the carbon dioxide concentration, usually from 330 to 660 ppm. Table 7.1 shows that soybean seed yields and biomass yields are predicted to increase 31 percent and 41 percent, respectively, from a doubling of carbon dioxide. Experimental studies have consistently showed a lower seed yield than biomass yield for soybean when grown under a doubled carbon dioxide concentration. If the harvest index—the ratio of seed yield to above-ground biomass yield (seed plus pod walls plus stems)—were 0.50 for soybean grown under a 330 ppm carbon dioxide concentration, then the harvest index would be 0.46 if the carbon dioxide concentration were doubled. This small decrease in soybean harvest index under elevated carbon dioxide conditions has been commonly observed (Allen, 1990; Jones et al., 1984). The relative midday maximum photosynthetic rates under carbon dioxide enrichment were consistently higher than relative biomass yields, probably because the photosynthetic response to elevated carbon dioxide levels is greater under high light conditions than it is under total daily solar irradiance conditions.

Transpiration Responses to Carbon Dioxide

The effect of carbon dioxide concentration on water use under field conditions has been discussed for many years. In the past, elevated carbon dioxide levels have been mentioned as the ideal antitranspirant. This conclusion seems reasonable, since elevated carbon dioxide has been observed to reduce stomatal conductance in numerous experiments. Morison (1987) reviewed 80 observations in the literature and found that a doubled carbon dioxide con-

TABLE 7.1 Percent increases of soybean midday photosynthesis rates, biomass yield, and seed yield predicted across selected carbon dioxide concentration ranges associated with relevant benchmark points in time.

Period of Time	CO_2 Concentration Initial	Final	Midday Photosynthesis	Biomass Yield	Seed Yield
	- - - - ppm - - - -		- - -% increase over initial CO_2- - -		
IA-1700[1]	200	270	38	33	24
1700-1973	270	330	19	16	12
1973-20??[2]	330	660	50	41	31

[1] IA, the Ice Age about 13,000 to 30,000 years before present. The atmospheric carbon dioxide concentrations that prevailed during the last Ice Age, and from the end of the glacial melt until preindustrial revolution times, were 200 and 270 ppm, respectively.

[2] The first world energy "crisis" occurred in 1973, when the carbon dioxide concentration was 330 ppm. This concentration is used as the basis for many carbon dioxide-doubling studies. The carbon dioxide concentration is expected to double sometime within the twenty-first century.

centration will reduce stomatal conductance of most plants by about 40 plus or minus 5 percent. Kimball and Idso (1983) calculated a 34 percent reduction in transpiration in response to a doubled carbon dioxide concentration in several short-term plant growth chamber experiments, which seems consistent with the review by Morison (1987). However, Morison and Gifford (1984) also showed that doubling carbon dioxide will cause a more rapid development of leaf area for many plants and hence an equal or greater transpiration rate in the early stages of plant growth, due to a more rapid development of transpiring surfaces. Therefore, increased rates of development of transpiring leaf surface offset the reduced stomatal conductance for water vapor.

Allen et al. (1985) and Allen (1990) also discussed the effect of reduction in stomatal conductance on foliage temperature. The cause-and-effect relationships can be summarized as follows: Any reduction in stomatal conductance due to increasing the carbon dioxide concentration will restrict transpiration rates per unit leaf area. A reduction in transpiration rates will result in less evaporative cooling of the leaves, and leaf temperatures will rise. As leaf temperatures rise, the vapor pressure inside the leaves will increase, and thus the leaf-to-air vapor pressure gradient, which is the driving force for transpiration, will increase. The increase in leaf vapor pressure will increase transpiration rates per unit leaf area; thus, the transpiration rates will be maintained at only slightly lower values than would exist at ambient environmental carbon dioxide concentrations. In effect, all of the energy balance factors involved in canopy foliage energy exchange—not just stomatal factors—must be considered.

Controlled environment studies of soybean at Gainesville, Florida, showed that canopy transpiration rate changes ranged from negative 2 percent (Jones et al., 1985a) to plus 11 percent (Jones et al., 1985b) for carbon dioxide treatments of 800 and 330 ppm with corresponding LAI values of 6.0 and 3.3. In another experiment in which differences in the LAI of soybean between the 330 ppm and the 660 ppm treatments were small (3.36 and 3.46, respectively), the seasonal cumulative water use decreased by 12 percent for the doubled carbon dioxide treatments (Jones et al., 1985c). Decreases were similar for both water-stressed and nonstressed treatments.

Field weighing lysimeter and neutron-probe water balance studies of cotton at Phoenix, Arizona, have shown evapotranspiration reductions due to elevated carbon dioxide levels ranging from 0 up to 9 percent (Kimball et al., 1983).

In conclusion, although stomatal conductance may be reduced by about 40 percent for doubled carbon dioxide concentrations, water use by C3 crop plants under field conditions will probably be reduced by only about 0 to 12 percent. If leaf area increases due to doubled carbon dioxide concentrations are small (or can be controlled), then the transpiration reductions may be meaningful, albeit small. If leaf area increases due to doubled carbon dioxide concentrations are large, then no reductions in transpiration are to be expected, and increases may be possible.

Streamflow Responses to Carbon Dioxide

Several attempts have been made to predict changes in streamflow due to an increase in carbon dioxide (Aston, 1984), changes in climate (Revelle and Waggoner, 1983), or both (Brazel and Idso, 1984). Aston (1984) modeled streamflow changes from a New South Wales, Australia, watershed over the course of a year based on reduction of stomatal conductance to one-half of current values. His model predicted a 40 to 90 percent increase in annual streamflow above the observed baseline of about 150 mm per year from the actual watershed. However, Aston (1984) did not consider any increase of the LAI, which is perhaps a justifiable assumption for C4 plants but probably not for C3 plants. Rosenberg et al. (1990) conducted a quantitative analysis of evapotranspiration sensitivity to several plant and environmental factors. Their analysis demonstrated that increasing the LAI could indeed partially offset the effects of decreasing stomatal conductance on transpiration.

For climate change only, Revelle and Waggoner (1983) predicted that western river streamflows could be reduced by about 40 to 76 percent from the combined effects of a 2°C rise in temperature and a 10 percent reduction in precipitation. Brazel and Idso (1984) considered that vegetation would reduce transpiration to about two-thirds of its current value with a doubling of the carbon dioxide concentration, which led to predictions of increasing Arizona streamflow from about 63 to 460 percent. When they included a temperature increase of 2°C and a precipitation decrease of 10 percent, the predictions were still a 4 to 326 percent increase in streamflow. However, Brazel and Idso's predictions did not include any likely increases in vegetation LAI due to increased carbon dioxide levels. Although efforts to relate carbon dioxide and climate change impacts on water resources are continuing (Waggoner, 1990), realistic integration of vegetation influences on the hydrologic cycle are lacking.

Changes in vegetation may be a moot point when streamflow depends largely on spring snowmelt from lower elevations and continuous warm season snowmelt from higher elevations in the mountains of the West. The combination of complex plant responses and complex terrain make accurate hydrologic modeling a difficult task.

Plant Water-Use Efficiency

Allen et al. (1985) compared water-use efficiencies of soybean canopies grown in outdoor, sunlit, controlled-environment chambers at 800 and 330 ppm carbon dioxide concentrations which had LAI values of 6.0 and 3.3, respectively. For each of these treatments (two replications), the exposure carbon dioxide levels were cross-switched for one day. The ratio of the WUE values (i.e., WUE at 800 ppm carbon dioxide exposure divided by the WUE at 330 ppm carbon dioxide exposure) averaged 2.33. The relative contributions of photosynthesis and transpiration to the ratio of WUE values were 73 and 27 percent, respectively. These comparisons are valid only for plant canopies with equal LAI values, because the same canopy was used for both carbon dioxide exposure levels. However, when the treatment and exposure levels of 800 ppm carbon dioxide were compared with the treatment and exposure levels of 330 ppm, the WUE ratio was 1.80, and the relative contributions to this ratio were 104 percent for photosynthesis and negative 4 percent for transpiration. The negative contribution of transpiration arises from the fact that canopy transpiration rates for the 800 ppm carbon dioxide treatment were slightly greater than the rates from the 330 ppm carbon dioxide treatment, due to the much larger LAI of the canopy exposed to the higher carbon dioxide treatment. Clearly, higher LAI values under elevated carbon dioxide concentrations can increase transpiration rates to the point where all of the improved WUE arises from increased photosynthetic rates and none from decreased water use.

Finally, it should be pointed out that increases in WUE in a world with higher carbon dioxide levels do not necessarily imply any reduction in crop water requirements per unit area of land. However, farmers should be able to achieve higher crop yields per unit land area with similar amounts of water. If temperatures rise, however, the overall WUE could actually decrease, because warmer climates have higher water requirements (as illustrated by Figure 7.2) and higher temperatures may cause yield reductions. The crop response scenarios that may affect hydrology and water resources management will be determined by the carbon dioxide and climate change scenarios and will differ depending on photosynthetic types (C4 versus C3) and species.

CLIMATE CHANGE EFFECTS ON PHOTOSYNTHESIS, GROWTH, AND TRANSPIRATION

Leaf photosynthetic rates are known to be sensitive to temperature. Figure 7.4 shows possible responses of leaf photosynthetic carbon dioxide uptake rates to temperature for C3 plants (bottom curve) and C4 plants (top curve) when grown at a 330 ppm carbon dioxide concentration and exposed to high light levels, such as would occur under midday summer conditions. This figure shows that C4 plants have a higher maximum photosynthetic carbon dioxide uptake rate and a higher temperature maximum than C3

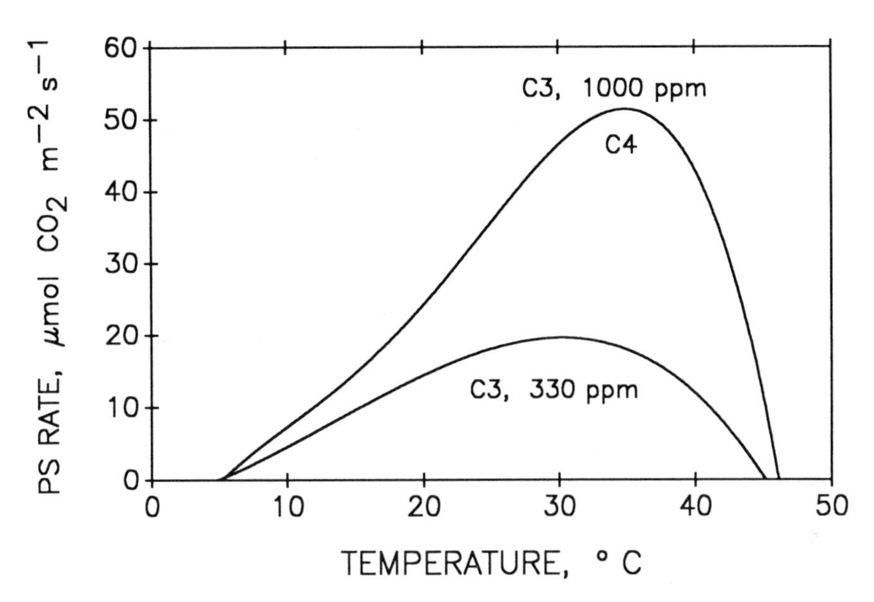

FIGURE 7.4 Examples of maximum photosynthetic (PS) rate responses to temperature of individual leaves of C3 plants under high light conditions when exposed to carbon dioxide concentrations of 330 ppm (lower curve) and 1,000 ppm or greater (upper curve). The upper curve is similar to the maximum PS response to temperature of C4 plant leaves, which have an internal mechanism for concentrating carbon dioxide for subsequent photosynthetic reactions. Various species differ widely, both in maximum leaf photosynthetic rates and in the distribution of leaf photosynthetic rates with temperature.
SOURCE: Modified and adapted from the example of Pearcy and Björkman (1983). See also Berry and Björkman (1980) and Penning de Vries et al. (1989) for further examples of the variability of response among species and experimental conditions.

plants. The relative differences are smaller at lower temperatures. These curves were drawn to represent active crop plants in temperate zones. The actual photosynthetic carbon dioxide uptake rates could be considerably different from those shown, and the temperature distribution of photosynthetic rates could be higher or lower, depending upon species, climate, or pretreatment temperature conditions (Berry and Björkman, 1980; Penning de Vries et al., 1989). In particular, the curves could be stretched to higher temperatures for species adapted to hot, desert environments (Pearcy and Björkman, 1983) or compressed to lower temperatures for species adapted to cool environments.

Nevertheless, from Figure 7.4 we can conclude that C4 plants could benefit more (or at least suffer less) than C3 plants from an increase in global temperatures. However, the differences for a whole canopy of leaves are somewhat reduced from the differences for individual leaves exposed perpendicularly to high light. First, a canopy of leaves generally has leaves oriented in all directions, so that much of the total leaf area is exposed to much less irradiance than in single leaf exposure systems. Under these conditions many of the individual leaves are limited by light, and the photosynthetic carbon dioxide uptake rates of the whole canopy become more similar. Second, solar irradiance levels are lower than midday values throughout much of the day. Nevertheless, the direction of the leaf-level differences, if not the magnitude, is maintained between C4 and C3 canopies.

Figure 7.4 also shows that C3 plant photosynthetic rates at elevated carbon dioxide levels may increase and resemble the rates of C4 plants (Pearcy and Björkman, 1983), but the extent of increase will vary widely among species. The photosynthetic rates of C3 plant leaves increase at elevated carbon dioxide levels because molecules of carbon dioxide compete more effectively with oxygen for binding sites on rubisco, the carboxylating enzyme (Bowes and Ogren, 1972).

When plants are well watered, leaf temperatures tend to rise more slowly than air temperatures throughout the daily cycle, so that foliage-to-air temperature differences become greater as air temperature rises (Idso et al., 1987; Allen, 1990). For soybean, Jones et al. (1985a) found no change in crop canopy photosynthetic rates across the air temperature set-point range of 28 °C to 35 °C. However, the transpiration rates increased 30 percent, which would lead to evaporative cooling of the leaves and larger foliage-to-air temperature differences. This 4 to 5 percent increase in transpiration rate per 1 °C rise in temperature is close to the 6 percent per 1 °C rise in saturation vapor pressure deficit over this temperature range.

Temperature affects growth of plants in several ways. The rate of development and expression of new nodes on plants increases with increasing temperature. Leaves expressed at new nodes will grow larger, in general, if there is no concurrent water stress. Thus, plant size increases at a more rapid rate, and solar radiation capture occurs earlier in crop development. Once full ground cover is achieved, at a LAI of about 2 to 3, light capture becomes limiting, and the overall temperature effects on growth are muted but not eliminated. The duration of each ontogenic phase of plant growth decreases with increasing temperature, which is the most important effect of temperature within the upper and lower limits of survival.

INTERACTIVE EFFECTS OF CARBON DIOXIDE AND CLIMATE CHANGE

Photosynthetic and Productivity Interactions

As explained above, Figure 7.4 shows leaf photosynthetic carbon dioxide uptake rate versus temperature responses typical of C4 plants and C3 plants at carbon dioxide concentrations of 330 ppm. The upper curve can also represent C3 plants at high carbon dioxide levels of (1,000 ppm or greater). These curves suggest that a combination of rising carbon dioxide concentration and rising temperature should lead to greater photosynthetic rates and hence greater biomass growth rates.

S. G. Allen et al. (1988; 1990a,b) conducted experiments in Phoenix, Arizona, on *Azolla*, water lily, and sorghum as seasonal temperatures were changing. They found that net photosynthetic rates were higher for *Azolla* and water lily during warmer times of year. Linear regressions on net photosynthetic rates for water lily versus air temperature at the time the measurements were taken showed a greater increase with temperature for plants grown at a 640 ppm carbon dioxide concentration than for those grown at a 340 ppm carbon dioxide concentration. However, there was also an interaction with solar radiation. The plants grown at 640 ppm of carbon dioxide also showed a much greater response to solar radiation than those grown at 340 ppm. Although the interactions among carbon dioxide treatment level, air temperature, and solar radiation were not resolved, the data show that all were interrelated in the carbon dioxide response. S. G. Allen et al. (1990a,b) also computed linear regressions of photosynthetic rate versus previous minimum air temperatures and previous maximum air temperatures for periods of 1, 3, 6, and 9 days. They found that net photosynthetic carbon dioxide uptake rates were more sensitive to previous minimum air temperatures than to previous maximum air temperatures. (Of course, maximum temperatures are closely related to minimum temperatures.) The slope of the regression increased with the number of previous days included in the calculation of air temperature. Most importantly, Allen et al., also found that leaf photosynthetic rates were more sensitive to previous minimum temperatures than to air temperature at the time the measurements were being made.

Photosynthetic rates of sorghum leaves increased only slightly when the leaves were grown at a high carbon dioxide concentration: the rates for a 640 ppm carbon dioxide concentration were about 10 percent higher than the rates for a 340 ppm temperature (S. G. Allen et al., 1990b). The effect of temperature on photosynthetic rates was also quite small—a response to be expected, since sorghum is a C4 plant. Leaves were about 1.0 to 1.5°C warmer under the 640 ppm carbon dioxide treatment.

Idso et al. (1987) compared growth rates (biomass accumulation rates) of carrot, radish, water hyacinth, *Azolla*, and cotton grown across a seasonal range of temperatures at carbon dioxide concentrations of 650 ppm with growth rates for these same crops grown at 350 ppm. They found that the ratio of biomass accumulation rates of plants grown at 650 ppm to plants grown at 350 ppm increased somewhat linearly with air temperature across the seasonal mean air temperature range of about 12°C to 36°C. The biomass growth ratio (biomass accumulation at specified elevated carbon dioxide concentration divided by biomass accumulation at a baseline carbon dioxide concentration) increased about 0.087 per 1°C over this temperature range and had a zero intercept at about 18.5°C mean daily air temperature. Under arid zone conditions at Phoenix, Arizona, daily minimum temperatures are 8 to 9°C lower than daily mean temperatures during the months of November to March, when mean air temperatures are well below 18.5°C. Low nocturnal temperatures, as well as low total solar irradiance, short photoperiod (day length), previous carbohydrate storage, and stage of growth of the plants may affect the biomass growth ratio during the winter months.

We may conclude that increasing both carbon dioxide concentrations and temperature will cause a greater increase in biomass productivity than increasing carbon dioxide levels alone. However, Baker et al. (1989) found different biomass growth ratios for both final harvest dry matter and seed yield for soybean grown at 330 and 660 ppm of carbon dioxide. Although early canopy vegetative growth rates suggest that the biomass growth ratio could increase with temperature, the final harvest data showed otherwise. The experiment was conducted with day/night air temperatures of 26/19, 31/24, and 36/29°C, which gave average air temperatures of about 22.8, 27.8, and 32.8°C under the 13/11 hour thermoperiod. The biomass growth ratios for final harvest dry matter were 1.50, 1.36, and 1.24 for the respective temperatures. The biomass growth ratios for final harvest seed yield were 1.46, 1.24, and 1.15 for the respective temperatures. The changes in the biomass growth ratio for dry matter and seed yield were -0.026 (r^2 = 0.98) and -0.031 (r^2 = 0.88) per 1°C, respectively. This cultivar of soybean, "Bragg," has a determinate growth habit that causes vegetative growth to nearly cease when flowering begins. Furthermore, elevated temperatures tend to hasten maturity and shorten the life cycle of this soybean crop. These factors were different from the Arizona study. Furthermore, the study of Baker et al. (1989) had identical light conditions for all treatments, and photoperiod interaction with temperature was minimized by two weeks of supplemental lighting at the beginning of the season.

In summary, the biomass growth ratio of plants grown at elevated carbon dioxide concentrations may increase with increasing temperature for vegetative growth, as suggested by Figure 7.4. However, this response may be reversed for seed grain crops that have a determinate growth habit, such as "Bragg" soybean.

Evapotranspiration

Evapotranspiration refers to the combination of plant transpiration and evaporation directly from the soil surface. Much of the following discussion of evapotranspiration will refer largely to the effects of carbon dioxide and climate on the plant component, which is in general much larger than the soil component except when the LAI is small.

The best modeling studies to date on the simulated effects of climate change and carbon dioxide concentration increase on plant canopy evapotranspiration were conducted by Rosenberg et al. (1990), using the Penman-Monteith model. A similar approach was used by Allen and Gichuki (1989) and Allen et al. (1991) to estimate effects of carbon dioxide-induced climate changes on evapotranspiration and irrigation water requirements in the Great Plains from Texas to Nebraska. Rosenberg et al. (1990) examined the effects of temperature, net radiation, air vapor pressure, stomatal resistance, and LAI on three types of plant canopies: wheat at Mead, Nebraska; grassland at Konza Prairie, Kansas; and forest at Oak Ridge, Tennessee. Increasing the temperature by 3°C gave a 6 to 8 percent increase in transpiration per 1°C. This compares reasonably well with the 4 to 5 percent increase in transpiration per 1°C measured experimentally in soybean across the 28 to 35°C range by Jones et al. (1985a). Rosenberg et al. (1990) also reported that evapotranspiration decreased 12 to 17 percent for a 40 percent increase in stomatal resistance. This corresponds closely to a 12 percent decrease in seasonal transpiration obtained experimentally for soybean grown in controlled-environment chambers by Jones et al. (1985c) for doubled carbon dioxide concentration conditions when leaf area index was very similar for both the ambient and doubled carbon dioxide treatments. In another study, Jones et al. (1985b) showed that exposure of soybean canopies to a level of 800 ppm carbon dioxide decreased daily total transpiration by 16 percent in comparison to an exposure level of 330 ppm. Jones et al. attributed this reduction to an increase in stomatal resistance.

Increases in LAI of 15 percent caused increases in predicted evapotranspiration of about 5 to 7 percent according to the model of Rosenberg et al. (1990). These values were comparable to those extracted from Jones et al. (1985b). Their data showed a 33 percent increase in measured daily transpiration for a change in LAI from 3.3 to 6.0 (an 82 percent increase in LAI). However, the effect of LAI may not be linear. By use of a soil-plant-atmosphere model, Shawcraft et al. (1974) showed that the effect of LAI on transpiration would be highly nonlinear across a LAI range of 0 to 8. Most of the effect of changing leaf area occurred across the LAI range of 0 to 4. However, these simulations were conducted with a moist soil surface (having a water potential of -60 MPa) and relatively high soil surface-to-air boundary-layer conductance. Thus, predicted evapotranspiration rates at a LAI of 2 were maintained at 85 percent or more of the rates at a LAI of 8. Nevertheless, the modeling results of Shawcraft et al. (1974) for three solar elevation angles and three leaf elevation angle classes showed that predicted plant transpiration increased by an average of 27 plus or minus 8 percent for a LAI increase from 3.3 to 6.0.

Net radiation could increase under climate change conditions from both greater downwelling thermal radiation and increased solar radiation (decreased cloudiness), or it could decrease from increased cloudiness. Rosenberg et al. (1990) showed that evapotranspiration should change about 0.6 to 0.7 percent for each 1 percent change in net radiation. Likewise, they showed that evapotranspiration should change about -0.4 to -0.8 percent for each 1 percent change in vapor pressure of the air. A combination of several factors gave changes in evapotranspiration ranging from 27 percent (for a case of increased net radiation and decreased vapor pressure) to negative 4 percent (for a case of decreased net radiation and increased vapor pressure. These factors were: a temperature increase of 3°C, net radiation changes of plus or minus 10 percent, vapor pressure changes of plus or minus 10 percent, stomatal resistance increase of 40 percent, and leaf area index increase of 15 percent. Each factor related to climate change and plant response to carbon dioxide affects the predicted evapotranspiration.

ESTIMATING YIELD AND WATER REQUIREMENTS OF CROPS UNDER TWO CURRENT CLIMATE CHANGE SCENARIOS

General Circulation Models

Climate changes under conditions of doubled atmospheric carbon dioxide levels have been predicted using five atmospheric general circulation models (GCMs). These models are the National Oceanic and Atmospheric Administration (NOAA) Geophysical Fluids Dynamics Laboratory (GFDL) model developed at Princeton University (Manabe and Wetherald, 1986, 1987), the NASA Goddard Institute for Space Studies (GISS) model developed at Columbia University (Hansen et al., 1984, 1988), the Community Climate Models (CCM) developed at the National Center for Atmospheric Research (NCAR) (Washington and Meehl 1983, 1984, 1986), the Oregon State University (OSU) model (Schlesinger, 1984), and the United Kingdom Meteorological Office (UKMO) model (Wilson and Mitchell, 1987; Mitchell, 1989). All of these models predict an increase of global average surface temperatures. The global mean surface temperature increases for recent modeling studies in which the carbon dioxide concentration was doubled were 2.8, 4.0, 4.0, 4.2 and 5.2°C for the OSU, CCM, GFDL, GISS, and UKMO models, respectively (Wilson and Mitchell, 1987). For a carbon dioxide concentration doubling, the global mean precipitation is also predicted to increase by 7.8, 7.1, 8.7, 11.0, and 15.0 percent for the above GCMs, respectively.

Grotch (1988) analyzed four of the GCM climate change scenarios for a climate with doubled carbon dioxide concentration.

Changes predicted across the United States were extracted and summarized, as were global, hemispherical, and other regional changes. The predicted June-July-August (JJA) median temperature increases for the United States were about 3.5, 3.0, 3.8, and 5.6°C for a carbon dioxide doubling from the OSU, CCM, GISS, and GFDL models, respectively. The predicted JJA changes in precipitation for the United States were 4, 10, 8, and -25 percent for the respective models. Thus, the GCM precipitation change scenarios are more variable for the United States than temperature change scenarios and may differ considerably among regions around the world. The UKMO model predictions for reduced precipitation for the United States are somewhat similar to the GFDL model scenario (Wilson and Mitchell, 1987). Higher JJA temperatures for the United States were associated with models with the lowest summer precipitation. As would be expected, both the GFDL model (Manabe and Wetherald, 1987) and the UKMO model (Wilson and Mitchell, 1987) predict serious decreases in soil wetness during the summer for the United States.

All of the GCMs predict a temperature increase for the Unites States for a doubling of atmospheric carbon dioxide. However, the predicted summer precipitation for the United States covers a range of 10 to -25 percent. The possibility of a significant reduction in summer precipitation, coupled with a temperature rise, could pose a serious problem for future agricultural productivity and water resources.

Modeling Crop Responses to Carbon Dioxide and Climate Changes

Many years of experimental observations on the interactions of carbon dioxide and climate factors would be required to provide complete information on responses of plants to climate change. However, plant growth models have been developed that are sensitive to environmental factors such as photoperiod, temperature, soil water availability, and light interception. These models can provide projections of crop response to future climate change scenarios in comparison with baseline climate records of the recent past.

Peart et al. (1989) and Curry et al. (1990a,b) used a soybean crop growth model, SOYGRO (Wilkerson et al., 1983; Jones et al., 1989), and a maize growth model, CERES-maize (Jones and Kiniry, 1986), for predicting growth and yield responses to doubled carbon dioxide climate change scenarios in the southeastern United States.

These simulations were conducted using 30 years of baseline weather data (1951 through 1980) from 19 sites in the southeastern United States—a region bounded by Virginia and Kentucky to the north and Arkansas and Louisiana to the west. Predicted climate changes within the appropriate grid cells of two GCMs, the GISS model (Hansen et al., 1988) and the GFDL model (Manabe and Wetherald, 1987), were used to change temperatures, precipitation, and solar radiation, month by month, for each of the baseline data sets at each site. These modified baseline data sets provided GISS and GFDL climatic scenarios (Smith and Tirpak, 1989). Monthly precipitation data of two sites for baseline climate and derived values for GISS and GFDL scenarios are given in Figures 7.5, 7.6, and 7.7, respectively, for Columbia, South Carolina, and Figures 7.8, 7.9, and 7.10, respectively, for Memphis, Tennessee. These two sites were essentially at the center of two GISS model grid cells and close to the center of two GFDL model grid cells, and thus should be appropriate sites for representing climate change scenarios within the grid cells of the two models. Monthly averages of July maximum daily temperatures for baseline, GISS, and GFDL scenarios were 33.27, 35.23, and 38.19°C, respectively, for Columbia, and 33.07, 35.44, and 36.03°C, respectively, for Memphis. Current planting dates, cultivars, and prevailing cropped soil types at each site were used in the simulations.

First, simulations were run based on rain-fed climate change conditions without direct fertilization effects of elevated carbon dioxide concentrations. Next, those simulations were repeated with optimum irrigation. Finally, simulations were run under both rain-fed and irrigated conditions for doubled atmospheric carbon dioxide conditions with a crop photosynthetic enhancement factor of 1.35 for soybean and 1.10 for maize.

Table 7.2 shows the soybean seed yield simulations from the SOYGRO model averaged over 19 sites and 30 years. First, under rain-fed conditions with climate change effects only, average soybean yield, compared to the baseline climate, was reduced by 71 percent under the GFDL scenario, but was reduced only 23 percent under the GISS scenario. The yields under the GFDL scenario were severely impacted because of the rainfall reductions predicted by this GCM (Figures 7.5 to 7.10).

Under optimum irrigation conditions, average soybean yields under both the GISS scenario and the GFDL scenario were reduced by 18 or 19 percent with respect to optimum irrigation under baseline climate conditions. However, in spite of higher temperatures, the irrigated yields under the GISS and GFDL scenarios were about

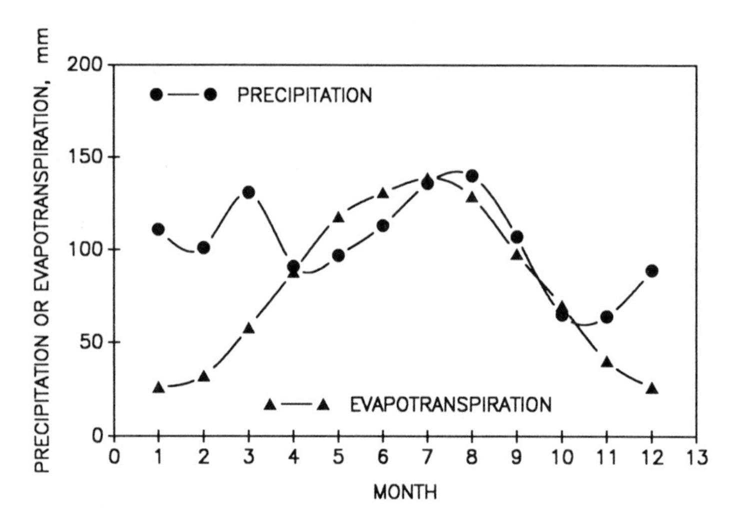

FIGURE 7.5 Average monthly precipitation and potential evapotranspiration for Columbia, South Carolina, for the 30-year base climate period, 1951 to 1980.
SOURCE: Precipitation data adapted from Table 6 of Peart et al., 1989.

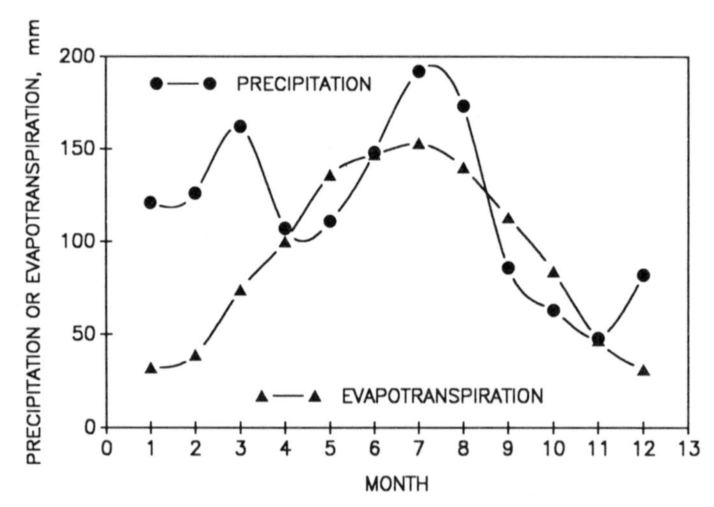

FIGURE 7.6 Derived average monthly precipitation and potential evapotranspiration for Columbia, South Carolina, from a GISS climate change scenario for doubled atmospheric carbon dioxide.
SOURCE: Precipitation values adapted from Table 6 of Peart et al., 1989.

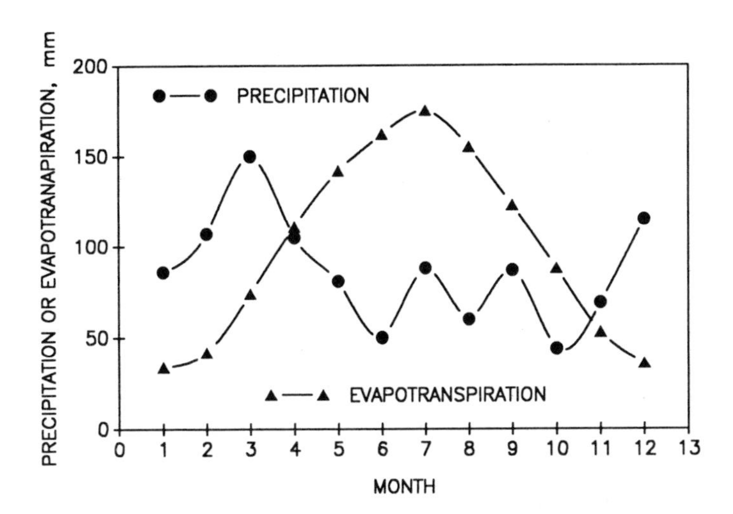

FIGURE 7.7 Derived average monthly precipitation and potential evapotranspiration for Columbia, South Carolina, from a GFDL climate change scenario for doubled atmospheric carbon dioxide.
SOURCE: Precipitation values adapted from Table 6 of Peart et al., 1989.

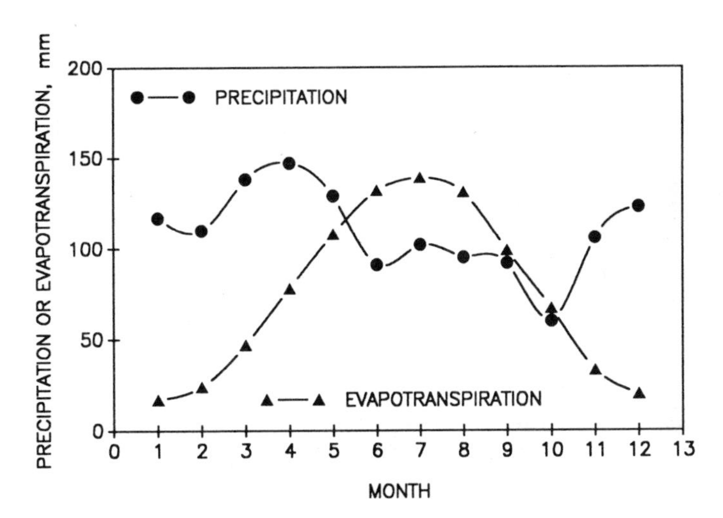

FIGURE 7.8 Average monthly precipitation and potential evapotranspiration for Memphis, Tennessee, for the 30-year base climate period, 1951 to 1980.
SOURCE: Precipitation data adapted from Table 6 of Peart et al., 1989.

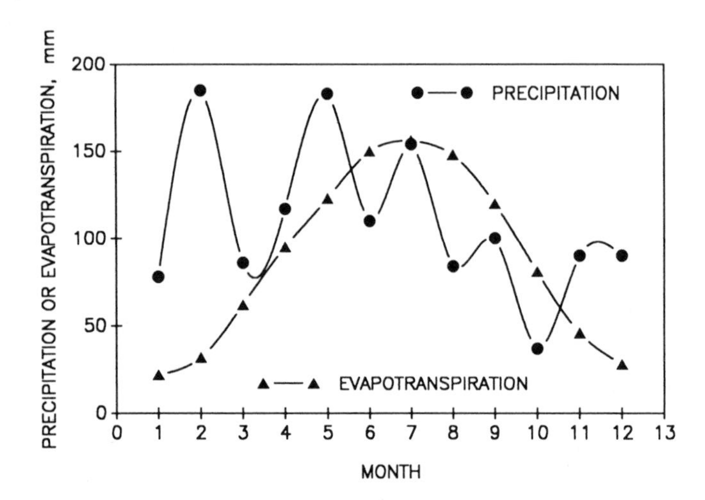

FIGURE 7.9 Derived average monthly precipitation and potential evapotranspiration for Memphis, Tennessee from a GISS climate change scenario for doubled atmospheric carbon dioxide.
SOURCE: Precipitation data values from Table 6 of Peart et al., 1989.

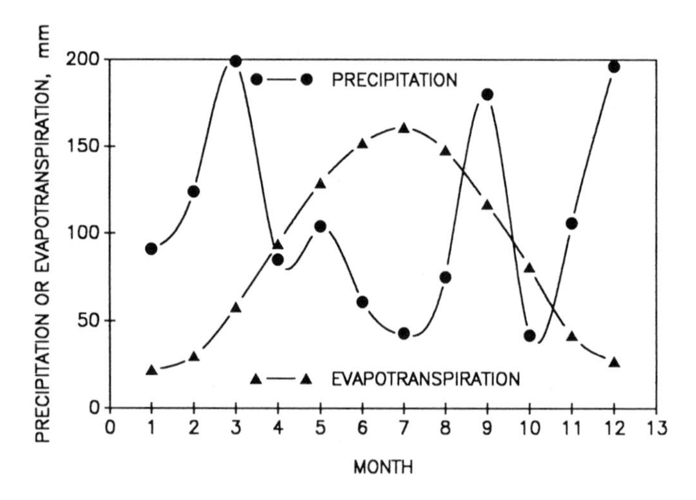

FIGURE 7.10 Derived average monthly precipitation and potential evapotranspiration for Memphis, Tennessee from a GFDL climate change scenario for doubled atmospheric carbon dioxide.
SOURCE: Precipitation values adapted from Table 6 of Peart et al., 1989.

TABLE 7.2 Effects of climate change only on simulated soybean yields (bushels/acre) for the southeastern United States when the atmospheric carbon dioxide concentration is doubled.

BASE	GISS Model		GFDL Model		Model
Yield	Yield	Diff.	Yield	Diff.	Diff.
Rainfed					
37	29	-23%	11	-71%	62%
Irrigated					
57	47	-18%	46	-19%	-2%

SOURCE: Adapted from Peart et al., 1989.

25 percent greater than the baseline climate scenario without irrigation.

When carbon dioxide fertilization plus climate change effects were simulated (Table 7.3), average soybean yields for the GISS climate scenario increased 11 percent under rain-fed conditions, whereas yields under the GFDL climate scenario still decreased (by 52 percent). Under optimum irrigation with carbon dioxide fertilization, yields under both GISS and GFDL scenarios were increased 13 to 14 percent relative to the irrigated baseline climate scenario.

Maize yields declined by only about 8 percent in the GISS climate scenario and by about 73 percent in the GFDL scenario when the effects of climate change due to the greenhouse effect were simulated (Table 7.4). Although irrigation increased predicted yields, the GISS and GFDL climate scenarios gave yield decreases of 18 and 27 percent, respectively, relative to the irrigated baseline weather crops. Including the direct effects of carbon dioxide fertilization plus climate change had little effect on the predicted yields of maize (Table 7.5), as expected because maize is a C4 plant.

Ritchie et al. (1989) found that of the climate changes that might occur because of increasing greenhouse gas concentrations, higher temperature had the greatest effect on predicted soybean and maize yields in the Great Lakes and corn belt regions. Increases in temperature caused a decrease in the duration of the crop life cycle. Yield reductions were greatest for the GFDL

TABLE 7.3 Direct carbon dioxide fertilization effects plus climate change effects on simulated soybean yields (bushels/acre) for the southeastern United States under a doubled carbon dioxide concentration.

BASE	GISS Model		GFDL Model		Model
Yield	Yield	Diff.	Yield	Diff.	Diff.
Rainfed					
37	41	+11%	18	-52%	-56%
Irrigated					
57	65	-13%	65	-14%	+1%

SOURCE: Adapted from Peart et al., 1989.

TABLE 7.4 The effects of climate change only on simulated maize yields (bushels/acre) when the atmospheric carbon dioxide concentration is doubled. Effects are averaged over four locations: Charlotte, Macon, Meridian, and Memphis.

BASE	GISS Model		GFDL Model		Model
Yield	Yield	Diff.	Yield	Diff.	Diff.
Rainfed					
137	126	-8%	37	-73%	-71%
Irrigated					
224	183	-18%	164	-27%	-10%

SOURCE: Adapted from Peart et al., 1989.

climate change scenario for the southernmost locations of this region. Predicted yields increased for the northernmost stations because temperatures and growing season duration became more favorable for these crops. Overall, irrigation water requirements

TABLE 7.5 Direct carbon dioxide fertilization effects plus climate change effects (for a doubled carbon dioxide concentration) on simulated maize yields (bushels/acre) averaged over four locations: Charlotte, Macon, Meridian, and Memphis.

BASE	GISS Model		GFDL Model		Model
Yield	Yield	Diff.	Yield	Diff.	Diff.
Rainfed					
137	130	-5%	35	-74%	-73%
Irrigated					
224	184	-18%	165	-26%	-10%

SOURCE: Adapted from Peart et al., 1989.

in this region increased about 90 percent for the GFDL scenario and decreased about 30 percent for the GISS scenario.

Rosenzweig (1989) modeled corn and wheat yields in the Great Plains under GISS and GFDL climate change scenarios. She found that corn and wheat yields decreases were most extreme in the southern Great Plains because higher temperatures shortened the life cycle of the crops. Where precipitation was predicted to decrease, irrigation requirements increased. Allen and Gichuki (1989) predicted a 15 percent overall irrigation requirement increase for this region, with greater requirements for alfalfa because its growing season was increased and lower requirements for corn and winter wheat because their growing seasons were decreased.

The direct effect of rising carbon dioxide concentrations offset the adverse effects of climate change at some, but not all, locations in the simulations of Ritchie et al. (1989) and Rosenzweig (1989).

The impact of climate change on California water resources and irrigated agriculture is of particular interest because of the wide range of vegetable, fruit, and nut crops produced there for the rest of the nation. Dudek (1989) predicted productivity changes for several of these crops in response to GISS and GFDL climate change scenarios in which the carbon dioxide concentration was doubled; the basis for Dudek's work was a United Nations Food and Agriculture Organization agro-ecological zone method

(Doorenbos and Kassam, 1979). This method is not as rigorous as crop climate models, but such models have not been developed for specialty crops. For climate change only, Dudek (1989) predicted productivity decreases of 3 to 40 percent, depending upon crop, region, and scenario. Depending upon crop, statewide average yield decreased by 8 to 34 percent for the GISS scenario and by 6 to 31 percent for the GFDL scenario. When carbon dioxide enrichment effects were introduced, based in part on information from Kimball (1983), predicted productivity changes ranged from plus 41 to minus 27 percent, depending upon crop, region, and scenario. However, ranges of statewide average yield changes were somewhat smaller. Depending upon crop, predicted yields ranged from about plus 17 to minus 12 percent for the GISS scenario and about plus 21 to minus 8 percent for the GFDL scenario.

Dudek (1989) used the California agriculture and resource model (CARM) to predict economic and market impacts of the productivity changes. Although production changes generally declined under the climate change only scenarios, commodity prices generally increased. Under the climate change plus carbon dioxide effects scenarios, the overall impacts were much smaller. Since predicted water resource supplies were reduced, especially for the San Joaquin Valley, predicted use of both ground water and surface water decreased.

Rosenberg et al. (1990) used the Penman-Monteith evapotranspiration model and GISS, GFDL, and NCAR scenarios for estimating the impact of predicted climate changes on summer evapotranspiration at Mead, Nebraska (for wheat), Konza Prairie, Kansas (for grassland), and Oak Ridge, Tennessee (for forest). They found that inclusion of other factors (net radiation, vapor pressure, wind speed, stomatal resistance, and leaf area index) reduced the impact of temperature increases on predicted evapotranspiration. For example, a 6.3°C rise in GFDL scenario temperature for Mead, Nebraska, resulted in a 42 percent increase in predicted evapotranspiration but only a 23 percent increase when all other climate change factors were also considered. Similarly, a GISS scenario temperature rise of 4.7°C for Konza Prairie, Kansas, gave a 28 percent predicted increase in evapotranspiration, but this increase was only 4 percent when all other climate change factors were included in the Penman-Monteith equation. On the basis of these computations, Rosenberg et al. (1990) caution against the presumption that evapotranspiration will increase in all scenarios of global climate change. When the decreased duration of the life cycle of plants (caused by increased temperature) is included in climate change scenarios, some crops (such as corn and winter wheat) may actually transpire less total water throughout their life cycles when grown on the Great Plains, as shown in model studies of Allen and Gichuki (1989) and Allen et al. (1991).

IMPLICATIONS OF CHANGING EVAPOTRANSPIRATION AND PRECIPITATION FOR WATER RESOURCES

Adaptations and Evapotranspiration Requirements

The crop yield simulations of the previous section demonstrate two main points: the adverse impact on crop production of inadequate rainfall under rising temperature scenarios and the importance of direct carbon dioxide effects in the face of elevated temperatures. However, the climate changes that would occur under the doubled carbon dioxide concentrations used in this simulation may occur at lower carbon dioxide concentrations if radiatively active trace gases other than carbon dioxide play a large role in the greenhouse effect. In that case, the direct carbon dioxide effects would be somewhat lower than those shown in the examples of Tables 7.2 through 7.5 for an equivalent climate change.

Much of the reduction in soybean yields reported by Peart et al. (1989) and Curry et al. (1990a,b) was due to decreases in the length of the grain filling period under higher temperatures. Changes in management practices may help restore part of the yield. Planting earlier or later in the season may help offset the higher temperatures. Selection of other cultivars could also help. In the future, it may be necessary to breed plants with new combinations of temperature tolerance and photoperiod responses, or perhaps to use growth regulators. Under conditions in which nonstructural carbohydrates accumulate in carbon dioxide-enriched plants, new germplasms that use photoassimilate more efficiently need to be developed.

Irrigation is not likely to be a panacea for climate change. In simulations by Peart et al. (1989) and Curry et al. (1990a,b), the average irrigation requirement increased by 33 percent under the GISS scenario and 134 percent under the GFDL scenario. However, with less summertime rainfall under the GFDL scenario (Figures 7.5 through 7.10), region-wide water resources would become scarce, and water may not be readily available for crops. Some areas of the United States may have to adapt by irrigating less land area and other areas by shifting from rain-fed to irrigated agriculture.

Increasing temperatures and decreasing precipitation for the nation as predicted by the GFDL model (and the UKMO model) would have serious negative overall impacts on agricultural productivity and society, although producers in some regions may benefit from higher prices brought on by scarcity (Adams et al., 1990).

Water Availability

Changes in crop yields and irrigation requirements caused by changed rainfall patterns and temperatures would influence ground water recharge, streamflow, and reservoir storage, as illustrated by Miller and Brock (1988; 1989) in a Tennessee Valley Authority modeling study of climate change. Climate change scenarios for Columbia, South Carolina, and Memphis, Tennessee for the GISS and GFDL models were selected to further illustrate the precipitation and temperature change effects on potential monthly and annual evapotranspiration and water deficits or excesses (Peart et al., 1989; Curry et al., 1990a,b).

Monthly potential evapotranspiration (PET) was calculated from the method of Jensen and Haise (1963) as modified by Jensen (1966). Cloud cover data were taken from Landers (1974) for Columbia and from Dickson (1974) for Memphis. Monthly solar radiation data were calculated from extraterrestrial solar irradiance and cloud cover data according to the procedure shown by Doorenbos and Pruitt (1977). The monthly Jensen-Haise PET computations were conducted based on computer programs by Zazueta and Smajstrla (1989). The summation of these monthly PET values was about 20 percent larger than values shown by Geraghty et al. (1973) in a water atlas of the United States. The method of Stephens and Stewart (1963), with coefficients applicable for Waynesville, North Carolina (Stephens, 1965), was applied to the Memphis baseline (1951 through 1980) climate data. Hand calculations using the Stephens-Stewart method gave an annual PET of 895 mm for Memphis, which was 84 percent of the computed Jensen-Haise value and was in close agreement with the water atlas value (Geraghty et al., 1973). Therefore, all monthly Jensen-Haise PET values were adjusted by multiplying them by 0.84, which gave annual PET values of 895 mm for Memphis and 955 mm for Columbia.

The average annual temperatures for the baseline period, the GISS scenario, and the GFDL scenario were 17.4, 19.9, and 20.6°C, respectively, for Columbia and 16.5, 19.9, and 19.3°C, respectively, for Memphis. Average monthly distributions of calculated PET, along with precipitation (PPT), for Columbia are shown in Figures 7.5, 7.6, and 7.7 for the baseline climate, the GISS scenario, and the GFDL scenario, respectively. On average, monthly PPT for the May through October period would be sufficient for crop production for both baseline climate and GISS scenario conditions (Table 7.6). However, PPT minus PET over the May through October period was -435 mm for the GFDL scenario, which indicates a severe water deficit during this period. Nevertheless, during the November through April period, PPT minus PET for the GFDL scenario was about 87 percent of the GISS scenario. Therefore, following dry summer months there is opportunity for sizable recharge and streamflow during the winter months despite the overall annual water deficit of about 153 cm for the GFDL scenario at Columbia (Table 7.6).

For the Memphis example, the baseline climate showed a slight tendency toward summer drought (Figure 7.8). The GISS scenario (Figure 7.9) and the GFDL scenario (Figure 7.10) showed more month-to-month variation for Memphis than for Columbia. However, the total annual PPT was similar for all three Memphis cases (Table 7.6). The PPT deficit was particularly severe for the GFDL scenario during the June through August period. Nevertheless, the potential for recharge and subsequent streamflow was great (528 mm) during the November through April period, so the GFDL scenario has the potential of providing more streamflow water resources than the GISS scenario, although the GISS scenario provides more rainfall for crops and other vegetation during the summer months (Table 7.6).

These examples at two locations in a humid region show that although climate change scenarios may imply severe rainfall shortages and soil water deficits during the growing period, PPT excesses over PET during the cooler parts of the year still have the potential for maintaining cool season streamflows and water storage in reservoirs.

Allen et al. (1982) showed that annual streamflow water yield of a Florida watershed was linearly related to annual PPT. Furthermore, the intercept on the PPT axis was 806 mm, which approached the average annual watershed evapotranspiration of 891 mm (Knisel et al., 1985) determined by water budget analysis. Knisel et al. (1985) further showed that streamflow in this watershed was generally linearly related to PPT during three 4-month periods over the annual cycle. The intercept on the PPT axis for each of the three periods was similar to the average

TABLE 7.6 Average annual precipitation (PPT), potential evapotranspiration (PET), PPT minus PET for a 6-month period of recharge and runoff (November through April), and PPT for a 6-month period of high PET (May through October) for baseline climate (1951 to 1980) and for GISS and GFDL scenarios at two locations: Columbia and Memphis.

	BASE (mm)	GISS (mm)	GFDL (mm)
Columbia, SC			
PPT	1245	1419	1042
PET	955	1096	1195
PPT-PET (Nov. - Apr.)	+317	+323	+282
PPT-PET (May - Oct.)	-27	0	-435
NET	+290	+323	-153
Memphis, TN			
PPT	1310	1314	1306
PET	895	1063	1061
PPT-PET (Nov. - Apr.)	+522	+361	+528
PPT-PET (May - Oct.)	-107	-110	-283
NET	+415	+251	+245

watershed evapotranspiration during each period. These studies demonstrated that, in general, evapotranspiration requirements had to be met before sizable amounts of runoff occurred. However, intense or long-lasting rainfall events and low soil infiltration and percolation would result in increased runoff and streamflow before the monthly PET requirements are met. Also, during cooler months with higher rainfall, some soil water recharge must occur before rainfall will produce streamflow.

The impact of several climate change scenarios on California water resources was summarized by King et al. (1989). In general, the GISS and GFDL models predicted annual temperature increases of about 4.6 and 4.4°C, respectively, whereas the OSU model predicted temperature increases of only 2.1°C. These predicted GISS and GFDL temperature changes would lead to greater winter runoff from snowmelt and rainfall in the mountains around the Central Valley, and the snowpack would melt earlier. Consequently, runoff in the late spring and early summer would be less. The predicted temperature changes of the OSU model led to only slight changes in runoff patterns, with peak runoff still occurring in May. The current reservoir storage and subsequent deliveries of the California State Water Project system could be reduced by 16, 14, and 7 percent based on the GISS, GFDL, and OSU climate change scenarios (King et al., 1989). According to the projections of the climate change scenarios, the Federal Central Valley Project would not be impacted as much.

Several assessments of climate change on water resources throughout the United States have been reported (e.g., Smith and Tirpak, 1989; Gleick, 1990; Waggoner, 1990). In general, watersheds in the West and Great Plains (except for those in the Pacific Northwest) have a high consumptive use to renewable supply ratio (averaging about 0.25) and a high total demand (the sum of consumptive use, ground water overdraft, water transfers, and evaporation) to renewable supply ratio (averaging about 0.35). Also, except for in the Pacific Northwest and the upper Colorado River basin, the ratio of ground water overdraft to ground water recharge is large (averaging about 0.42). These factors illustrate the vulnerability of the western states' water resources to climate change scenarios that include reduced precipitation. The impacts of reduced precipitation on native ecosystems, rain-fed agriculture, and water supply systems could be equally severe for eastern states. Variability of precipitation causes water shortages under present conditions, as demonstrated by the current California drought. Therefore, not only climate changes but also climate variability must be considered in managing water resources for sustained agricultural production to ensure an adequate food supply. Studies of interannual and daily variability of temperatures and precipitation show variations, in turn, among GCM models and among climate scenarios driven by both current and future projected levels of greenhouse gases (Mearns, 1989). Whether we should anticipate the consequences of a hotter and drier climate or a hotter and wetter climate, with more, less, or no change in variability, remains to be seen.

SUMMARY

Elevated carbon dioxide levels appear to increase the size and dry weight of most plants and plant components. Relatively more photoassimilate appears to be partitioned into structural components (stems and petioles) during vegetative development in order to support the light-harvesting apparatus (leaves). This observation may be a manifestation of plant size rather than a unique carbon dioxide effect. In general, the harvest index tends to decrease with increasing carbon dioxide concentration and with increasing temperature. Selection of plants that would use more photoassimilates for reproductive growth seems a useful goal for future research as the atmospheric carbon dioxide concentration increases. Plant growth regulators may also play an important role. Research efforts should be directed not only toward assessing the impacts of climate change on agriculture but also toward exploring biological adaptations and management systems for reducing the impacts of climate change. Whether we should anticipate a hotter and drier climate or a hotter and wetter climate remains to be seen.

The key topics discussed in this paper are summarized below:

- The biomass and seed yield production of crops is linearly related to water use. The ratio of dry matter produced (measured as, for example, total biomass, or seed yield) to crop water used is called water-use efficiency (WUE). More water is required per unit biomass produced (in other words, the WUE is lower) in hot, arid environments than in cool, humid environments (which have a higher WUE). The water requirement also varies with species. Under current carbon dioxide levels, plants with the C4 photosynthetic mechanism require less water per unit biomass produced (they have a higher WUE) than plants with the C3 photosynthetic mechanism (which have a lower WUE).
- The carbon dioxide concentration was as low as 180 to 200 ppm during the coldest part of the last ice age. As the ice melted, the concentration rose to about 270 ppm, where it remained stable until the advent of the industrial revolution and expansion of human population. The carbon dioxide concentration increased to 315 ppm by 1958, when continuous measurements were begun at Mauna Loa, Hawaii. The atmospheric carbon dioxide concentration rose to 330 ppm by 1973, the year of the first "energy crisis," and today has reached about 355 ppm. Based on soybean experiments, the increase from 270 to 330 ppm should have increased seed yields by 12 percent. If carbon dioxide levels double from 330 to 660

ppm, seed yield should increase by about 31 percent, which is in agreement with a 33 plus or minus 6 percent increase in plant growth and yields reported in a literature survey by Kimball (1983).

- Doubling the carbon dioxide concentration will cause stomatal conductance of leaves to decrease by about 40 percent. However, whole crop cumulative transpiration over the course of a season may only be reduced by 0 to 12 percent, for two reasons. First, when stomata partially close and restrict vapor from leaves, the foliage temperature rises. This raises the vapor pressure of water in the intercellular air space of leaves and increases the leaf-to-air vapor pressure difference. Thus, the reduction of stomatal conductance is substantially overcome by the driving force for evaporation. Secondly, elevated carbon dioxide levels promote growth of a greater amount of leaf area, so that a larger surface area for transpiration exists. In conclusion, a 40 percent reduction in stomatal conductance probably provides only a 10 to 15 percent reduction in transpirational water use. The greater amount of leaf area under elevated carbon dioxide conditions could eliminate this small difference in crop water use.
- Increases in WUE of carbon dioxide-enriched crops is due largely to sizable increases in photosynthesis, growth, and yield. Decreases in water use are small and contribute very little, if anything, to increases in WUE.
- Plants exposed to elevated carbon dioxide concentrations (650 ppm in comparison with 350 ppm) have shown greater response to carbon dioxide at high average daily temperatures than at low average daily temperatures during vegetative growth across the range of 12 to 35°C. This response is in agreement with single leaf photosynthetic measurements of responses to elevated carbon dioxide across a range of temperatures. One could conclude that all crops will respond more to the combination of increasing temperature with elevated carbon dioxide than to elevated carbon dioxide levels alone. However, during reproductive growth of soybean plants, the opposite trend was found. The complex responses of various kinds of plants to interactions of carbon dioxide, temperature, water supply, light, and photoperiod (day length) need further research.
- Increasing temperatures across the range of 28 to 35°C appears to increase the transpiration rate by about 4 to 5 percent per 1°C, as shown in both experimental and modeling studies. This is in close agreement with the rise in saturation vapor pressure of about 6 percent per 1°C.

- Outputs from two GCMs, the GISS and GFDL models, for a doubled carbon dioxide concentration were used as examples of predictions of soybean crop yield in the southeastern United States. The GISS model predicted about 25 percent more June through August rainfall than baseline 1951 through 1980 values, and the GFDL model predicted about 40 percent less rainfall. Under rain-fed conditions and considering only the effects of climate change, the soybean crop model predicted a decrease in yields of 23 and 71 percent for the GISS and GFDL scenarios, respectively, compared with baseline weather conditions. When the direct effect of carbon dioxide on plant growth was included, the predictions were plus 11 percent for the GISS model and minus 52 percent for the GFDL model. When the simulated crops were irrigated, there was little difference between the predicted yields (plus 13 to 14 percent) of the two climate change scenarios. These data illustrate the critical importance of temperature and especially rainfall in any climate change scenario. Predicted crop yields for other regions (the Great Lakes, corn belt, Great Plains, and California) in response to carbon dioxide and climate change were generally similar to those of the Southeast.
- Under the soybean simulations for the southeastern United States, the average requirement for irrigation increased by 33 percent and 134 percent above the baseline for the GISS and GFDL scenarios, respectively.
- Rising carbon dioxide levels will cause an increase of photosynthetic rates, growth, yield, and water use efficiency for C3 crop plants. Water use per unit land area will not change much unless temperatures increase. Elevated temperatures may reduce seed crop yields in most areas of production, and potential rainfall decreases could cause serious reductions in rain-fed agriculture. Under irrigated agriculture conditions, reductions of precipitation could limit the number of acres available for irrigation and could lead to serious competition for water resources among various users.
- Changes in precipitation amounts and distribution are the most serious climate change factors, from the standpoint of both crop productivity and water resources. However, low rainfall during the summer growing season does not necessarily mean a large reduction in overall water yield of a basin. If high rainfall should occur during the cool seasons, annual runoff, streamflow, and reservoir storage may not be impacted as much as if there were no change in seasonal patterns. Thus, water resources for nonagricultural uses may not be impacted as much as water for agriculture. Models that incorporate sufficient details of the hydro-

logic cycle, as well as vegetation and energy balance factors, should be developed to provide a more informed physical basis for managing water resources. This information could be used in assessing long-range ecological, food supply, economic, and societal consequences of water management decisions under conditions of climate uncertainty.

ACKNOWLEDGMENTS

This work was supported in part by the U.S. Department of Energy Interagency Agreements DE-AI05-88ER69014 and DE-AI01-81ER60001 with the U.S. Department of Agriculture, Agricultural Research Service. This work was conducted in cooperation with the University of Florida at Gainesville. Florida Agricultural Experiment Station Journal Series number R-01423.

REFERENCES

Acock, B., and L. H. Allen, Jr. 1985. Crop responses to elevated carbon dioxide concentration. Pp. 53-97 in B. R. Strain and J. D. Cure, eds., Direct Effects of Increasing Carbon Dioxide on Vegetation. Report DOE/ER-0238. Washington, D.C.: U.S. Department of Energy, Carbon Dioxide Research Division.

Adams, R. M., C. Rosenzweig, R. M. Peart, J. T. Ritchie, B. A. McCarl, J. D. Glyer, R. B. Curry, J. W. Jones, K. J. Boote, and L. H. Allen, Jr. 1990. Global climate change and U.S. agriculture. Nature 345:219-224.

Allen, L. H., Jr. 1990. Plant responses to rising carbon dioxide and potential interactions with air pollutants. J. Environ. Qual. 19:15-34.

Allen, L. H., Jr., K. J. Boote, J. W. Jones, P. H. Jones, R. R. Valle, B. Acock, H. H. Rogers, and R. C. Dahlman. 1987. Response of vegetation to rising carbon dioxide: Photosynthesis, biomass, and seed yield of soybean. Global Biogeochemical Cycles 1:1-14.

Allen, L. H., Jr., P. Jones, and J. W. Jones. 1985. Rising atmospheric CO_2 and evapotranspiration. Pp. 13-27 in Advances in Evapotranspiration. ASAE Pub. 14-85. St. Joseph, Michigan: American Society of Agricultural Engineers.

Allen, L. H., Jr., W. G. Knisel, and P. Yates. 1982. Evapotranspiration, rainfall, and water yield in south Florida research watersheds. Soil Crop Sci. Soc. Fla. Proc. 41:127-139.

Allen, R. G., and F. N. Gichuki. 1989. Effects of projected CO_2-induced climatic changes on irrigation water requirements in the Great Plains States (Texas, Oklahoma, Kansas, and Nebraska). Appendix C (Agriculture), Vol. 1, Chapter 6 in J. B. Smith and D. A. Tirpak, eds., The Potential Effects of Global Climate Change on the United States. Report EPA-230-05-89-053. Washington, D.C.: U.S. Environmental Protection Agency.

Allen, R. G., F. N. Gichuki, and C. Rosenzweig. 1991. CO_2-induced climatic changes on irrigation water requirements. J. Water Resources Planning and Management 117:157-178.

Allen, S. G., S. B. Idso, B. A. Kimball, and M. G. Anderson. 1988. Interactive effects of CO_2 and environment on photosynthesis of *Azolla*. Agric. For. Meteorol. 42:209-217.

Allen, S. G., S. B. Idso, and B. A. Kimball. 1990a. Interactive effects of CO_2 and environment on net photosynthesis of water lily. Agric., Ecosystems, and Environ. 30:81-88.

Allen, S. G., S. B. Idso, B. A. Kimball, J. T. Baker, L. H. Allen, Jr., J. R. Mauney, J. W. Radin, and M. G. Anderson. 1990b. Effects of Air Temperature on Atmospheric CO_2-Plant Growth Relationships—TR048. Report DOE/ER-0450T. Washington, D.C.: U.S. Department of Energy and U.S. Department of Agriculture.

Allison, F.E., E. M. Roller, and W. A. Raney. 1958. Relationship between evapotranspiration and yield of crops grown in lysimeters receiving natural rainfall. Agron. J. 50:506-511.

Arkley, R. J. 1963. Relationship between plant growth and transpiration. Hilgardia 34:559-584.

Aston, A. R. 1984. The effect of doubling atmospheric CO_2 on streamflow: a simulation. Jour. Hydrol. 67:273-280.

Baker, J. T., L. H. Allen, Jr., K. J. Boote, P. Jones, and J. W. Jones. 1989. Response of soybean to air temperature and carbon dioxide concentration. Crop Sci. 29:98-105.

Barnola, J. M., D. Raynaud, Y. S. Korotkevich, and C. Lorius. 1987. Vostok ice core provides 160,000-year record of atmospheric CO_2. Nature 329:408-414.

Bennett, O. l., B. D. Doss, D. A. Ashley, V. J. Kilmer, and E. C. Richardson. 1964. Effects of soil moisture regime on yield, nutrient content, and evapotranspiration for three annual forage species. Agron. J. 56:195-198.

Berry, J., and O. Björkman. 1980. Photosynthetic response and adaptation to temperature in higher plants. Ann. Rev. Plant Physiol. 31:491-543.

Bowes, G., and W. L. Ogren. 1972. Oxygen inhibition and other properties of soybean ribulose-1,5-diphosphate carboxylase. J. Biol. Chem. 247:2171-2176.

Brazel, A. J., and S. B. Idso. 1984. Implications of the Rapidly Rising CO_2 Content of the Earth's Atmosphere for Water Resources in Arizona. Scientific Paper No. 19, Laboratory of Climatology. Tempe, Arizona: Arizona State University.

Briggs, L. J., and H. L. Shantz. 1913a. The Water Requirements of Plants: I. Investigations in the Great Plains in 1910 and 1911. Bureau of Plant Industry Bull. 284. Washington, D.C.: U.S. Department of Agriculture.

Briggs, L. J., and H. L. Shantz. 1913b. The Water Requirements of Plants: II. A Review of the Literature. Bureau of Plant Industry Bull. 285. Washington, D.C.: U.S. Department of Agriculture.

Briggs, L. J., and H. L. Shantz. 1914. Relative water requirements of plants. J. Agric. Res. 3:1-63.

Chang, J. H. 1968. Climate and Agriculture, An Ecological Survey. Chicago: Aldine Publishing Co.

Curry, R. B., R. M. Peart, J. W. Jones, K. J. Boote, and L. H. Allen, Jr. 1990a. Simulation as a tool for analyzing crop response to climate change. Trans. ASAE 33:981-990.

Curry, R. B., R. M. Peart, J. W. Jones, K. J. Boote, and L. H. Allen, Jr. 1990b. Response of crop yield to predicted changes in climate and atmospheric CO_2 using simulation. Trans. ASAE 33:1381-1390.

De Wit, C. T. 1958. Transpiration and Crop Yields. Versl. Landbouwk. Onderz. (Agr. Res. Rep.) 64.6. Wageningen, Netherlands: Centre for Agricultural Publication and Documentation (Pudoc).

Dickson, R. R. 1974. The climate of Tennessee. Pp. 370-384 in Climate of the States, Vol. 1. Port Washington, New York: Water Information Center, Inc.

Doorenbos, J., and A. H. Kassam. 1979. Yield Responses to Water. FAO Irrigation and Drainage Paper No. 33. Rome, Italy: Food and Agriculture Organization of the United Nations.

Doorenbos, J., and W. O. Pruitt. 1977. Crop Water Requirements. FAO Irrigation and Drainage Paper No. 24. Rome, Italy: Food and Agriculture Organization of the United Nations.

Dudek, D. J. 1989. Climate change impacts upon agriculture and resources: a case study of California. Appendix C (Agriculture), Vol. 1, Chapter 5 in The Potential Effects of Global Climate Change on the United States. Report EPA-230-05-89-053. Washington, D.C.: U.S. Environmental Protection Agency.

Enoch, H. Z., and B. A. Kimball. 1986. Carbon Dioxide Enrichment of Greenhouse Crops, Vols. I and II. Boca Raton, Florida: CRC Press, Inc.

Geraghty, J. J., D. W. Miller, F. van der Leeder, F. L. Troise, M. Pinther, and R. S. Collins. 1973. Water Atlas of the United States. Port Washington, New York: Water Information Center, Inc.

Gleick, P. H. 1990. Vulnerability of water systems. Pp. 223-240 in P. E. Waggoner, ed., Climate Change and U.S. Water Resources. New York: John Wiley and Sons.

Grotch, S. L. 1988. Regional Intercomparisons of General Circulation Model Prediction and Historical Climate Data—TR041. Report DOE/NBB-0884. Washington, D.C.: U.S. Department of Energy, Carbon Dioxide Research Division.

Guinn, G., and J. R. Mauney. 1980. Analysis of CO_2 exchange assumptions: feedback control. Pp. 1-16 in J. D. Hesketh and J. W. Jones, eds., Predicting Photosynthesis for Ecosystem Models, Vol. II. Boca Raton, Florida: CRC Press, Inc.

Hanks, R. J., H. R. Gardner, and R. L. Florian. 1969. Plant growth-evapotranspiration relationships for several crops in the Great Plains. Agron. J. 61:30-34.

Hansen, J., A. Lacis, D. Rind, G. Russell, P. Stone, I. Fund, R. Ruedy, and J. Lerner. 1984. Climate sensitivity: analysis of feedback mechanisms. Pp. 130-163 in J. E. Hansen and T. Takahashi, eds., Climate Processes and Climate Sensitivity. Geophys. Monog. Ser., Vol. 29. Washington, D.C.: American Geophysical Union.

Hansen, J., I. Fung, A. Lacis, S. Lebedeff, D. Rind, R. Ruedy, G. Russell, and P. Stone. 1988. Global climate change as forecast by the GISS 3-D model. J. Geophys. Res. 93:9341-9364.

Idso, S. B., B. A. Kimball, M. G. Anderson, and J. R. Mauney. 1987. Effects of atmospheric CO_2 enrichment on plant growth: the interactive role of air temperature. Agri. Ecosystems Environ. 20:1-10.

Jensen, M. E. 1966. Empirical methods of estimating or predicting evapotranspiration using radiation. Pp. 57-61, 64 in Proceedings of a Conference on Evapotranspiration and Its Role in Water Resources Management. St. Joseph, Michigan: American Society of Agricultural Engineers.

Jensen, M. E. 1974. Consumptive Use of Water and Irrigation Water Requirements. New York: Technical Committee on Irrigation Water Requirements, Irrigation and Drainage Division, American Society of Civil Engineers.

Jensen, M. E., and H. R. Haise. 1963. Estimating evapotranspiration from solar radiation. Journal of the Irrigation and Drainage Division, American Society of Civil Engineers 89:15-41.

Jones, C. A., and J. R. Kiniry, eds. 1986. CERES-Maize: A Simulation Model of Maize Growth and Development. College Station: Texas A & M University Press.

Jones, J. W., K. J. Boote, S. S. Jagtap, and J. W. Mishoe. 1989. Soybean development. Chapter 5 in R. J. Hanks and J. T. Ritchie, eds., Modeling Plant and Soil Systems. Madison, Wisconsin: American Society of Agronomy.

Jones, P., L. H. Allen, Jr., J. W. Jones, K. J. Boote, and W. J. Campbell. 1984. Soybean canopy growth, photosynthesis, and transpiration responses to whole-season carbon dioxide enrichment. Agron. J. 76:633-637.

Jones, P., L. H. Allen, Jr., and J. W. Jones. 1985a. Responses of soybean canopy photosynthesis and transpiration to whole-day temperature changes in different CO_2 environments. Agron. J. 77:242-249.

Jones, P., L. H. Allen, Jr., J. W. Jones, and R. R. Valle. 1985b. Photosynthesis and transpiration responses of soybean canopies to short- and long-term CO_2 treatments. Agron. J. 77:119-126.

Jones, P., J. W. Jones, and L. H. Allen, Jr. 1985c. Seasonal canopy CO_2 exchange, water use, and yield components in soybean grown under differing CO_2 and water stress conditions. Trans. ASAE 28:2021-2028.

Jouzel, J., C. Lorius, J. R. Petit, C. Genthon, N. I. Barkov, V. M. Kotlyakov, and V. M. Petrov. 1987. Vostok ice core: a continuous isotope temperature record over the last climatic cycle (160,000 years). Nature 329:403-407.

Keeling, C. D., R. B. Bacastow, A. F. Carter, S. C. Piper, T. P. Whorf, M. Heinmann, W. G. Mook, and H. Roeloffzen. 1989. A three dimensional model of atmospheric CO_2 transport based on observed winds: Analysis of data. Pp. 165-234 in D. H. Peterson, ed., Aspects of Climate Variability in the Pacific and the Western Americas. Geophysical Monograph 55. Washington, D.C.: American Geophysical Union.

Kimball, B. A. 1983. Carbon dioxide and agricultural yield: an assemblage and analysis of 430 prior observations. Agron. J. 75:779-788.

Kimball, B. A., and S. B. Idso. 1983. Increasing atmospheric CO_2: effects on crop yield, water use, and climate. Agric. Water Management 7:55-72.

Kimball, B. A., J. R. Mauney, G. Guinn, F. S. Nakayama, P. J. Pinter, Jr., K. L. Clawson, R. J. Reginato, and S. B. Idso. 1983. Response of Vegetation to Carbon Dioxide, Ser. 021: Effects of Increasing Atmospheric CO_2 on the Yield and Water Use of

Crops. Joint program of the U.S. Department of Energy and the U.S. Department of Agriculture, U.S. Water Conservation Lab, and U.S. Western Cotton Research Lab. Phoenix, Arizona: U.S. Department of Agriculture, Agricultural Research Service.

King, G. A., R. L. DeVelice, R. P. Neilson, and R. C. Worrest. 1989. Pp. 251-285 (Chapter 4) in The Potential Effects of Global Climate Change on the United States. Report EPA-230-05-89-050. Washington, D.C.: U.S. Environmental Protection Agency.

Knisel, W. G., P. Yates, J. M. Sheridan, T. K. Woody, L. H. Allen, Jr., and L. E. Asmussen. 1985. Hydrology and Hydrogeology of Upper Taylor Creek Watershed, Okeechobee County, Florida: Data and Analysis. Report ARS-25. Washington, D.C.: U.S. Department of Agriculture, Agricultural Research Service.

Landers, H. 1974. The climate of South Carolina. Pp. 353-369 in Climate of the States, Vol. 1. Port Washington, New York: Water Information Center, Inc.

Lorius, C., J. Jouzel, D. Raynaud, J. Hansen, and H. Le Treut. 1990. The ice core record: climate sensitivity and future greenhouse warming. Nature 347:139-145.

Manabe, S., and R. T. Wetherald. 1986. Reduction in summer soil wetness induced by an increase in atmospheric carbon dioxide. Science 232:626-628.

Manabe, S., and R. T. Wetherald. 1987. Large-scale changes of soil wetness induced by an increase in atmospheric carbon dioxide. J. Atmos. Sci. 44:1211-1235.

Mearns, L. O. 1989. Climate variability. Pp. 29-55 (Chapter 3) in The Potential Effects of Global Climate Change on the United States. Report EPA-230-05-89-050. Washington, D.C.: U.S. Environmental Protection Agency.

Miller, Barbara A., and W. G. Brock. 1988. Sensitivity of the Tennessee Valley Authority Reservoir System to Global Climate Change. Report No. WR28-1-680-101. Norris, Tennessee: Tennessee Valley Authority.

Miller, Barbara., and W. G. Brock. 1989. Global climate change: implications for the Tennessee Valley Authority reservoir system. Pp. 493-500 in J. C. Topping, Jr., ed., Coping with Climate Change: Proceedings of the Second North American Conference on Preparing for Climate Change. Washington, D.C.: The Climate Institute.

Mitchell, J. F. B. 1989. The "greenhouse" effect and climate change. Reviews of Geophysics 27:115-139.

Morison, J. I. L. 1987. Intercellular CO_2 concentration and stomatal response to CO_2. Pp. 229-252 in E. Zeiger, G. D.

Farquhar, and I. R. Cowan, eds., Stomatal Function. Stanford, California: Stanford University.

Morison, J. I. L., and R. A. Gifford. 1984. Plant growth and water use with limited water supply in high CO_2 concentrations: I. leaf area, water use, and transpiration. Australian J. Plant Physiol. 11:361-374.

Pearcy, R. W., and O. Björkman. 1983. Physiological effects. Pp. 65-105 in E. R. Lemon, ed., Carbon Dioxide and Plants: The Response of Plants to Rising Levels of Atmospheric Carbon Dioxide. American Association for the Advancement of Science Selected Symposium 84. Boulder, Colorado: Westview Press.

Peart, R. M., J. W. Jones, R. B. Curry, K. J. Boote, and L. H. Allen, Jr. 1989. Impact of climate change on crop yield in the Southeastern USA: a simulation study. Appendix C (Agriculture), Vol. 1, Chapter 2 in J. B. Smith and D. A. Tirpak, eds., The Potential Effects of Global Climate Change on the United States. Report EPA-230-05-89-053. Washington, D.C.: U.S. Environmental Protection Agency.

Penning de Vries, F. W. T., D. M. Jansen, H. F. M. ten Berge, and A. Bakema. 1989. Simulation of Ecophysiological Processes of Growth in Several Annual Crops. Wageningen, Netherlands: Centre for Agricultural Publication and Documentation (Pudoc).

Revelle, R., and P. W. Waggoner. 1983. Effects of a carbon dioxide induced climatic change on water supplies in the Western United States. Pp. 419-432 in Changing Climate: Report of the Carbon Dioxide Assessment Committee. Washington, D.C.: National Academy Press.

Ritchie, J. T., B. D. Baer, and T. Y. Chou. 1989. Effect of global climate change on agriculture: Great Lakes region. Appendix C (Agriculture), Vol. 1, Chapter 1 in J. B. Smith and D. A. Tirpak, eds., The Potential Effects of Global Climate Change on the United States. Report EPA-230-05-89-053. Washington, D.C.: U.S. Environmental Protection Agency.

Rosenberg, N. J., B. A. Kimball, P. Martin, and C. F. Cooper. 1990. From climate and CO_2 enrichment to evapotranspiration. Pp. 151-175 in P. E. Waggoner, ed., Climate Changes and U.S. Water Resources. New York: John Wiley and Sons.

Rosenzweig, C. 1989. Potential effects of climate change on agricultural production in the Great Plains: a simulation study. Appendix C (Agriculture), Vol. 1, Chapter 3 in J. B. Smith and

D. A. Tirpak, eds., The Potential Effects of Global Climate Change on the United States. Report EPA-230-05-89-053. Washington, D.C.: U.S. Environmental Protection Agency.

Schlesinger, M. E. 1984. Climate model simulation of CO_2-induced climate change. Pp. 141-235 in B. Saltzman, ed., Advances in Geophysics, Vol. 26. New York: Academic Press.

Shawcraft, R. W., E. R. Lemon, L. H. Allen, Jr., D. W. Stewart, and S. E. Jensen. 1974. The soil-plant-atmosphere model and some of its predictions. Agric. Meteorol. 14:287-307.

Smith, J. B., and D. A. Tirpak, eds. 1989. The potential effects of global climate change on the United States. Report EPA-230-05-89-050. Washington, D.C.: U.S. Environmental Protection Agency.

Stanhill, G. 1960. The relationship between climate and the transpiration and growth of pastures. Pp. 293-296 in P. J. Boyle, and L. W. Raymond, eds., Proceedings, Eighth International Grasslands Congress, University of Reading, Berkshire, England. Oxford, England: Alden Press.

Stephens, J. C. 1965. Estimating evaporation from insolation. J. Hydr. Div., Am. Soc. Civ. Engr. 91(HY5):171-182.

Stephens, J. C., and E. H. Stewart. 1963. A Comparison of Procedures for Computing Evaporation and Evapotranspiration. Pub. No. 62, Intern. Assoc. of Sci. Hydrol. Berkeley, California: Trans. Intern. Union Geodesy and Geophysics.

Tanner, C. B., and T. R. Sinclair. 1983. Efficient water use in crop production: research or research? Pp. 1-27 in H. M. Taylor, W. R. Jordan, and T. R. Sinclair, eds., Limitations to Efficient Water Use in Crop Production. Madison, Wisconsin: American Society of Agronomy, Crop Science Society of America, and Soil Science Society of America.

Tolbert, N. E., and I. Zelitch. 1983. Carbon metabolism. Pp. 21-64 in E. R. Lemon, ed., CO_2 and Plants: The Response of Plants to Rising Levels of Atmospheric Carbon Dioxide. American Association for the Advancement of Science Selected Symposium 84. Boulder, Colorado: Westview Press.

Waggoner, P. E. 1990. Climate Change and U.S. Water Resources. New York: John Wiley & Sons.

Warrick, R. A., R. M. Gifford, and M. L. Parry. 1986. CO_2, climate change, and agriculture. Pp. 393-473 in B. Bolin, B. R. Doos, J. Jager, and R. A. Warrick, eds., The Greenhouse Effect, Climate Change, and Ecosystems (SCOPE 29). New York: John Wiley and Sons.

Washington, W. M., and G. A. Meehl. 1983. General circulation model experiments on the climatic effects due to doubling and quadrupling of carbon dioxide concentrations. J. Geophys. Res. 88:6600-6610.

Washington, W. M., and G. A. Meehl. 1984. Seasonal cycle experiment on the climate sensitivity due to a doubling of CO_2 with an atmospheric general circulation model coupled to a simple mixed layer ocean model. J. Geophys. Res. 89:9475-9503.

Washington, W. M., and G. A. Meehl. 1986. General circulation model CO_2 sensitivity experiments: snow-sea ice albedo parameterizations and globally averaged surface air temperature. Clim. Change 8:231-241.

Wilkerson, G. G., J. W. Jones, K. J. Boote, K. T. Ingram, and J. W. Mishoe. 1983. Modeling soybean growth for crop management. Trans. ASAE 26:53-73.

Wilson, C. A., and J. F. B. Mitchell. 1987. A doubled CO_2 climate sensitivity experiment with a GCM including a simple ocean. J. Geophys. Res. 92:13315-13343.

Zazueta, F. S., and A. G. Smajstrla. 1989. Water Management Utilities. Pub. 89-1, Fla. Coop. Ext. Serv., Institute of Food and Agricultural Sciences. Gainesville: University of Florida.

8

Hydrologic Implications of Climate Uncertainty in the Western United States

Marshall E. Moss
U.S. Geological Survey
Tucson, Arizona

By now, a vast majority of the inhabitants of the United States of America are aware of and accept the fact that levels of greenhouse gases, such as carbon dioxide, are increasing in the earth's atmosphere. Probably, most people with even a minimal education in science also accept that increased atmospheric greenhouse gas concentrations will cause additional energy retention in the earth's immediate environment—an increased greenhouse effect. However, drawing conclusions about changes in hydrologic phenomena brought about by the augmented green-house effect is not a straightforward exercise. One might logically conclude that a part of the additional energy would accelerate the hydrologic cycle: that is, there would be more precipitation, more infiltration, more evapotranspiration, and more runoff. But, heretical as it may seem, the hydrologic cycle as depicted in most basic texts does not exist. There is a myriad of paths by which a molecule of water can transit about the globe, and the likelihood of it doing so in a cyclic manner is infinitesimally small. Thus, the simplified concept of a hydrologic cycle offers little insight into the hydrologic impacts of potential climate change.

Instead of a cycle as the conceptual analog for hydrology, a random walk seems more appropriate. In the hydrologic random walk, a water molecule's passage through one of the reservoirs of the earth's hydrosphere or its transition from one reservoir to another is controlled probabilistically by the distribution of energy and mass within and among adjacent reservoirs. Hydrology is the science of understanding the aggregations of a great many molecules of water passing through and among the reservoirs. To predict the hydrologic impacts of increased atmospheric concentrations of greenhouse gases, knowledge of the partitioning of the incremental energy within the various reservoirs is required. This paper explores our current ability to define this partitioning and draws conclusions about the resulting implications on the hydrology of the western United States.

To illustrate the complexity of the hydrologic random walk, it is useful to consider the concept of teleconnections (Namias, 1981), which is a statistical approach relating the magnitudes of weather or hydrologic events that occur at great distances from each other. For teleconnections to be more than a statistical oddity, they must be the result of seasonally preferred paths through the hydrologic random walk. The relation of weather patterns around the globe to an aperiodic anomalous warming of the eastern Pacific Ocean, the El Niño Southern Oscillation (ENSO), is a teleconnection that has been much explored recently. For example, as shown in Figure 8.1, Ropelewski and Halpert (1987) demonstrated a positive correlation between ENSO events and precipitation magnitudes over much of the Colorado River basin. This correlation implies that an energy exchange between the atmosphere and the Pacific Ocean can alter the probabilities of precipitation in the western United States during ENSO events.

A more explicit depiction of a preferred path of a similar or even greater spatial scale is found in the work of Koster as reported by Eagleson (1986). Figure 8.2 shows the regions where water that is evaporated in the month of March from a grid cell of 10 degrees longitude by 8 degrees latitude located in southeastern Asia is first redeposited on earth. Most of the land area under the mandate of the Bureau of Reclamation receives moisture from this cell, as does most of eastern Asia and the northern Pacific. Thus, evidence indicates a very complex system of reservoirs of moisture and energy in the oceans, in the atmosphere, and on the land that interacts with itself to define the existing climate and hydrology in the western United States.

What do we know about the response of this complex system to an increased greenhouse effect? Probably, we know best the physics of the transport of mass and energy in most reservoirs of the system. However, we know the physics only at spatial scales that are not fully compatible with the data bases and computing facilities that are available today. Climatologists and oceanographers have attempted to bridge this incompatibility by constructing mathematical general circulation models (GCMs) of the earth's atmosphere and oceans. GCMs are, at best, compromises between the sophistication of the description of the physics and the temporal and spatial scales at which transport is computed; thus,

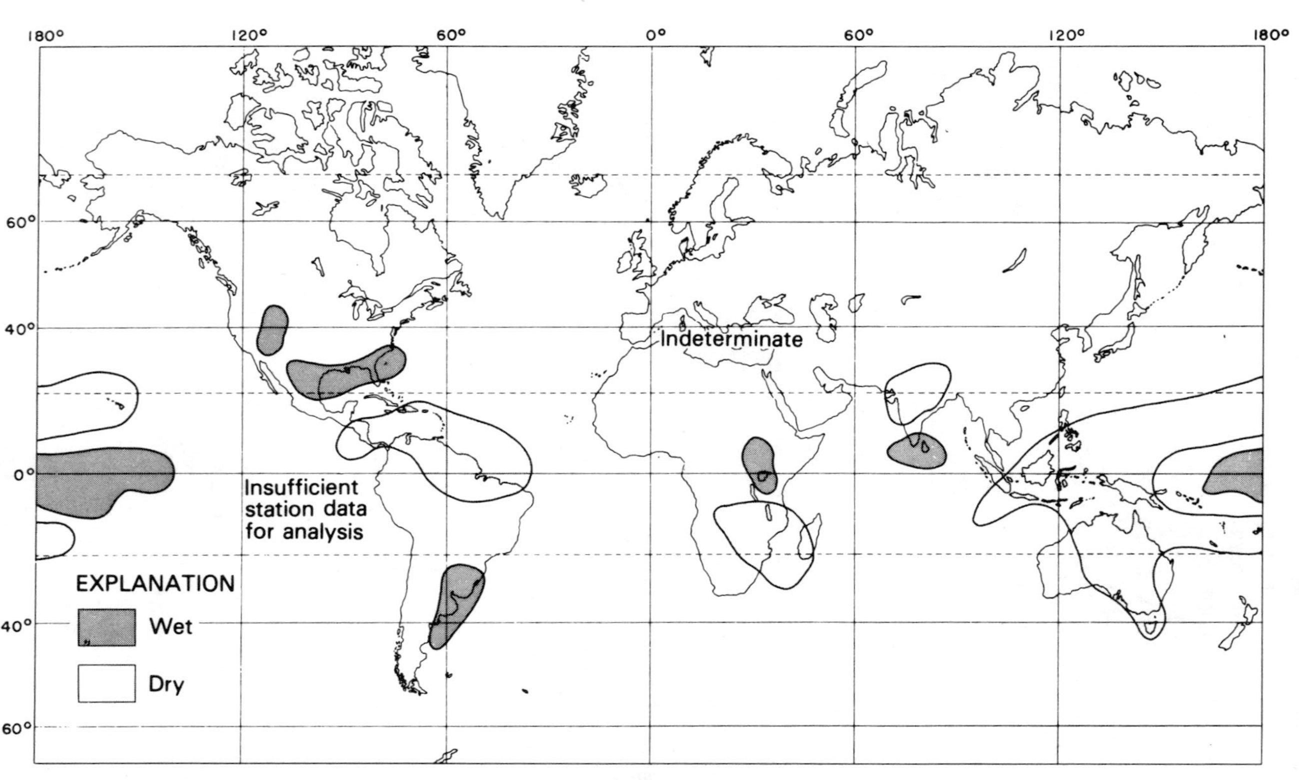

FIGURE 8.1 Regions exhibiting a consistent precipitation response to El Niño-Southern Oscillation episodes. **SOURCE:** Reprinted, by permission, from Ropelewski and Halpert (1987). Copyright ©1987 by Monthly Weather Review.

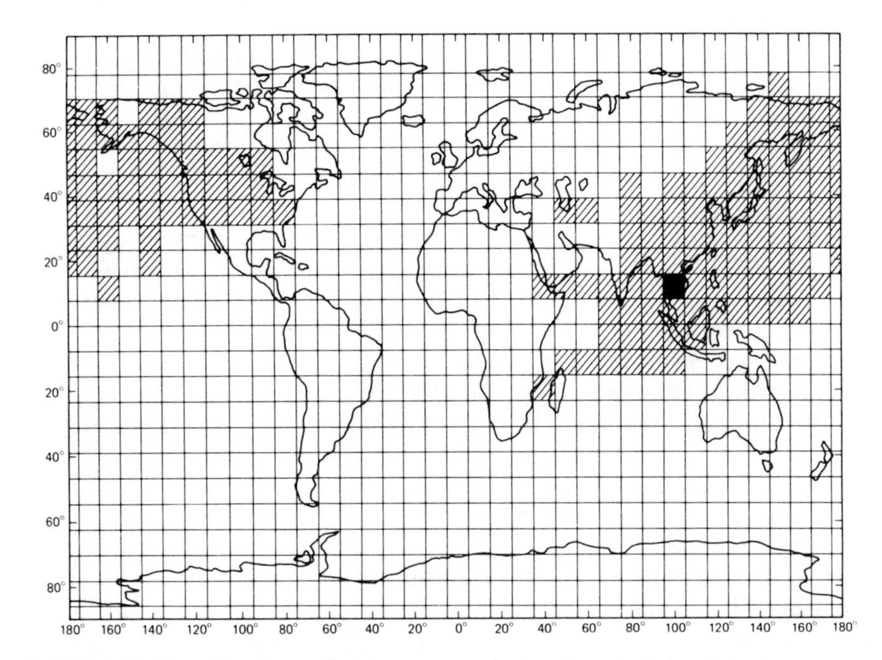

FIGURE 8.2 Region of influence of Southeast Asia March evaporation.
SOURCE: Reprinted, by permission, from Eagleson (1986). Copyright © 1986 by the American Geophysical Union.

they have computational grids that are several degrees in both latitude and longitude.

Information about hydrologic processes on and beneath the land surfaces of earth is poorly served by these compromises. For example, a common GCM representation of the land surface within a grid is that of a uniform soil of constant depth, the so-called bucket model, from which runoff is generated when the soil is saturated and precipitation exceeds evapotranspiration during any time step of computation. Runoff so defined is quite a different phenomenon from that recognized under the same name in hydrology. Furthermore, the spatial averaging that takes place over a grid cell yields variables that have little relevance to most problems of interest to traditional hydrologists and little applicability in water-resources decisionmaking.

To extract relevant hydrologic information from GCMs, two requirements must be met. First, outputs from the GCMs must be hydrologically meaningful. Second, there must be significant information contained in those outputs. Hydrologic models that convert GCM runoff to a meaningful variable do not currently exist; thus, GCM runoff fails the first criterion.

Other outputs from GCMs, such as precipitation, temperature, and relative humidity, might be hydrologically meaningful except for the discrepancies between their spatial scales of aggregation and the spatial resolution needed for these variables in existing hydrologic models. One approach that could resolve such discrepancies is statistical disaggregation (Valencia and Schaake, 1973). If this approach can be applied with some degree of confidence to spatial disaggregation of such GCM outputs, the outputs could meet the first criterion, whereas GCM runoff could not.

Another approach for the resolution of the scale discrepancy is the use of models of a finer scale for selected geographical regions of interest nested within a GCM (Giorgi, 1990). This approach shows great promise at the meteorological mesoscale, which is pertinent for many of the larger-scale hydrologic problems. It is conceivable that nested models and disaggregation models could be combined to address hydrologic problems of an even smaller scale.

With respect to the second criterion, quite a large body of literature exists that describes qualitatively the uncertainties inherent in climate modeling. A comprehensive review of this literature was done recently by Dickinson (1989). However, in the only attempt to date to quantify the information derived from GCMs, Moss (1991) has found that, for the grid cell highlighted in Figure 8.3, information from the Community Climate Model of the National Center for Atmospheric Research about July precipitation under current climatic conditions is limited to less than 20 percent of the information contained in 30 years of actual records. For January precipitation, the model output is limited to about 15 percent of the 30-year record. Extrapolations required to estimate future climate changes would undoubtedly degrade the resulting information below these limits. Thus, there may be some useful hydrologic information in GCMs, but our current ability to extract it is very limited.

Because of the paucity of hydrologic information that can be extracted from climate models, hydrologists generally have opted for scenario analysis (Lave and Epple, 1985) as a means to investigate the sensitivity of hydrologic systems to climate change. For example,

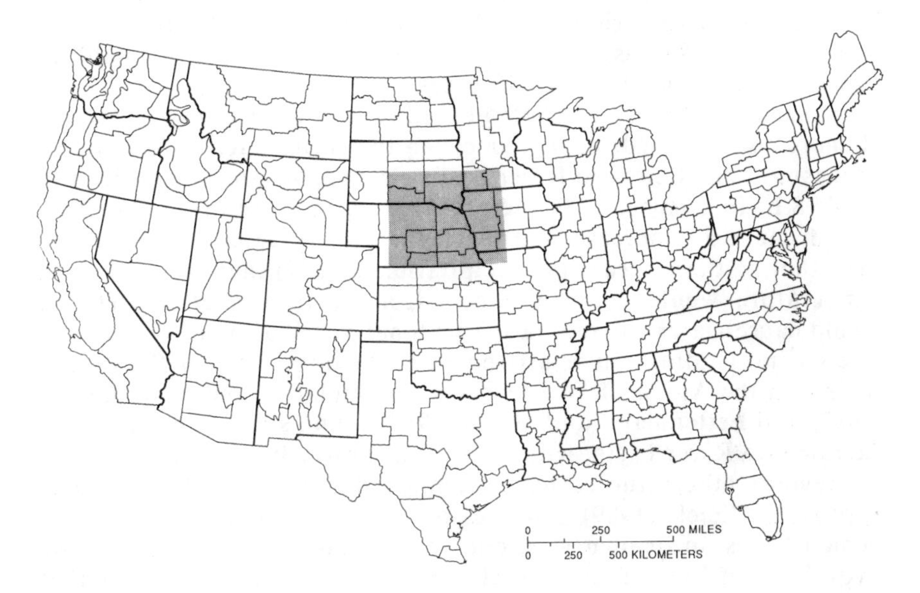

FIGURE 8.3 A GCM cell superimposed on the map of the State Climate Divisions.

Revelle and Waggoner (1983) assumed a scenario of a 10 percent decrease in precipitation and an increase of 2°C and used the empirical relations from Langbein and others (1949) to explore the impacts on runoff in the Colorado River basin. They found a potential decrease in average annual runoff of approximately 50 percent. However, the data used by Langbein were collected in the first half of this century, when carbon dioxide in the atmosphere was not at the levels contemplated in the climate-change scenarios. Most plants respond to increased levels of carbon dioxide by decreasing their rates of transpiration, and this feedback is not included in the work of Revelle and Waggoner (1983). Idso and Brazel (1984) estimated the vegetation effect and found that the 50 percent decrease in runoff reported earlier would become a 50 percent increase instead. Stockton (1975) has estimated the annual runoff of the upper Colorado River basin for the period 1520 to 1961 using dendrochronology and has found the mean annual runoff for this period to be approximately

13.5 million acre-feet. Figure 8.4 shows the decadal averages for the reconstructed record and the range of plus and minus 50 percent of the long-term average. It can be seen that the level of hydrologic uncertainty as depicted by the mean annual runoff, given the climate-change scenario of Revelle and Waggoner (1983) and of Idso and Brazel (1984), is greater than the decadal runoff variability experienced during at least 440 years.

Schaake (1990) has attempted to reduce this uncertainty by the use of more complex hydrologic models on the Animas River, which is a subbasin of the Colorado River basin. His results are summarized in Table 8.1. In essence, he found: (1) that an increase in precipitation would cause an increase in runoff that was greater in percentage than that of the increase in precipitation, and (2) that an increase in temperature would cause a minor decrease in annual runoff but would cause major changes in the seasonal distribution of the runoff. The second finding can be attributed to the dominance of the runoff regime in the Animas River by snow accumulation and melt. Gleick (1987) and Lettenmaier and others (1989) found similar results in the Sacramento River basin, which also is a snowmelt-dominated system.

Several other studies have been conducted using the scenario approach; Gleick (1989) provides a recent review of these. Each demonstrates, in its own way, one or more possible outcomes for the hydrologic effects of climate change. It should be reiterated that each is subject to its own inherent assumptions and should be considered as a measure of the system's sensitivity to those assumptions and not necessarily as a likely outcome of climate change. At this time, the plausibility of each scenario can only be determined subjectively, because the probability of any climate scenario cannot yet be determined.

Thus, today's state of understanding concerning the hydrologic implications of climate change is best characterized as one of uncertainty in which the level of uncertainty itself is uncertain. Because the validity of water resources decisions is very sensitive to hydrologic uncertainty, one of the first priorities of hydrologic research should be the quantification of the added uncertainty caused by an enhanced greenhouse effect. In other words, water resources planners and decisionmakers should encourage the research community to assess the level of hydrologic uncertainty while concurrently reducing it.

There are two paths for the reduction of uncertainty: (1) data collection, and (2) research. Traditionally, data collection has played the primary role in uncertainty reduction in hydrology. However, in the nonstationary world caused by climate change, the dominant role

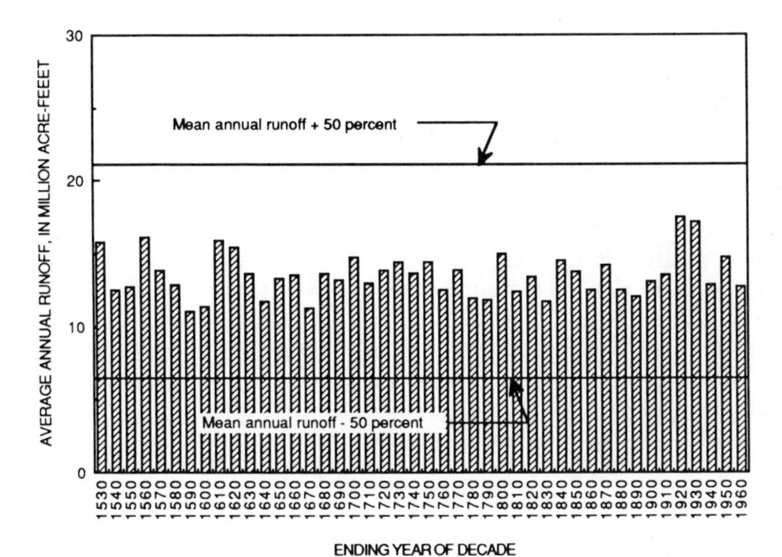

FIGURE 8.4 Reconstructed runoff for the Colorado River at Lee's Ferry.

TABLE 8.1 Sensitivity of runoff to climate change for the Animas River at Durango, Colorado.

	Period		
	Annual Average	January	June
Runoff (thousands of acre-feet)	545	12	163
Percentage runoff change for 10% increase in precipitation	19.7	17.1	12.1
Percentage runoff change for 10% increase in ETP	-7.0	3.8	8.9
Percentage runoff change for 2°C temperature increase	-2.1	37.4	-26.8
Percentage runoff change for 10% increase in ETP and 2°C temperature increase	-8.6	22.9	-30.4

SOURCE: Schaake, 1990.

of data must be reduced (Moss and Lins, 1989), because the information content of the raw data degrades with time. In other words, the hydrologic processes operational at the time that the data were collected are subsequently modified by the climate changes. Thus, yesterday's data lose their relevance to today's circumstances unless sufficient understanding of the climate-hydrology interactions is available to capture the information in yesterday's data and apply it to today's situation. The current level of understanding of hydroclimatology is not sufficient to perform this act of information retention. Therefore, increased support for research in hydroclimatology is a prerequisite to uncertainty reduction. Nevertheless, data collection is a necessity as well. Research without supporting data is a tenuous approach at best; furthermore, the data will be valuable in their own right once sufficient research is done so that they can be properly interpreted.

REFERENCES

Dickinson, R. E. 1989. Uncertainties of estimates of climatic change: a review. Climatic Change 15:5-13.

Eagleson, P. S. 1986. The emergence of global-scale hydrology. Water Resources Research 22(9):6s-14s.

Giorgi, F. 1990. Simulation of regional climate using a limited area model nested in a general circulation model. Journal of Climate 3:941-963.

Gleick, P. H. 1987. Regional hydrologic consequences of increases in atmospheric $C0_2$ and other trace gases. Climatic Change 10:137-161.

Gleick, P. H. 1989. Climate change, hydrology, and water resources. Reviews in Geophysics 27:329-344.

Idso, S. B., and A. J. Brazel. 1984. Rising atmospheric carbon dioxide concentrations may increase streamflow. Nature 312:51-53.

Langbein, W. B., et al. 1949. Annual Runoff in the United States. Circular 52. Washington, D.C.: U.S. Geological Survey.

Lave, L. B., and D. Epple. 1985. Scenario Analysis, Climate Impact Analysis. New York: John Wiley & Sons.

Lettenmaier, D. P., T. Y. Gan, and D. R. Dawdy. 1989. Interpretation of hydrologic effects of climate change in the Sacramento-San Joaquin river basin, California. Appendix A, Pp. 1-1 to 1-52 in J. B. Smith, and D. A. Tirpak, eds., Potential Effects of Global Climate Change on the United States. Washington, D.C.: U.S. Environmental Protection Agency.

Moss, M. E. 1991. Bayesian relative information measure—a tool for analyzing the outputs of general circulation models. Journal of Geophysical Research, in press.

Moss, M. E., and H. F. Lins. 1989. Water Resources in the Twenty-first Century: A Study of the Implications of Climate Uncertainty. Circular 1030. Washington, D.C.: U.S. Geological Survey.

Namias, J. 1981. Teleconnections of 700 mb Height Anomalies for the Northern Hemisphere. CALCOFI Atlas No. 29. La Jolla, California: Scripps Institution of Oceanography.

Revelle, R. R., and P. E. Waggoner. 1983. Effects of carbon dioxide-induced climate change on water supplies in the western United States. Pp. 419-432 in Changing Climate. Washington, D.C.: National Academy Press.

Ropelewski, C. W., and M. S. Halpert. 1987. Global and regional scale precipitation patterns associated with the El Niño/Southern Oscillation. Monthly Weather Review 115:1606-1626.

Schaake, J. C. 1990. From climate to flow. Pp. 177-206 in Climate Change and U.S. Water Resources. New York: John Wiley and Sons.

Stockton, C. W. 1975. Long-term Streamflow Records from Tree Rings. Tucson: University of Arizona Press.

Valencia R. D., and J. C. Schaake. 1973. Disaggregation processes in stochastic hydrology. Water Resources Research 9:580-585.

9

The Implications of Climatic Change for Streamflow and Water Supply in the Colorado Basin

Linda Nash
Pacific Institute for Studies in Development, Environment, and Security
Berkeley, California

INTRODUCTION

The Colorado River is one of the most important river systems in the United States. Although not a large river even by North American standards, the Colorado flows through some of the most arid regions of the country and is the sole source of water for a region with extensive agriculture, large cities, and a diverse ecosystem. Existing global models suggest that climatic changes will have dramatic impacts on water resources. Water availability, quality, and demand will be affected by higher temperatures, new precipitation patterns, rising sea level, and changes in storm frequency and intensity. These changes will be important to the Colorado River basin because of their effect on water supply and water management—issues already hotly contested in the region. Moreover, potential climatic impacts will have significant ramifications for decisions about water allocations and water rights that are likely to be made in the coming decade.

Despite recent advances in modeling the atmosphere, large uncertainties remain about the details of regional hydrologic changes. Until current climate models improve both their spatial resolutions and their hydrologic parameterizations, information on the hydrologic effects of global climatic changes can best be obtained using regional hydrologic models. At this time such studies are limited to sensitivity analyses that describe the vulnerability of hydrologic basins to a range of plausible climate scenarios. While these scenarios cannot be regarded as reliable predictions of future conditions, they do provide insights into regional vulnerabilities.

A variety of techniques are available for studying the hydrologic impacts of climatic change, including stochastic methods that rely primarily on statistical techniques and deterministic or conceptual models that use physically based, mathematical descriptions of hydrologic phenomena. Because climatic changes are expected to alter underlying statistical relationships among variables, stochastic models are probably inappropriate for climate-impact studies. In contrast, deterministic techniques should be more robust so long as the modeled processes (such as percolation, soil-moisture storage, and snow melt) are not expected to change significantly under a carbon dioxide-altered climate. To date, climate-impact studies on the Colorado River basin have been limited to stochastic methods (Revelle and Waggoner, 1983; Stockton and Boggess, 1979). In contrast, this project used a conceptual hydrologic model to study the sensitivity of the basin to greenhouse warming.

The large size of the Colorado basin complicates the development of a hydrologic model. This study used a conceptual model developed and operated by the National Weather Service River Forecasting Service (NWSRFS) in Salt Lake City, Utah (Burnash et al., 1974; Anderson, 1973). This model has advantages and limitations, described in detail in Nash and Gleick (1991), but its success as a forecasting tool provides reason for believing that the model has the capability to simulate the effects of changes in temperature and precipitation. The NWSRFS models the upper Colorado basin as a series of approximately 50 small subbasins that are linked together. In addition, an aggregated model has been developed that divides the entire upper Colorado basin into two elevation zones and uses a limited number of meteorological stations to predict inflow into Lake Powell. In addition to this two-elevation model, we selected three subbasins that were known to make a substantial contribution to basin flow: the White River at Meeker, the East River at Almont, and the Animas River at Durango.

To assess the potential impacts of climatic change on runoff in the Colorado River basin, scenarios of changes in temperature and precipitation were used as inputs into the NWSRFS model. For this study, we relied on purely hypothetical scenarios as well as scenarios derived from the outputs of general circulation models (GCMs). These scenarios are listed in Table 9.1. The baseline data for the model consist of six-hourly data for the years 1949 through 1983, inclusive.

Using the results of the NWSRFS model, we chose a range of plausible runoff scenarios with which to assess the sensitivity of the basin to changes in water supply. The impacts of changes in runoff on water deliveries, hydropower production, reservoir levels, and several other variables were studied with the U.S. Bureau of Reclamation's Colorado River Simulation System (CRSS). The CRSS is a reservoir-system simulation model that tracks streamflow and water supply throughout the Colorado River basin. The CRSS incorporates the legal and administrative requirements that affect water supply in the basin and is the most detailed simulation model currently available for the Colorado River. It serves as the Bureau of Reclamation's primary tool for studying the operation of the river and the impacts of operational changes and projected developments in the basin. The model is documented in the U.S. Department of the Interior (DOI) publication Colorado River Simulation System Overview (DOI, 1987).

The CRSS uses historical streamflow data (a 78-year record, extending from 1906 to 1983) to analyze possible future conditions. The input to the model is "natural" streamflow—defined as historical flow data adjusted to remove the effects of human development—at 29 gaging stations throughout the basin. The output from the model is actual streamflow, reservoir levels, hydropower production, reservoir spills, salinity, and water deliveries. The model incorporates a large set of supply constraints and decision rules that determine reservoir operation, including the allocation of shortages and surpluses among various water users. In no sense does the model predict future shortages or surpluses, but it does portray the sensitivity of those outcomes to particular inputs or operating parameters.

For the purposes of this study, operating procedures (such as rule curves and target storages) were held constant, and the natural flow database was uniformly altered by plus or minus 5, 10, and 20 percent. These scenarios are consistent with results generated by the NWSRFS model. Although it is likely that operational parameters would be adjusted over time to increase the system's efficiency with respect to changed hydrologic conditions, such changes would be implemented slowly and only after a general acknowledgment of changed conditions. At this point, it is difficult to estimate to what extent changes in operations might mitigate the impacts of changes in flow; this is an area for further research. Our results summarize a model run of 78 years in which natural flows were altered by the specified percentage at all 29 input stations. Reservoir evaporation rates were unchanged, even though they would be expected to increase under conditions of higher temperature. The demand data used in these runs were the

TABLE 9.1 Climate-Change Scenarios Used in the NWSRFS Model.

	Two-Elevation	White River	East River	Animas River
Hypothetical				
T+2°C, P-20%	- -	X	X	X
T+2°C, P-10%	X	X	X	X
T+2°C, P+0	X	X	X	X
T+2°C, P+10%	X	X	X	X
T+2°C, P+20%	- -	X	X	X
T+4°C, P-20	X	X	X	X
T+4°C, P-10%	X	X	X	X
T+4°C, P+0	X	X	X	X
T+4°C, P+10%	X	X	X	X
T+4°C, P+20%	X	X	X	X
GCM[1]				
GISS 1: T+4.8°C, P+20%	- -	X	- -	- -
GISS 2: T+4.9°C, P+10%	X	- -	X	X
GFDL: T+4.7°C, P+0	X	X	X	
UKMO 1: T+6.8°C, P+30%	X	X	- -	- -
UKMO 2: T+6.9°C, P+10%	X	X	X	X

NOTE: [1] All GCM scenarios represent annual average changes for an equilibrium (2xCO_2) run.

Bureau of Reclamation's projections for the year 2040 and were held constant for the period analyzed. For the model runs presented in this paper, the total amount of reservoir storage at the beginning of the run was approximately 36.5 million acre-feet (maf), or about 60 percent of the system's total storage capacity.

RESULTS

Hydrologic (NWSRFS) Model

Large changes in the magnitude of annual flow in the Colorado basin may result from plausible climatic changes (Tables 9.2 through 9.5). A 2°C rise in temperature corresponds to a decrease in runoff of 4 percent on the White River, 9 percent on the East River, and 7 percent on the Animas River. For the two-elevation model, a tem-

TABLE 9.2 Annual In-Flow (taf) Into Lake Powell (Two-Elevation Model) for All Scenarios.

Scenario	Mean[1]		SD	CV	Minimum		Maximum	
Base	10940		2983	0.27	4481		17040	
T+2° P-10%	8386	(-23.3%)	2418	0.29	3357	(-25.1%)	12940	(-24.1%)
T+2° P+0	9656	(-11.7%)	2727	0.28	3924	(-12.4%)	14330	(-15.5%)
T+2° P+10%	11000	(0.6%)	3046	0.28	4504	(0.5%)	16350	(-4.0%)
T+4° P-20%	6447	(-41.0%)	1970	0.31	2520	(-43.8%)	11480	(-32.6%)
T+4° P-10%	7522	(-31.2%)	2260	0.30	2892	(-35.5%)	12480	(-26.8%)
T+4° P+0	8668	(-20.7%)	2554	0.30	3373	(-24.0%)	13490	(-20.8%)
T+4° P+10%	9879	(-9.7%)	2854	0.29	3911	(-12.7%)	14530	(-14.8%)
T+4° P+20%	11150	(2.0%)	3162	0.28	4443	(-0.9%)	16180	(5.1%)
GISS 2	9444	(-13.6%)	2804	0.30	3624	(-19.1%)	14220	(-16.5%)
GFDL	8369	(-23.5%)	2514	0.30	3180	(-29.0%)	13270	(-22.1%)
UKMO 1	10950	(0.2%)	3240	0.30	4107	(-8.3%)	16070	(-5.7%)
UKMO 2	8639	(-21.0%)	2693	0.31	3173	(-29.2%)	13296	(-18.3%)

NOTE: [1] Numbers in parentheses represent percent change from the base case.

TABLE 9.3 Annual Flow (taf) of the White River for All Scenarios.

Scenario	Mean[1]	SD	CV	Minimum	Maximum
Base	434.9	104.5	0.24	242.8	670.5
T+2° P-20%	335.1 (-22.9%)	70.6	0.21	193.6 (-20.3%)	474.7 (-29.2%)
T+2° P-10%	374.6 (-13.9%)	82.9	0.22	214.6 (-11.6%)	541.1 (-19.3%)
T+2° P+0	417.0 (-4.1%)	97.5	0.23	234.7 (-3.4%)	608.7 (-9.2%)
T+2° P+10%	465.1 (-7.0%)	114.8	0.25	255.0 (-5.0%)	697.1 (4.0%)
T+2° P+20%	515.7 (18.6%)	132.9	0.26	279.0 (14.9%)	788.6 (17.6%)
T+4° P-20%	320.9 (-26.2%)	70.0	0.22	180.7 (-25.6%)	468 (-30.2%)
T+4° P-10%	357.6 (-17.8%)	80.6	0.23	200.6 (-17.4%)	532.4 (-20.6%)
T+4° P+0	396.9 (-8.7%)	92.9	0.23	221.5 (-8.8%)	599.7 (-10.6%)
T+4° P+10%	440.4 (-1.3%)	107.9	0.24	241.7 (-0.5%)	666.9 (-0.5%)
T+4° P+20%	487.9 (12.2%)	126.2	0.26	264.0 (8.7%)	756.2 (12.8%)
GISS 1	476.2 (9.6%)	122.9	0.26	252.9 (4.2%)	746.2 (11.3%)
GFDL	389.7 (-10.4%)	91.7	0.24	214.1 (-11.8%)	599.7 (-10.6%)
UKMO 1	488.5 (-12.3%)	128.3	0.26	250.2 (3.0%)	790.1 (17.8%)
UKMO 2	401.3 (-7.7%)	97.4	0.24	211.8 (-12.8%)	640.4 (-4.5%)

NOTE: [1] Numbers in parentheses represent percent change from the base case.

TABLE 9.4 Annual Flow (taf) of the East River for All Scenarios.

Scenario	Mean[1]	SD	CV	Minimum	Maximum
Base	230.7	84.9	0.37	76.9	477.0
T+2° P-20%	165.8 (-27.6%)	60.6	0.36	60.2 (-22.8%)	358.6 (-24.8%)
T+2° P-10%	186.9 (-18.7%)	69.1	0 37	66.4 (-14.0%)	401.8 (-15.8%)
T+2° P+0	209.4 (-9.1%)	77.8	0.37	72.5 (-5.8%)	446.1 (-6.5%)
T+2° P+10%	233.5 (1.3%)	86.2	0.37	79.1 (2.8%)	490.5 (2.8%)
T+2° P+20%	258.7 (12.3%)	94.3	0.36	86.4 (12.2%)	535.0 (12.2%)
T+4° P-20%	153.8 (-33.1%)	58.8	0.38	54.4 (-29.3%)	348.9 (-26.8%)
T+4° P-10%	172.8 (-25.0%)	66.9	0.39	61.6 (-19.9%)	388.4 (-18.6%)
T+4° P+0	192.8 (-16.5%)	74.9	0.39	68.8 (-10.6%)	428.6 (-10.2%)
T+4° P+10%	223.4 (-3.4%)	86.3	0.37	77.6 (0.8%)	487.0 (2.1%)
T+4° P+20%	246.4 (6.6%)	93.8	0.38	84.7 (10.1%)	528.3 (10.8%)
GISS 2	205.6 (-11.2%)	80.9	0.39	70.2 (-8.8%)	456.2 (-4.4%)
GFDL	187.0 (-19.1%)	73.4	0.39	64.6 (-16.1%)	420.2 (-11.9%)
UKMO 2	187.6 (-19.0%)	76.2	0.41	64.2 (-16.6%)	438.9 (-8.0%)

NOTE: [1] Numbers in parentheses represent percent change from the base case.

TABLE 9.5 Annual Flow (taf) of the Animas River for All Scenarios.

Scenario	Mean[1]		SD	CV	Minimum		Maximum	
Base	550.6		192.5	0.35	240.4		941.7	
T+2° P-20%	406.6	(-26.1%)	143.5	0.35	165.9	(-31.0%)	682.6	(-27.5%)
T+2° P-10%	458.6	(-16.7%)	162.3	0 35	188.8	(-21.5%)	762.2	(-19.1%)
T+2° P+0	512.3	(-7.0%)	181.6	0.35	212.3	(-11.7%)	853.0	(-9.4%)
T+2° P+10%	568.4	(3.2%)	200.8	0.35	238.0	(-1.0%)	947.8	(0.6%)
T+2° P+20%	628.2	(14.1%)	220.5	0.35	264.4	(1.0%)	1051.5	(11.7%)
T+4° P-20%	376.8	(-31.5%)	133.2	0.35	150.5	(-37.4%)	640.1	(-32.0%)
T+4° P-10%	424.3	(-22.9%)	150.8	0.36	170.6	(-29.0%)	715.8	(-24.0%)
T+4° P+0	473.3	(-14.1%)	168.8	0.36	191.5	(-20.3%)	791.8	(-15.9%)
T+4° P+10%	525.0	(-4.7%)	187.1	0.36	214.6	(-10.7%)	874.2	(-7.2%)
T+4° P+20%	578.9	(5.1%)	205.5	0.35	240.2	(-0.1%)	961.5	(2.0%)
GISS 2	505.5	(-8.4%)	182.4	0.36	205.0	(-14.7%)	847.2	(-10.0%)
GFDL	459.3	(-16.7%)	165.7	0.36	184.8	(-23.1%)	775.1	(-17.7%)
UKMO 2	465.3	(-15.7%)	169.2	0.36	182.1	(-24.2%)	798.8	(-15.2%)

NOTE: [1] Numbers in parentheses represent percent change from the base case.

perature increase of 2°C reduces runoff by 12 percent, excluding the effect of higher temperatures on reservoir evaporation, and an increase of 4°C decreases runoff by between 9 and 21 percent. Increases and decreases in precipitation of 10 percent and 20 percent lead to equivalent changes (10 to 20 percent) in runoff. All relationships between runoff and precipitation are nearly linear for the range of scenarios studied (Figure 9.1), with the exception of the scenarios in which the temperature increased 4°C on the East River. In the latter case, runoff increases more slowly than precipitation. Overall, runoff in the White River is slightly less affected by temperature increases than is runoff in the Animas and East rivers.

Annual average flows are normally distributed on the East River and approximately lognormally distributed on the White and Animas rivers. Temperature increases of 2°C and 4°C strongly skew these distributions toward the right for the Animas and East rivers, indicating a greater frequency of occurrence of low-flow years. This shift is evident for the White River as well but is not nearly as pronounced. In all cases, the climate change scenarios result in distributions of annual flow that are approximately lognormal. As expected, percentage changes in runoff are significantly greater for low-flow years, while absolute effects are greater for high-flow years.

Temperature increases cause peak runoff to occur earlier in the year. A temperature increase of 2°C shifts peak runoff from June to May for the White and Animas rivers. For the East River, peak runoff still occurs in June, although it is not nearly as exaggerated. For all three basins, the 2°C rise creates a double peak, with runoff in May and June nearly equal. When temperature is increased by 4°C, the East River also undergoes a distinct shift in the timing of peak runoff—from June to May. The United Kingdom Meteorological Office scenario for the Animas and White rivers shifts peak runoff from June to April, which reflects the larger 6.8°C temperature rise for this GCM.

Figure 9.2 illustrates mean flow as it varies between high- and low-flow seasons on the White River. Spring flow is averaged over the three highest flow months (April, May, and June) and fall flow over the three lowest flow months (October, November, and December). These results suggest less extreme seasonal flows as a result of climate change in most cases. In the Animas River model, climate scenarios tend to diminish the differences between spring and fall flows, and spring flows decrease in all scenarios. In the White and East river models, spring flows do not decrease as dramatically, and scenarios that incorporate precipitation increases of 20 percent augment spring flow substantially. Because spring flows are already

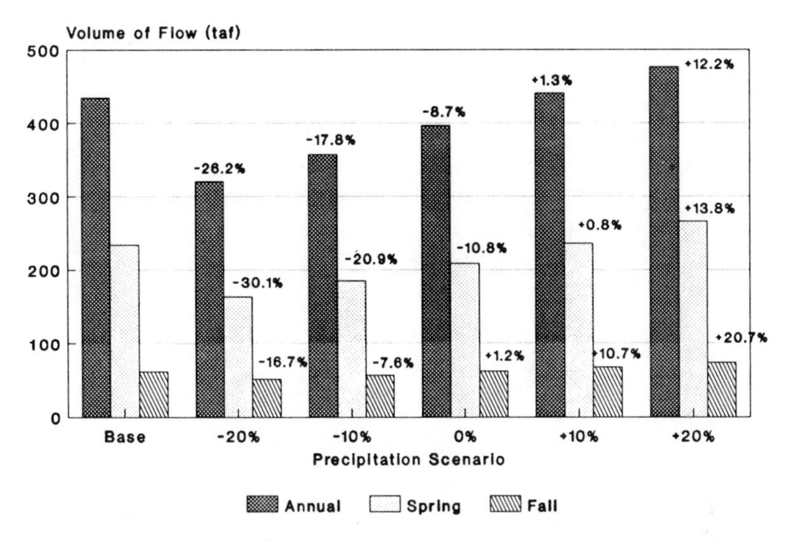

FIGURE 9.1 Effect of hypothetical changes in precipitation on streamflow in the White River.

very high in these basins, significant increases in runoff would increase the risk of flooding.

CRSS Model Results

Changes in natural flow produce uneven changes in actual (modeled) flow (Figure 9.3). For example, increases in natural flow have a relatively greater effect on actual flow at the compact point (the dividing point between the upper and lower Colorado River basins, which occurs at Lee's Ferry) than do corresponding decreases in natural flow. This reflects the fact that under conditions of decreased flow, reservoir releases are being increased. Accordingly, a 10 percent decrease in natural flow produces a 30 percent decrease in upper basin storage and only a 12 percent decrease in flow at the compact point. A similar effect can be seen for actual flow below Imperial Dam, where a 10 percent increase in natural flow produces an 11 percent increase in actual flow, whereas a 10 percent increase in natural flow results in only a 7 percent decrease. At an upstream point—

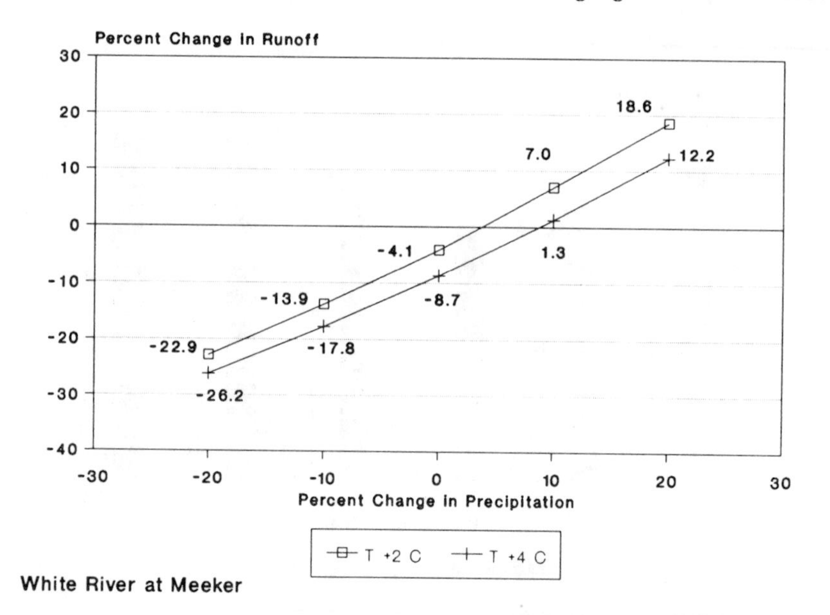

FIGURE 9.2 Mean annual, spring (April through June), and fall (October through December) flows for the White River. The base case and the T+4°C scenarios are shown. Percentage numbers indicate the change in flow compared to the base case.

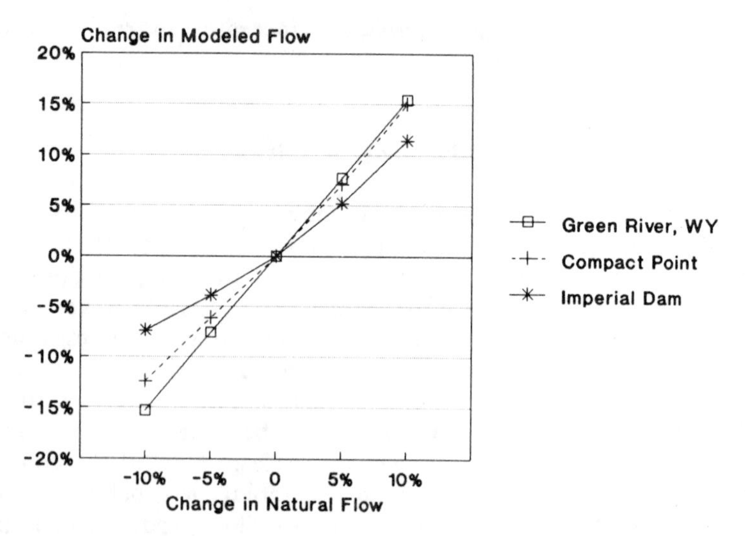

FIGURE 9.3 Effect of changes in natural flow on changes in actual flow at three stations.

Green River at Green River, Wyoming—these discrepancies are not discernable. A 10 percent change in natural flow (equivalent to an absolute change of 150 trillion acre-feet [taf]) in either direction causes a 15 percent change in actual flow (also equivalent to 150 taf).

The impact of changes in natural flow on storage can be seen in Figure 9.4. When natural flow is increased by 10 percent, storage in the upper basin increases by 19 percent, to roughly 85 percent of total capacity. Storage changes rapidly because the reservoir system is designed and operated for historic flow conditions. The buffering capacity of the current system is relatively ineffective in the face of 10 percent changes in flow over the long term.

Hydroelectric power production is also strongly affected by relatively moderate changes in flow. A 5 percent decrease in natural flow corresponds to a 4 to 8 percent decrease in actual flow but results in a 15 percent decrease in hydropower production. A 10 percent increase in natural flow creates a 10 to 15 percent increase in actual flow and a 20 percent increase in annual average hydropower production.

The impact of changes in natural flow on depletions (consumptive use) is shown in Figure 9.5. Overall deliveries and depletions do not decline as much as decreases in flow might suggest, because reservoir storage is being dramatically reduced so water-supply requirements can be met. The minus 20 percent scenario causes an 11 percent decline in average annual consumptive use, while the plus 20 percent scenario causes an 8 percent rise in average annual consumptive use. It is difficult to assess the relative impacts of extreme shortages on different users because allocations and operating procedures would be subject to many political, technical, and economic variables. In the CRSS model, shortages are allocated based on the assumption that they would not ultimately exceed those that occurred during the most critical period on record (1953 through 1964). Consequently, the model imposes reductions in the upper basin that equal only about 5 percent of total demand. Additional shortages are then passed on to the lower basin. Using this shortage strategy, upper and lower basin depletions decline by 2 and 9 percent, respectively, in the minus 10 percent scenario. Compact requirements were violated 33 percent of the time under the minus 5 percent scenario and 61 percent of the time under the minus 20 percent scenario.

It must be emphasized, however, that this shortage strategy results in an unlikely apportionment of river flows. In fact, legally, the upper basin should suffer relatively greater shortages than the lower basin (Getches, 1991, pp. 15-16). Under a strict interpretation of the Colorado River Compact, upper basin consumption could be limited

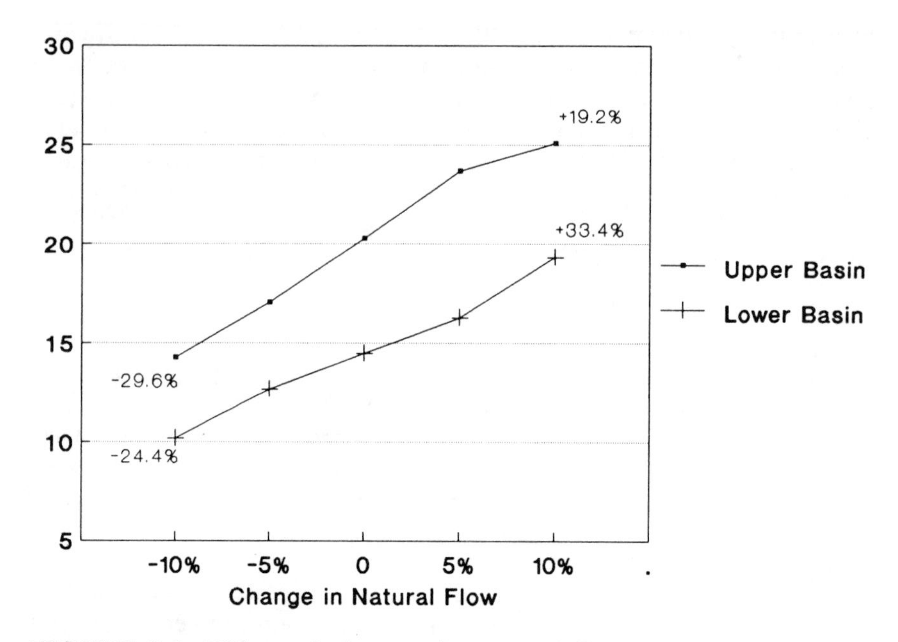

FIGURE 9.4 Effect of changes in natural flow on average storage on August 1.

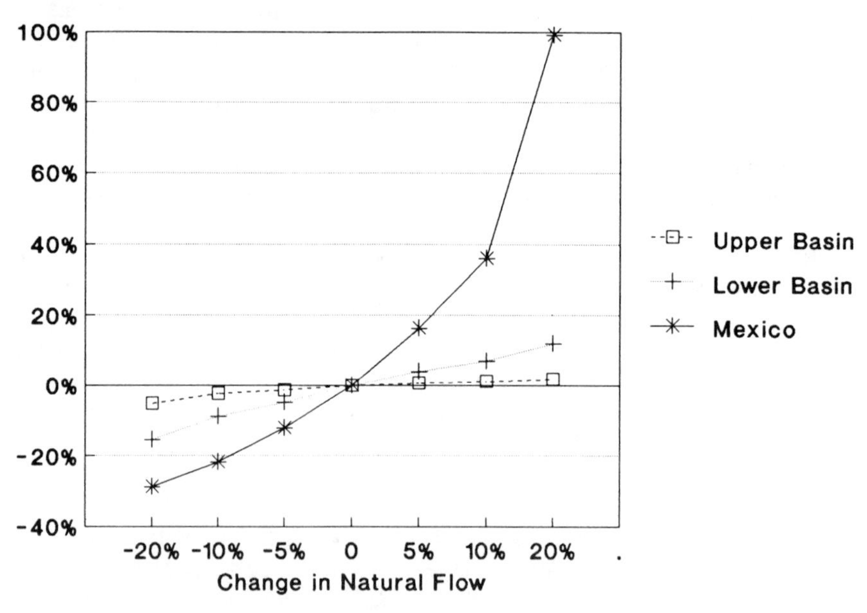

FIGURE 9.5 Effect of changes in natural flow on average annual consumptive water use.

to 3 maf or less roughly one-third of the time even under the base-case scenario. This would be a cutback of more than 20 percent over current levels. Under the minus 20 percent scenario, the upper basin might never receive more than 3 maf per year and could receive as little as 2 maf in about 10 percent of the years.

Not surprisingly, the greatest vulnerability of the system is in the area of water quality (salinity). Under almost no circumstances can existing water-quality standards be met given projected demands and operating constraints. Our results suggest that at least a 20 percent increase in natural flow would be needed to bring the salinity levels in the lower basin into compliance with existing standards (Figure 9.6). Although the scenarios presented here result in only moderate changes in salinity, the problem is already so severe in the base case that even moderate declines in water quality are of particular concern.

DISCUSSION AND CONCLUSIONS

In the first study to analyze the impacts of climatic change on the Colorado River, Stockton and Boggess (1979) used Langbein's relationships (Langbein et al., 1949) to estimate the effects of a 2°C temperature rise and a 10 percent decrease in precipitation. They found that streamflow in the upper basin would decline by about 44 percent. Following up on that work, Revelle and Waggoner (1983) developed a linear regression model of runoff using precipitation and temperature as independent variables. Their model predicted that a 2°C temperature increase would decrease mean annual flow by 29 percent, while a 10 percent decrease in precipitation would decrease runoff by about 11 percent. In combination, these changes would result in a 40 percent decrease in runoff, in close agreement with Stockton and Boggess's earlier result.

In contrast, our studies with the NWSRFS model suggest less severe impacts on runoff and a greater sensitivity of annual runoff to precipitation rather than temperature changes. A 2°C temperature rise combined with a decrease in precipitation of 10 percent would decrease runoff by 14 to 23 percent. While these results are lower than results of the earlier statistical studies, they still represent dramatic decreases in water availability in the Colorado basin. These results are comparable to similar studies of arid and semi-arid basins that used conceptual hydrologic models (Gleick, 1987; Flaschka et al., 1987), supporting Karl and Riebsame's (1989) conclusion that the Langbein relationships overstate the role of evaporation. For the

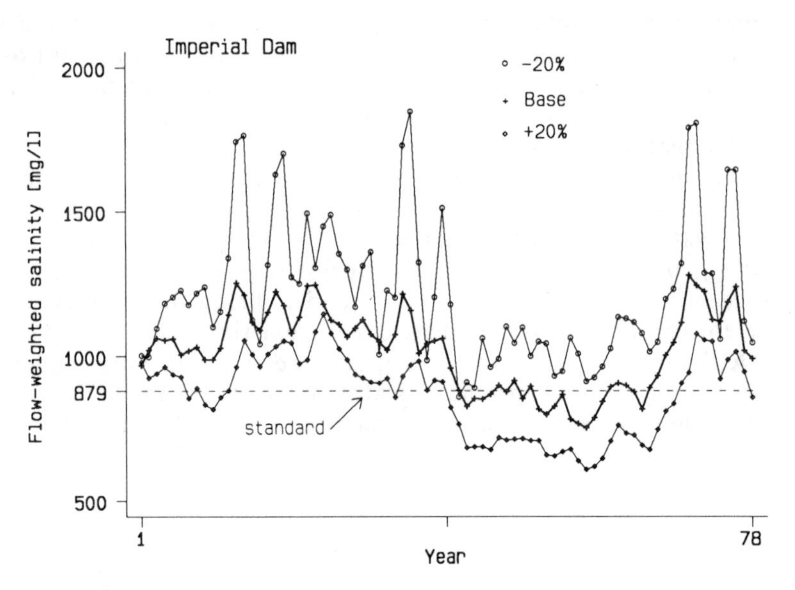

FIGURE 9.6 Salinity as a function of year at Imperial Dam. The base case and the +20 percent scenarios are shown. Water-quality standards are continually violated in all but the +20 percent scenario.

range of scenarios presented here, mean annual runoff changes nearly linearly with precipitation, although this relationship begins to break down as precipitation increases by 20 percent, at which point runoff begins to increase more quickly.

Our analysis suggests that variations in mean annual runoff of 30 percent are possible as a result of climatic change, with even greater changes likely in the most arid subbasins, but that precipitation changes of more than 10 percent would be necessary before changes in annual runoff would be significantly different from the historic flow series (Nash and Gleick, 1991). This does not imply that the impacts of climatic change are insignificant, but it does suggest the difficulty inherent in detecting the effects of climatic change given a relatively short and variable streamflow record. The results also suggest that increases in precipitation would be needed to balance the effects of higher temperatures on runoff. If precipitation stays the same or decreases, substantial decreases in water availability may result.

An increase in temperature shifts the seasonality of runoff as well, with peak runoff occurring earlier in the spring. This change reflects the fact that under higher temperatures, more precipitation falls as rain rather than snow, and snowmelt runoff occurs earlier in the year. Because this seasonal result is induced by changes in temperature, rather than by less-certain changes in precipitation, we believe it is fairly robust.

Table 9.6 summarizes the sensitivity of several water supply variables to changes in natural flow. Looking back at the hydrologic modeling discussed above, we can relate climate scenarios to changes in water supply. A temperature increase of 2°C and a decrease in precipitation of 10 to 20 percent corresponds, more or less, to a decrease in runoff of 20 percent. This, in turn, would cause reductions in storage of 60 to 70 percent, reductions in power generation of 60 percent, and an increase in salinity of 15 to 20 percent. A temperature increase of 2°C accompanied by an increase in precipitation of 20 percent corresponds roughly to a 20 percent increase in runoff, a 30 to 60 percent increase in storage, a 40 percent increase in power production, and a 13 to 15 percent decrease in salinity. A temperature increase of 4°C coupled with a precipitation decrease of 20 percent would result in approximately a 30 percent decrease in runoff, which is more extreme than any of the scenarios modeled with the CRSS.

The magnitude of these changes suggests that the Colorado basin would be highly vulnerable to the level of climatic changes contemplated in this study. Without any change in precipitation, these temperature changes alone imply decreases in runoff of 5 to 10 percent. The water supply impacts of such a change would be severely felt throughout the basin. The upper basin is likely to suffer given current institutional arrangements, but the lower basin will not be immune either because of its heavy (and increasing) dependence on Colorado River water. Although the Law of the River places the burden of shortages on the upper basin, the volume of lower basin demands, combined with the high priority given to supplying them on an annual basis, removes water management flexibility. Moreover, decreases in flow would exacerbate an already severe salinity problem in the lower basin. Should precipitation increase, some or all of these impacts might be offset, but should precipitation decrease, the impacts may exceed even those presented here.

It should be borne in mind, however, that these results reflect flow changes of 5 to 20 percent imposed on the hydrology of the last 80 years. The results would be different if a different hydrologic record had been used. For instance, the hydrology of the last 400

TABLE 9.6 Sensitivity of water-supply variables to changes in natural flow in the Colorado River Basin.[1]

Change in Natural Flow (%)	Change in Actual Flow (%)	Change in Storage (%)	Change in Power Generation (%)	Change in Depletions (%)	Change in Salinity (%)
-20	(10-30)	(61)	(57)	(11)	15-20
-10	(7-15)	(30)	(31)	(6)	6-7
-5	(4-7)	(14)	(15)	(3)	3
5	5-7	14	11	3	(3)
10	11-16	28	21	5	(6-7)
20	30	38	39	8	(13-15)

NOTE: [1] Numbers in parentheses represent decreases.

years suggests that much more severe and sustained droughts have occurred in the past. If this hydrology were used as a basis for a similar study, decreases in flow would have still greater impacts on the basin.

In summary, this research points out that significant changes in streamflow are likely if scenarios of future climatic changes prove to be accurate. Most hydrologic records are quite short and have been subject to other complicating effects (e.g., dams, diversions, and changes in vegetative water use). Thus, changes in annual runoff may not be distinguishable from natural variability until mean runoff has changed by more than 10 percent. This suggests that prudent management policies should not wait for verifiable change. While the Colorado basin is prepared to handle very large short-term fluctuations in streamflow, it may nonetheless be vulnerable to long-term increases or decreases in streamflow and water supply.

REFERENCES

Anderson, E. A. 1973. National Weather Service River Forecast System: Snow Accumulation and Ablation Model. National Oceanic and Atmospheric Administration Technical Memorandum NWS HYDRO-17. Silver Spring, Md.: U.S. Department of Commerce.

Burnash, R. J., R. L. Ferral, and R. A. McGuire. 1974. A Generalized Streamflow Simulation System: Conceptual Modeling for Digital Computers. Sacramento, Calif.: Joint Federal-State River Forecast Center.

Flaschka, I. M., C. W. Stockton, and W. R. Boggess. 1987. Climatic variation and surface water resources in the great basin region. Water Resources Bulletin 3:47-57.

Getches, D. H. 1991. Water Allocation During Drought in Arizona and Southern California: Legal and Institutional Responses. Boulder, Colo.: University of Colorado Natural Resources Law Center.

Gleick, P. H. 1987. Regional hydrologic consequences of increases in atmospheric CO_2 and other trace gases. Climatic Change 10:137-161.

Karl, T. R., and W. E. Riebsame. 1989. The impact of decadal fluctuations in mean precipitation. Climatic Change 15:423-448.

Klemes, V. 1985. Sensitivity of Water Resource Systems to Climate Variations. World Climate Applications Programme WCP-98. Geneva: World Meteorological Organization.

Langbein, W. B., et al. 1949. Annual Runoff in the United States. U.S. Geological Survey Circular 5. Washington, D.C.: U.S. Department of the Interior.

Nash, L. L., and P. H. Gleick. 1991. The sensitivity of streamflow in the Colorado basin to climatic changes. Journal of Hydrology 125:221-241.

Nemec, J., and J. Schaake. 1982. Sensitivity of water resource systems to climate variation. Hydrological Sciences 27:327-343.

Revelle, R. R., and P. E. Waggoner. 1983. Effects of a carbon dioxide-induced climatic change on water supplies in the western United States. Pp. 419-432 in Changing Climate. Washington, D.C.: National Academy Press.

Stockton, C. W., and W. R. Boggess. 1979. Geohydrological Implications of Climate Change on Water Resource Development. Fort Belvoir, Va.: U.S. Army Coastal Engineering Research Center.

U.S. Department of the Interior (DOI). 1987. Colorado River Simulation System: System Overview. Denver, Colo.: U.S. Department of the Interior.

10

Climate Uncertainty: Implications for Operations of Water Control Systems

John A. Dracup and Donald R. Kendall
University of California, Los Angeles
Loyola Marymount University, Los Angeles

INTRODUCTION

The key question we are addressing here is how water resource systems will be operated under conditions of uncertain future climate variability and change. This question in turn raises the question of whether or not a current water resource system is adequately designed for climate variability and change. We attempt to answer these two questions using the method of climate variability analogies as suggested by Glantz (1988; 1989). Traditionally, water resources design, such as reservoir sizing, has been predicated on an assumption that the statistical parameters of historic streamflow series are stationary: that is, the parameters that characterize the historical streamflow series do not change with time. This is not to say that traditional water resource system designs assume that climate is actually stationary; traditional designs account for changing climates by considering the variance of streamflow (σ^2) about its mean ($\overline{Q}$). Stationarity assumes, however, that future variations in climate, as expressed in streamflows, will be similar to those observed in the past.

Future scenarios of climate variability of interest to water resource managers are those of long-term extreme high or low streamflows. The scenarios developed here will be based on historical analogies. We will use the high flow event that occurred on the Colorado River during the spring of 1983 as an analogy of climate variability or change causing significant increased streamflow in the Colorado River. We will use low flow events evident from tree-ring studies in the Colorado River basin as analogies of climate variability or change causing significant decreased streamflow in the Colorado River.

The idea of forecasting by the use of historical climate analogies has been used by Glantz (1988; 1989), who poses four caveats in their use. First, the use of the climate analogy should be made clear so that it is not misleading when viewed in a different context. Second, the climate analogy should not be overextended to support unjustified conclusions. Third, one should recognize that the analogy may be inappropriate for cultural or historical reasons. Fourth, one should note that analogies can lead to the development of possible but inconsistent scenarios.

The use of historical climate analogies provides a different look into the future than that provided by large-scale computer models. Furthermore, climate analogies provide an insight into how the responsible agencies actually dealt with the climate variability as it occurred.

The topic of the operation of water resource systems under climatic stress addressed here is, of course, not new. One of the best discussions of this can be found in the National Research Council's *Climate, Climate Change, and Water Supply* (1977) study. This study emphasizes that "unless the exact sequence of future flows can be predicted with certainty there may be little benefit to hydrologic system design." The study proposes designing water resource systems with robustness (the ability to perform reasonably well under a variety of possible climates) and resilience (the ability of a system designed for one climate and set of conditions to be modified in response to persistent new climates or conditions) (Matalas and Fiering, 1977). The question to be considered here, then, is whether climate analogies can be used to determine whether or not an existing water resource system exhibits the properties of robustness and resilience.

Forecasting by analogy using the Colorado River as an example also has been studied by Brown (1988). However, she focused on the Colorado River Compact rather than on floods and droughts along the river.

THE COLORADO RIVER BASIN

If the streamflow in the Colorado River somehow could be equated with the number of words written about it, the river would constantly flow as a torrent (Dracup, 1977; Dracup et al., 1985; Hundley, 1975; Rhodes et al., 1984).

The Colorado River dominates water resource development in the seven states of the southwestern United States (see Figure

10.1). It also is one of the most carefully managed river systems in the world. Its multipurpose uses of water supply, hydroelectric power, and water-based recreation compete with its management priority of flood control. With the exception of the deserts of the Great Basin, this 243,000-square-mile basin has the greatest water deficiency (average precipitation less potential evapotranspiration) of any basin in the contiguous United States. Yet, more water is exported from the Colorado River basin than from any other basin in the United States.

The basin has been divided by the Colorado River Compact into the upper Colorado basin and lower Colorado basin for purposes of interstate administration. The upper basin drainage includes the areas of Arizona, Colorado, New Mexico, Wyoming, and Utah that drain into the Colorado River above Lee's Ferry, Arizona. It is bounded on the east and north by mountains forming the Continental Divide, and on the south it opens to the lower Colorado region. The lower basin drainage includes most of Arizona, parts of southeastern Nevada, southeastern Utah, southeastern California, and western New Mexico.

WATER AVAILABILITY ESTIMATES

A wide range of climates occur in the Colorado River basin because of differences in altitude, latitude, and topographic features. In the north, summers are short and warm and winters are long and cold. In the south, the summers are longer and the winters are moderate at low altitude, but colder temperatures occur in the mountains.

About 83 percent of the water that flows in the Colorado River basin comes from the upper basin. The average annual precipitation throughout the entire upper basin is about 16 inches (40.6 cm), which amounts to 93,440,000 acre-feet per year (115 x 10^9 m^3). Approximately 15 percent of the precipitation runs off, and most is lost to evapotranspiration.

One of the most famous and controversial hydrologic records in the United States is that of the virgin flow of the Colorado River at Lee's Ferry, Arizona. Lee's Ferry is defined as a point on the Colorado River one mile below the mouth of the Paria River. Estimates of virgin flow have been made there for the upper basin since 1896; however, runoff has been measured and recorded only since the first gaging station was established at Lee's Ferry during the summer of 1921. (The Bureau of Reclamation now uses natural

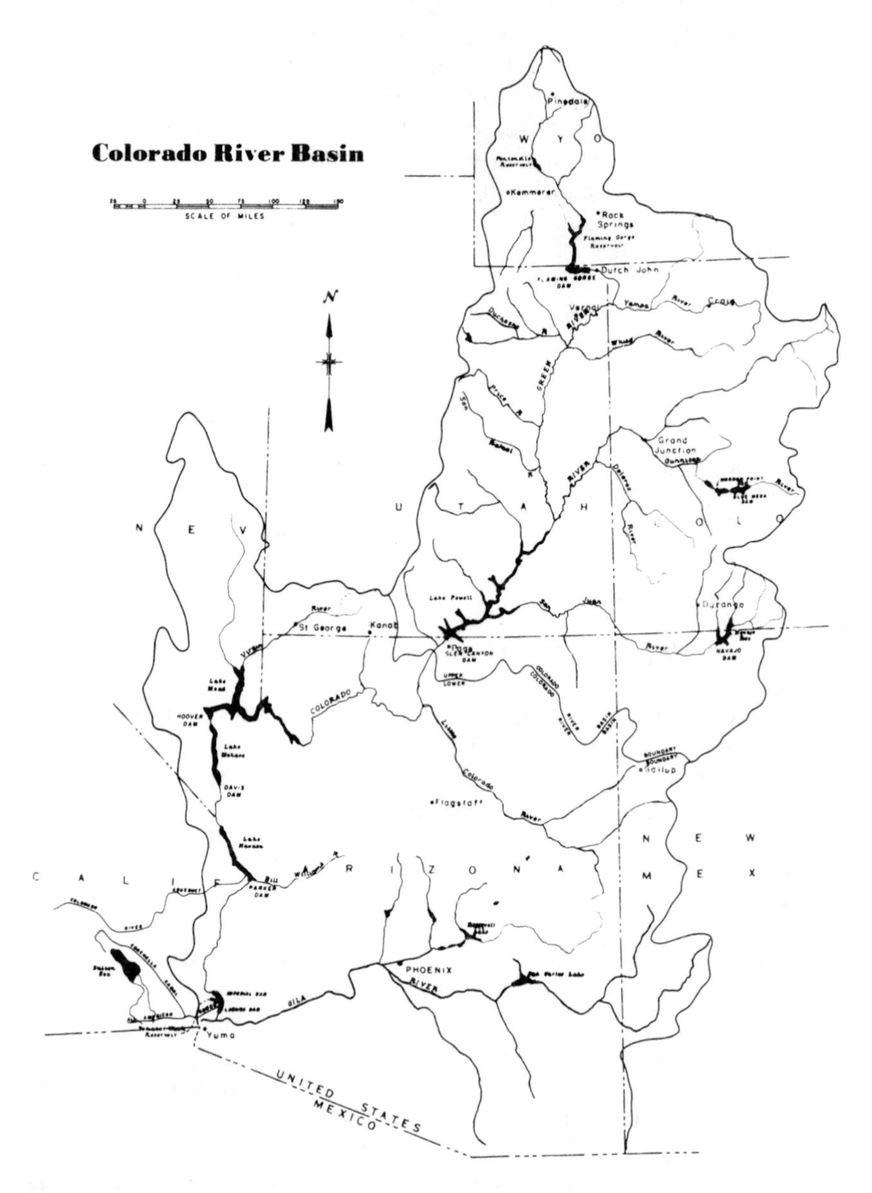

FIGURE 10.1 Colorado River basin upstream of the inflow to Mexico.

flow, not virgin flow, for its operation studies.) The importance of this flow is accentuated by the Colorado River Compact, which requires that the upper basin deliver 75 million acre-feet (maf) (92.5×10^9 m^3) at Lee's Ferry each 10 years. Estimates of the long-term annual average flow vary from 11.8 to 16.8 maf (14.5 to 20.7×10^9 m^3) depending on the time period selected (Colorado River Board of California, 1969). Recent tree-ring analysis dating back to 1512 has indicated the long-term mean to be approximately 13.5 maf (16.6×10^9 m^3) (Stockton, 1977).

The current estimates of available surface-water supply within the upper basin are less than those at the time the Colorado River Compact was negotiated. This is because of the abnormally wet period that occurred during the early part of this century. The range of annual natural flow at Lee's Ferry has varied from a low of 5.0 maf (6.2×10^9 m^3) in 1977 to a high of 24.0 maf (29.6×10^9 m^3) in 1917. The average natural flow from 1931 through 1989 of 14.2 maf (17.5×10^9 m^3) may be closer to the long-term mean.

The laws governing the Colorado River have been presented in detail by Meyers (1966) and Hundley (1975; 1983). Only a brief summary of the major treaties, laws, and compacts will be presented here.

The allocation of Colorado River water is based on the concept of beneficial consumptive use. The allocation system operates at four levels: international, interregional, interstate, and intrastate (Weatherford and Jacoby, 1975).

The international allocation was accomplished by the Mexican Treaty of 1944. Mexico was guaranteed an annual amount of 1.5 maf (1.8×10^9 m^3) except in times of extreme shortage. However, this treaty contained no provision for water quality. Thus, joint agreements in 1965 and 1973 called for a temporary agricultural drainage water bypass and eventually a desalting plant to improve the quality of water crossing the border.

The interregional allocation was achieved when Congress approved the Colorado River Compact, which became effective in June 1929. Sectional rivalry has caused the states included in the drainage basin to agree to an equal apportionment in the use of the Colorado River system waters between the states of the upper basin and the states of the lower basin (an agreement set forth in Articles III (b) and III (d) of the Colorado River Compact).

Traditionally, the fertile lowland valleys in the lower basin states have developed economically more rapidly than have the mountain headwater "areas of origin" in the upper basin states. The upper basin states insisted that an equitable apportionment of the river be made to them prior to the expenditure of large sums of federal money, which might result in a modification of equities adverse to the upper basin states. This is in essence what was achieved in the Colorado River Compact.

The intent of this landmark document was to give each basin the perpetual right to the "exclusive beneficial use of 7.5 maf (9.25 x 10^9 m^3) of water per annum." However, the lower basin was assured that depletion in the upper basin would allow at least 75 maf (92.5 x 10^9 m^3) of flow to the lower basin at Lee's Ferry in each successive 10-year period. Thus, the lower basin received a guaranteed 10-year, not annual, minimum flow, and the upper basin assumed the burden of any deficiency caused by a hydrologic dry cycle. It is important to note that the division of the use of water between the upper and lower basins is a fixed amount rather than a proportional amount (such as one-half of a 10-year moving average).

CURRENT COLORADO RIVER MANAGEMENT TECHNIQUES

The joint operation of Lakes Powell and Mead is subject to the following criteria, according to the Law of the River (Nathanson, 1978):

- a Lake Powell minimum objective release of 8.2 maf per year;
- additional releases from Lake Powell to equalize end-of-year active storages in Lakes Powell and Mead if Lake Powell would otherwise contain more water in storage; and
- sufficient storage in the upper basin reservoirs to assure future deliveries to the lower basin without impairing annual consumptive use in the upper basin (called 602(a) storage under the Colorado River Basin Project Act of 1968).

As a result of the Law of the River, each basin currently possesses many storage facilities, including a large linchpin reservoir: Lake Powell for the upper basin and Lake Mead for the lower basin. The storage in these two reservoirs totals 51.0 maf (62.9 x 10^9 m^3), or 85 percent of the total storage in the entire Colorado River basin. The reservoir system now stores about four times the annual flow of the river. This volume of water in storage reflects the determination of the basin states to conserve as much water as possible, providing a margin of safety in the event that a run of dry years occurs. The upper basin states prefer that releases to the lower basin be the absolute minimum required by law. Furthermore, current operational regulations require that the river's manager, the U.S. Bureau of Reclamation, maintain Lakes Mead and Powell at or near equal volumes of water in storage at the end of each operating year (which occurs on September 30) (Nathanson, 1978). Each basin is thus assured "equal ownership" of the river.

Flood control protection is provided to residents, farms, and businesses below Hoover Dam. Flood control operations rest on two central elements:

1. scheduled dedicated water storage space made available to catch the spring runoff in Lake Mead; and
2. a forecast of how much water will enter Lake Powell from April 1 through July 31, produced by the National Weather Service (NWS) Colorado Basin River Forecast Center in Salt Lake City, Utah.

The objective of the flood control procedure, in effect, is to create enough storage space, through reservoir releases from August through January, to catch the predicted April through July runoff. The plan, in action since 1968 and slightly modified in 1982, uses a monthly streamflow forecast generated for the period January through July that predicts the spring inflow to Lake Powell. Adjustments in the forecasted storage space for the inflow can then be made to keep downstream releases below damaging levels and at the same time conserve as much water as possible.

The NWS Colorado Basin River Forecast Center uses monthly estimates to arrive at forecasts of the maximum probable and minimum probable April through July runoff. To meet these runoff estimates, the Bureau of Reclamation increases flood control space in Lake Mead starting on August 1 of each year to have 5.35 maf (6.6×10^9 m^3) available by January 1. According to the flood control plan for the Colorado River, Lake Mead is the only major basin reservoir with an explicit flood control space schedule. Prior to the construction of Glen Canyon Dam, the standard flood control procedure was to have 5.8 maf (7.1×10^9 m^3) of storage available on January 1, as recommended by Debler (1930). This storage requirement was increased each month until a maximum requirement of 9.5 maf (11.7×10^9 m^3) was reached on April 1. These procedures were formalized by the U.S. of Army Corps of Engineers in 1955 and were continued until 1968.

The Bureau of Reclamation's scheduled outflow release rates through the dams and the storage space availability based on the

NWS inflow forecasts have worked well in recent decades in minimizing water lost through unneeded anticipatory releases and potential flooding below Hoover Dam. This has helped to maximize hydroelectric generation, water conservation, storage, and flood control. However, the conditions in the Colorado basin in the early 1980s were radically different from those of the 1960s and the 1970s. The changes in those conditions were significant contributors to the flooding that occurred in the spring and summer of 1983.

OPERATION DURING HIGH STREAMFLOWS: THE COLORADO RIVER SPRING FLOODS OF 1983

On January 1, 1983, there were 6.6 maf (8.1×10^9 m^3) of storage space available in Lake Mead and upstream—more than the required January target of 5.35 maf (6.6×10^9 m^3). Yet even with the surplus storage space available, the reservoir system was overwhelmed by the magnitude of the spring inflow to Lake Powell. Because of late precipitation and cool weather throughout the upper basin, snowpack continued to increase during April and May. Figure 10.2 shows the rapid and unusual changes in the forecasted inflow to Lake Powell from January through June 1983.

Figure 10.3 illustrates the relationship between Lake Powell inflows, Lake Powell outflows, and Hoover Dam releases from April through July 1983. Because of the massive influx of water into Lake Powell, the Bureau had to increase outflows from Glen Canyon Dam. This, in turn, obliged the Bureau to raise outflows from Hoover Dam. The releases at Hoover Dam, which historically had been held to approximately 25,000 cubic-feet per second (cfs) (708 m^3 s^{-1}), were elevated to over 40,000 cfs (1,132 m^3 s^{-1}) in July. This is a critical point, for 40,000 cfs (1,132 m^3 s^{-1}) was the targeted maximum outflow rate from Lake Mead under the 1968 revised flood control procedures. The Bureau of Reclamation operators successfully limited Hoover release rates to 40,000 cfs (1,132 m^3 s^{-1}) except for the month of July, during which Hoover releases averaged 41,854 cfs (1,184 m^3 s^{-1}). However, flooding downstream of Hoover Dam begins when the flow exceeds 19,900 cfs (538 m^3 s^{-1}).

The rapid sequence of meteorological events occurring late in the spring, coupled with the problem of attempting to move massive amounts of water through Lakes Powell and Mead in a short period of time, resulted in streamflows greater than those

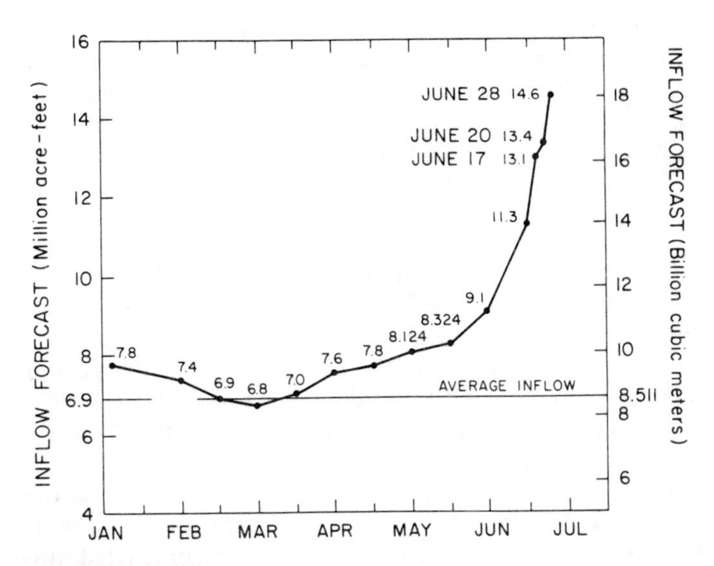

FIGURE 10.2 Forecasted inflow to Lake Powell from January through June 1983.
SOURCE: Dozier and Brown, 1983.

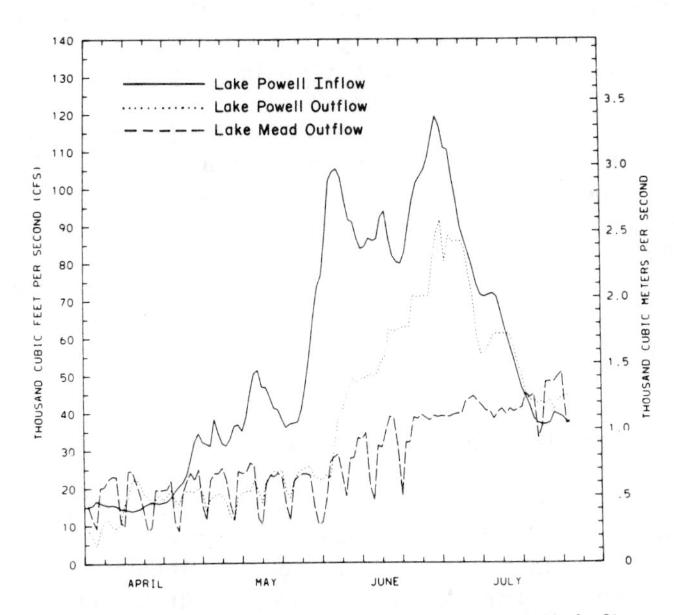

FIGURE 10.3 Relationship among Lake Powell inflows, Lake Powell outflows, and Hoover Dam releases from April through July 1983.

experienced during the previous two decades by lower basin residents and businesses. The unusual and unexpected flooding along the Colorado River during 1983 was the result of three converging factors: the sudden required operation of a full river system, an encroachment into the downstream flood plain, and climate variability in the basin. Each of these is discussed below.

The Full River System

The filling of Lake Powell behind Glen Canyon Dam began in 1963 and was completed in 1980. During this 17-year period, there were virtually no required flood control operations on the Colorado River. Runoff in excess of downstream water supply and hydropower generation was easily stored.

However, in 1980 Lake Powell became full, which required that the river now be operated in a careful, prudent manner; there became little room for forecast error. The forecasted inflow to Lake Powell had to be not only accurate, but also carefully monitored on a real-time basis. Monitoring inflows allows corrective management responses if conditions permit. However, in 1983, Lake Powell inflows rose so rapidly that there was no time for mitigating responses. For example, on May 24, 1983, the unregulated inflow to Lake Powell was approximately 37,000 cfs (1,047 m^3 s^{-1}). Eight days later, on June 1, the unregulated inflow was 102,000 cfs (2,887 m^3 s^{-1}) (U.S. Bureau of Reclamation, 1983).

Even the availability of real-time data may not have been sufficient to manage a wet year such as 1983, since it takes substantial time to move water through dams with structurally limited release rates. The 1983 April through July inflow into Lake Powell was more than 14 maf (17.3 X 10^9 m^3). Approximately 140 days would be required to discharge that quantity at a rate of 50,000 cfs (1,415 m^3 s^{-1}).

Physical Encroachment into the Flood Plain

Physical encroachment into the lower basin flood plain is a function of the defined flood plain boundaries, the relative stability in the annual streamflows, and societal decisions. Thus, encroachment into the lower basin flood plain, which would not have been possible in the absence of the upstream storages, was encouraged by a combination of technological fixes and lax zoning practices in counties bordering on the Colorado River (Arizona Republic, June 22, 1983; July 1, 1983).

The two major dams on the Colorado River have performed as planned in controlling the variability of streamflow rates (Figure 10.4). Glen Canyon and Hoover dams have consequently provided substantial protection to the lower Colorado flood plain in terms of their ability to reduce the river's meanderings and the flooding associated with high spring streamflows. Even in 1983, releases at Hoover Dam did not significantly exceed 50,000 cfs (1,415 $m^3 s^{-1}$).

The history of development in the flood plain roughly began with the construction of earthen levees in the area around Yuma, Arizona. The levee system was constructed to protect agricultural land (fertile flood plain soil) from the annual rush of spring snowmelt (U.S. Army Corps of Engineers, 1982). With the completion of Hoover Dam and the subsequent decrease in spring streamflow variability, more flood plain acreage became available for development. This resulted in the construction of residential and commercial structures in these areas. As the 1982 review of flood control operating procedures notes (U.S. Army Corps of Engineers, 1982):

> Few, if any, structures were located in the 40,000 cfs (1,132 $m^3 s^{-1}$) flood plain in the lower Colorado River at the time of the closure of Hoover Dam (1935) and for some years thereafter. For many years the flood control operation plan for Hoover Dam has incorporated a "target maximum" flood control release of 40,000 cfs (1,132 $m^3 s^{-1}$). With the completion of Glen Canyon Dam in 1962, streamflow variability was sharply narrowed. This coincided with the period of the greatest physical encroachment into the flood plain, including construction and development within the streamflow profile of less than 28,000 cfs (792 $m^3 s^{-1}$).

The period when the Colorado reservoir system was filling with water constituted a time during which true exposure to climatic impacts, such as precipitation variability, did not exist. It was not representative of a new climatic regime in the basin, but only of anthropogenic interference with the flow of the river. The encroachment into the flood plain was possible because water was in storage upstream and also because the filling of Lake Powell was drawn out for almost two decades. Two decades are more than sufficient to affect societal perceptions of climate stability.

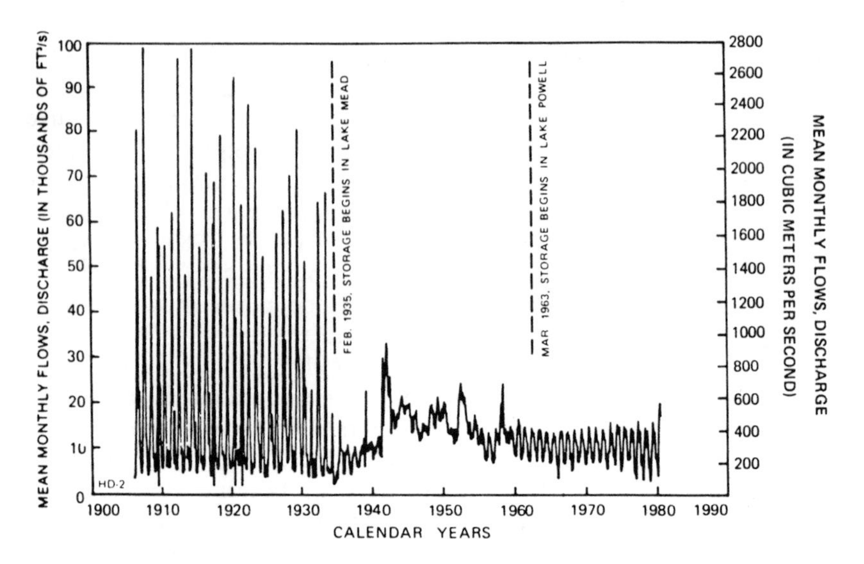

FIGURE 10.4 Mean monthly flows downstream of Hoover Dam in the Colorado River.

Climate Variability

Perceptions of climate stability in the Colorado River basin are not borne out by historical data. In fact, the third factor that contributed to the 1983 lower basin flooding was the variability of the climate in the arid American Southwest. As Figure 10.5 demonstrates, the variability of the river's streamflow is substantial. As recently as 1977, the western United States was hit with a severe drought that resulted in a significant drop—to 5.0 maf (6.2×10^9 m^3)—in the estimated virgin flow of the Colorado (Upper Colorado River Commission, 1982).

It is interesting to note that the 1977 drought also played a key role in the decision among the Colorado River basin states to defer any action regarding revision of the river management scheme (Broadbent, 1983). The concern in the Colorado River basin was again reinforced in terms of the major adverse climatic impact anticipated: i.e., drought is of concern, while a wet year is not. Because of the effect of a full system and a variable climate, a

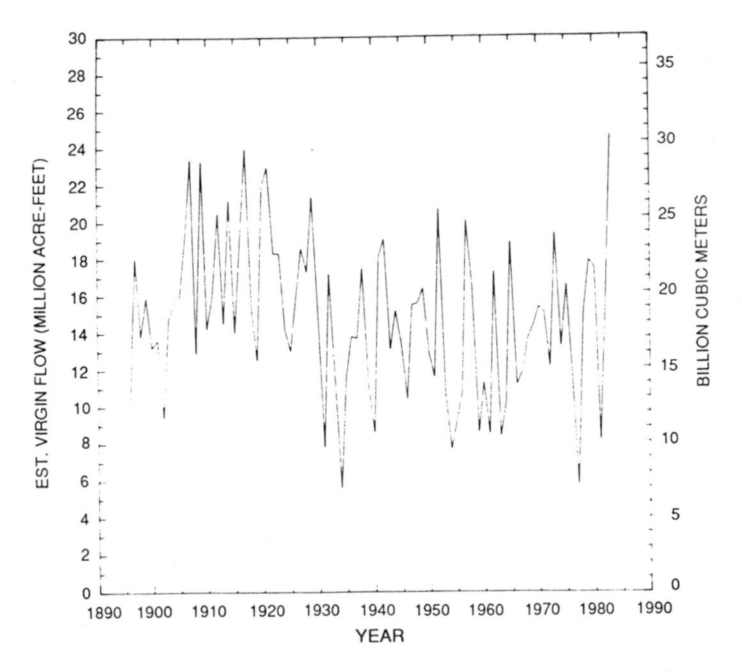

FIGURE 10.5 Estimated virgin flow in the Colorado River.

Bureau of Reclamation official had noted in 1979 that "the present operation strategy . . . . involves an 85 percent risk that damaging floodflows will occur between 1980 and 1984" (Freeny, 1981). Furthermore, because of climate variability, it has been estimated in a preliminary study by the U.S. General Accounting Office that if the river continues to be managed by keeping reservoirs full or nearly full, controlled flooding similar to that which occurred in 1983 can be expected to be repeated once every 10 to 15 years (Arizona Republic, Sept. 3, 1983).

The flooding on the Colorado River during the spring of 1983 also was exacerbated by a late spring snow, rapid warming, and rain on snow. Furthermore, the antecedent moisture conditions were higher than predicted (J. Lease, Bureau of Reclamation, personal communication, November 1990).

Thus, one can argue that it is not the streamflow variability in itself that caused the 1983 spring flooding and the associated damages. Flood plain encroachment and a full reservoir system, in conjunction with streamflow variability, converged to create appropriate conditions for the events of 1983.

Conclusions for River Basin Management Under High Flow Conditions

The issue of dedicated flood control storage space is a critical point in the management of the Colorado River. Accepting the Bureau of Reclamation's assumption that the Central Arizona Project will help to alleviate the flooding problem by providing an additional water diversion point, a temporary change in the dedicated flood control space could help to protect property in the lower basin (Broadbent, 1983; Freeny, 1981). However, since the precipitation and runoff of 1983 were so abnormal compared with recent years, proposals for increasing storage space will not necessarily be received warmly (Broadbent, 1983).

The response to the 1977 to 1978 drought indicated that there is substantial support for maintaining full reservoirs upstream and that the flood control for lower basin residents and other interests is not given high priority by the other beneficiaries of the river. Water resources in the American Southwest are managed for dry years, not for extremely wet years such as 1983. This is precisely the justification for the reservoir system to be maintained at a nearly full level.

Actions have been taken by the United States government since 1983 to reduce the potential for and the impact of flooding. A Colorado River flood warning system was developed through congressional appropriations in 1984 and 1985. A broad band of uncertainty was incorporated into the runoff forecasts in determining reservoir flood control space. The Federal-State Colorado River Management Work Group developed annual operating plans that provide for more effective river management under full and near full reservoir conditions.

Public Law 99-450, the Colorado River Floodway Protection Act, was enacted by Congress and approved by the President in 1986. The act established a floodway from Davis Dam to the Mexican border. The floodway was designed to accommodate a once-in-a-hundred-year river flow consisting of controlled releases and tributary inflows, or a flow of 40,000 cfs (1,132 $m^3 s^{-1}$). Except for limited purposes, no new expenditures or new federal financial assistance can be made available for construction within the floodway. No new flood insurance may be provided for new construction within the floodway. No new leases of lands within the floodway can be granted that are inconsistent with the operation and maintenance of the floodway. A floodway task force was formed to advise the Secretary of the Interior and the Congress on the restoration and maintenance of the floodway.

During the past eight years the natural flow of the Colorado River continued to be highly variable. The 1983 to 1986 recorded streamflows were greater than 20 maf (24.7 x 10^9 m^3) per year. The 1988 to 1990 streamflows were the lowest three consecutive streamflows on record (Colorado River Board of California, 1990).

SEVERE AND SUSTAINED DROUGHT IN THE COLORADO RIVER BASIN

The second climate analogy we can use as a climate change scenario is the supposition of a severe and sustained drought in the Colorado River basin. Since the Colorado River has a storage capacity of more than four times the average annual natural runoff and an elaborate system of laws governing water allocation and system operation, some entities believe that the basin has been drought-proofed. However, a close examination of the hydrologic reconstructions of unimpaired flows at Lee's Ferry based on tree-ring analysis indicates that droughts may be much more severe than analysis of conventional streamflow data indicates.

A study of severe, sustained drought affecting the Southwest involves establishing three factors: (1) streamflow characteristics and probabilities of drought events, (2) the effects of storage systems on water availability at key points in the system over time given streamflow characteristics during drought, and (3) relationships between available supplies and projected demands at key points during the period of analysis.

The Analytical Model

Severe and sustained drought along the Colorado River will be buffered by the massive storage provided in the Colorado River basin. For this reason, this drought analysis focuses on annual operation of the two major reservoirs in the upper and lower basins—Lakes Powell and Mead—and related effects on water availability relative to aggregate demand. Unimpaired measured streamflow at Lee's Ferry was taken to be the "base case" hydrology and compared against reconstructed sequences of equal length from tree-ring data dating back to 1520 (D. Meko, University of Arizona, personal communication, 1989).

The Colorado River Annual System Regulation Model (CRASR) (Metropolitan Water District of Southern California, 1980) was used to simulate the basin operation. It was substantially modified and rewritten by the authors to reflect current operating regimes. The model simulates annual regulation of Lake Powell and Lake Mead. Seasonal operation is not critical in this study because of the large amount of available storage capacity compared to annual runoff.

Nature of the Deficit

Streamflow is a useful hydrologic variable for drought studies at the basin scale, because it integrates a variety of processes over a watershed, including runoff, soil moisture, and evapotranspiration. Changes in climate are thus reflected as changes in streamflow volumes over a specified time period. Typically in the western United States, the record consists of a period of streamflow measurements dating back 60 to 80 years. At Lee's Ferry, the Colorado River streamflow records date back to 1906. In contrast, streamflow records at Lee's Ferry based on tree-ring studies have been reconstructed back to 1520, offering a record of greater variability than the measured record.

Under conditions of climate change, the magnitude of the change in the statistical parameters is unknown. However, we do know that climate will cause a change not only in the mean ($\bar{Q}$), but also in the variance (σ^2) and the skew (γ). Rather than performing a sensitivity analysis with these parameters and creating what-if scenarios, streamflow reconstructions provided by tree-ring analyses from the basin's recent past are used as a basis for this climate analogy study of droughts.

A comparison of some reconstructed sequences with hydrologic streamflow records of equal length is shown in Figure 10.6. What is most striking is the smaller mean annual flow for the reconstructed sequences—which ranges from 12.97 to 13.78 maf (16.0 to 17.0 x 10^9 m^3)—compared to the 1906 to 1985 record of 15.06 maf (18.6 x 10^9 m^3). Stochastic flows generated from an autoregressive model of order 1 (AR(1)) were also analyzed in this drought study (Kendall and Dracup, 1991). It has been shown that hydrologic sequences provided by tree-ring reconstructions were the most appropriate choice for this type of analysis (Kendall and Dracup, 1990).

Time Interval and Threshold Level of the Data

Drought analyses are typically considered on a monthly or annual time scale. For this analysis, a time scale of one-year

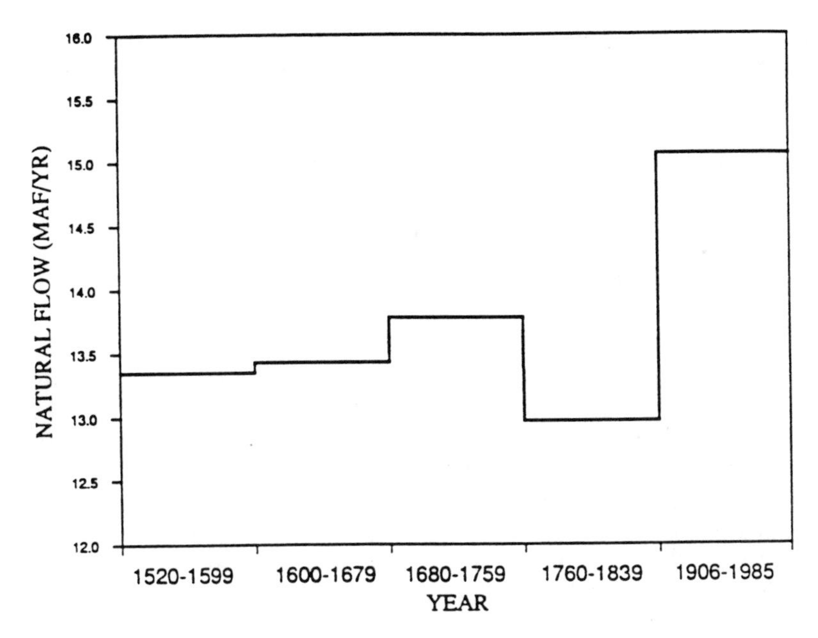

FIGURE 10.6 Eighty-year mean flows for the Colorado River at Lee's Ferry are nonstationary.
SOURCE: Adapted from data obtained from D. Meko, personal communication, October 1990.

intervals was chosen, as this is consistent with the time resolution period of tree-ring reconstructions.

Threshold levels in drought analyses are typically taken to be a streamflow's mean or median. Other appropriate thresholds may include average annual demand. For this analysis, the threshold level was taken to be the mean annual measured unimpaired flow at Lee's Ferry, important because comparisons using this threshold level will be made with annual unimpaired flows at Lee's Ferry reconstructed from tree-ring data.

Drought Parameters for the Colorado River

A study of the Colorado River using streamflow indicates that there is a tendency for low flows to follow low flows and for high flows to follow high flows. One possible reason for this behavior is that the process that governs streamflow has a nonstationary mean. Incorporation of tree-rings into this type of analysis gives a measure of support to this idea. An anthropogenically-induced climate change would exacerbate this process. Evidence of nonstationarity in the mean is shown in Figure 10.6.

Previous analyses performed by a panel on water and climate for the National Research Council (1977) argued against a stationarity-independent or short-memory process for the Colorado River streamflows. Furthermore, the panel estimated that the long-term mean for flows at Lee's Ferry was 13.5 maf (16.7×10^9 m^3) (Dracup, 1977).

A drought may be characterized as any year or series of consecutive years during which average annual streamflow is continuously below some specified threshold level, which is typically taken to be the long-term mean (Yevjevich, 1967; Dracup et al., 1980a). A drought event is considered to be composed of three defining attributes: duration (D); severity (S), which is the cumulative deficit; and magnitude (M), which is the average water deficit, such that $S = M \times D$. The parameters are interrelated; two are necessary and sufficient to completely define a single drought event (Dracup et al., 1980b). Duration and severity are the most correlated parameters and may be considered the two primary parameters dependent on streamflow values. Magnitude is a secondary parameter; duration and magnitude are weakly correlated.

Twenty hydrologic droughts occurred between 1906 and 1985 (see Table 10.1). Of these, four had a severity of over 15 maf (18.5×10^9 m^3), with a duration ranging from 3 to 5 years. It is apparent that a severe and sustained drought in the Colorado River basin is not necessarily produced by a single continuous drought event but may arise from a series of events separated by 1 or 2 years. For the period of record, a series of droughts, which account for the critical period, began in 1954 and continued through 1971.

Two sets of 80-year sequences from tree-ring reconstructed streamflows at Lee's Ferry that exhibited different types of severe droughts were chosen for the purpose of comparison. The periods selected are 1520 to 1599 (Figure 10.7) and 1600 to 1679 (Figure 10.8). These periods display different types of droughts with regard to their number, duration, and severity. The 1520 to 1599 sequence displays a series of droughts that contribute collectively toward severe and sustained drought conditions, while the 1600 to 1679 sequence displays a single severe drought.

TABLE 10.1 Historic hydrologic droughts on the Colorado River at Lee's Ferry, 1906 to 1985.

Drought No.	Duration (Years)	Starting Year	Severity (maf)	Magnitude (maf/Year)
1	1	1909	2.954	2.954
2	2	1911	0.665	0.333
3	1	1914	0.769	0.769
4	1	1916	1.103	1.103
5	1	1920	2.290	2.290
6	2	1925	2.432	1.216
7	1	1932	6.609	6.609
8	5	1934	16.520	3.304
9	2	1940	9.384	4.692
10	1	1944	1.617	1.617
11	2	1946	5.506	2.753
12	2	1951	4.678	2.339
13	4	1954	20.092	5.023
14	3	1960	15.267	5.089
15	2	1964	10.422	5.211
16	4	1967	9.176	2.294
17	2	1972	2.971	1.486
18	1	1974	1.945	1.945
19	3	1977	15.351	5.117
20	1	1982	7.410	7.410

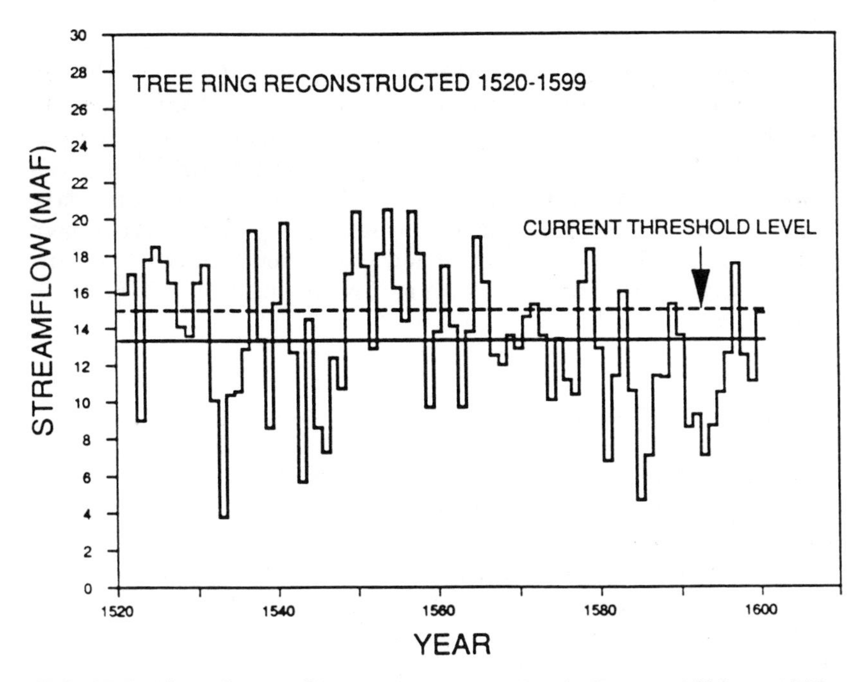

FIGURE 10.7 Streamflow sequence at Lee's Ferry, 1520 to 1599.

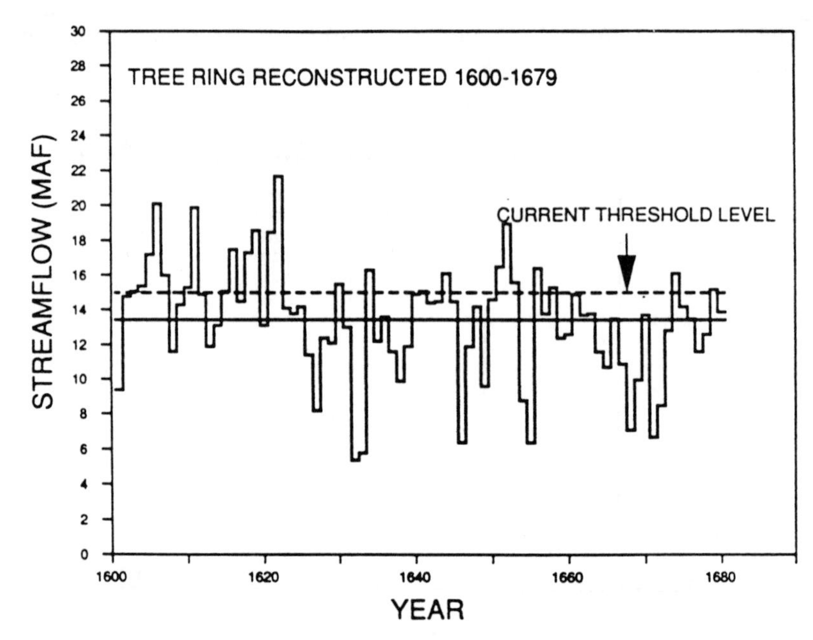

FIGURE 10.8 Streamflow sequence at Lee's Ferry, 1600 to 1679.

Maintaining the same threshold level of 15 maf (18.5 x 10^9 m^3) (the long-term mean from 1906 through 1985) for all sequences, 15 drought events were identified for both tree-ring reconstructed streamflow records. The drought characteristics differ significantly. The period from 1520 through 1599 exhibits four droughts, beginning in 1573 and extending through 1597. These are shown in Table 10.2. Conversely, the period from 1600 through 1679 displays a single severe drought with a 15-year duration, as shown in Table 10.3.

Index-Sequential Hydrologic Sequences

In order to develop probability estimates of water supply, many U.S. agencies use the index-sequential method to create a set of realistically probable streamflows (Kendall and Dracup, 1991; U.S. Bureau of Reclamation, 1985; Metropolitan Water District of Southern District, 1989). Each sequence differs from the next only by the fact that the first streamflow year is incremented by one. For example, in hydrologic sequence one, the first historic streamflow

TABLE 10.2 Hydrologic droughts on the Colorado River at Lee's Ferry, Arizona, from 1520 through 1599, as reconstructed from tree-ring analysis.

Drought No.	Duration (Years)	Starting Year	Severity (maf)	Magnitude (maf/Year)
1	1	1523	6.000	6.000
2	2	1528	2.300	2.300
3	5	1532	27.20	5.440
4	2	1538	8.000	4.000
5	7	1542	33.10	4.729
6	1	1552	2.100	2.100
7	1	1556	0.600	0.600
8	2	1559	6.500	3.250
9	3	1562	7.400	2.467
10	5	1567	9.400	1.880
11	5	1573	16.30	3.260
12	3	1580	13.90	4.633
13	5	1584	29.90	5.980
14	7	1590	34.60	4.943
15	3	1598	6.600	2.200

TABLE 10.3 Hydrologic droughts on the Colorado River at Lee's Ferry, Arizona, from 1600 through 1679, as reconstructed from tree-ring analysis.

Drought No.	Duration (Years)	Starting Year	Severity (maf)	Magnitude (maf/Year)
1	2	1601	5.800	2.900
2	2	1608	4.100	2.050
3	3	1612	5.100	1.700
4	1	1617	0.500	0.500
5	1	1620	1.900	1.900
6	7	1623	18.80	2.686
7	3	1631	20.80	6.933
8	6	1635	15.90	2.650
9	2	1642	1.100	0.550
10	6	1645	18.80	3.133
11	2	1654	14.80	7.400
12	1	1657	1.200	1.200
13	15	1659	52.10	3.773
14	4	1675	8.100	2.025
15	1	1680	1.100	1.100

year is assumed to occur in the first demand year of a study. In sequence two, the second historic streamflow occurs in the first demand year, with the first historic streamflow being wrapped to the bottom of the deck, so to speak. Consequently, there are as many sequences as there are years of hydrologic data. In order to develop probability distribution functions, this approach was used to generate pseudo-likely hydrologic traces of both the measured and tree-ring reconstructed streamflows.

Matalas and Fiering (1977) noted that a bias exists in geochronological records such as tree rings when used to estimate streamflows. Long-term mean estimates based on streamflows constructed from tree rings are reasonable, but tree-ring indices are more normally distributed and more highly correlated than recorded historical streamflows. However, it has never been determined whether these differences in statistical properties are significant enough to offset the utility of the reconstructed flows in water resource systems analyses.

Demand Assumptions

System regulation studies to determine future water supplies are performed over a series of years to accommodate growing demands. For this study, a demand period from 1989 to 2020, or 32 years, was simulated to accommodate the large drought periods observed in the tree-ring reconstructed hydrologic sequences. Demand data are from data used by the Bureau of Reclamation as the base case for its Colorado River System Simulation model (U.S. Bureau of Reclamation, 1985). Upper and lower basin demand schedules used in the analysis are shown in Tables 10.4 and 10.5.

The model output is directly related to the hydrologic sequence input. For water system design purposes, streamflow is recognized as a random process, of which the historic hydrologic record is one expression.

The hydrologic sequences comprise what can be taken to be year-to-year water supply. Coupled with this are year-to-year water demands. Lower basin demands include all uses from Lake Mead and downstream. The demands include those for California, Arizona, Nevada, and Mexico. Water losses due to evaporation or seepage from the conveyance system can be regarded as an additional demand.

TABLE 10.4 Upper Colorado River Basin Water Demands (maf).

Water Year	Moderate Demand
1988-89	3.540 acre feet
1989-90	3.550
1990-92	4.040
1991-92	4.050
1992-93	4.060
1993-94	4.070
1994-95	4.080
1995-96	4.090
1996-97	4.200
1997-98	4.310
1998-99	4.410
1999-2000	4.520
2000-01	4.630
2001-02	4.640
2002-03	4.660
2003-04	4.680
2004-05	4.700
2005-06	4.720
2006-07	4.740
2007-08	4.760
2008-09	4.780
2009-10	4.800
2010-11	4.830
2011-12	4.840
2012-13	4.860
2013-14	4.860
2014-15	4.890
2015-16	4.900
2016-17	4.910
2017-18	4.930
2018-19	4.940
2019-20	4.960

SOURCE: U.S. Bureau of Reclamation, 1985.

TABLE 10.5 Normal Lower Colorado River Basin Water Demands (maf).

Water Year	Normal Demands (inc. Mexico and losses)
1988-89	9.370
1989-90	9.460
1990-92	9.480
1991-92	9.490
1992-93	9.490
1993-94	9.500
1994-95	9.500
1995-96	9.510
1996-97	9.510
1997-98	9.520
1998-99	9.530
1999-2000	9.540
2000-01	9.550
2001-02	9.550
2002-03	9.550
2003-04	9.550
2004-05	9.550
2005-06	9.550
2006-07	9.550
2007-08	9.550
2008-09	9.550
2009-10	9.550
2010-11	9.550
2011-12	9.550
2012-13	9.560
2013-14	9.560
2014-15	9.560
2015-16	9.570
2016-17	9.570
2017-18	9.570
2018-19	9.580
2019-20	9.580

SOURCE: U.S. Bureau of Reclamation, 1985.

Comparison of Severe, Sustained Drought Simulations

Simulations were made for the hydrologic periods 1906 to 1985 (the base case), 1520 to 1599, and 1600 to 1679. Each simulation produced results from 80 index-sequential hydrologic sequences and included reservoir storages, releases, and imposed shortages in the upper and lower basins. These results were ranked using a Weibull plotting position in order to form empirical cumulative distribution functions. Average storages, releases, and shortages were computed. The simulations were based on one interpretation of the Law of the River and the 602(a) storage requirement.

Lake Powell

Results were derived from simulating 80 index-sequential sequences for each of the three 80-year hydrologic periods studied. Each simulation included streamflow at Lee's Ferry and reservoir storages and releases given water demands as specified in Tables 10.4 and 10.5 for periods beginning in the water year 1988-1989.

Exceedance probabilities developed for Lake Powell indicate that a repeat of a hydrologic period like that of 1520 to 1599 or 1600 to 1679 would be significant. Figure 10.9 indicates that at about the 90 percent exceedance level (i.e., the level at which there is a 90 percent probability that the reservoir storage will be equal to or greater than the indicated amount), the difference between the base-case period 1906 to 1985 and the period 1520 to 1599 is about 11 maf (13.6×10^9 m^3) for the year 2000. Conversely, there is a 10 percent probability that the reservoir storage will be equal to or less than the indicated amount. Furthermore, Lake Powell is at minimum power pool level at about the 90 percent exceedance level, as shown in Figure 10.9.

For the hydrologic period 1520 to 1599, the minimum power pool probability exceedance decreases to about 80 percent by the year 2010, as shown in Figure 10.10, and to about 75 percent by the year 2020, as shown in Figure 10.11.

A comparison of the two drought scenarios shows that they have similar probability curves with maximum differences at the higher exceedance levels. While both sequences exhibit quite different types of droughts, preceding or subsequent high flow periods tended to produce an equalizing effect between them, when considering the entire 80-year sequence.

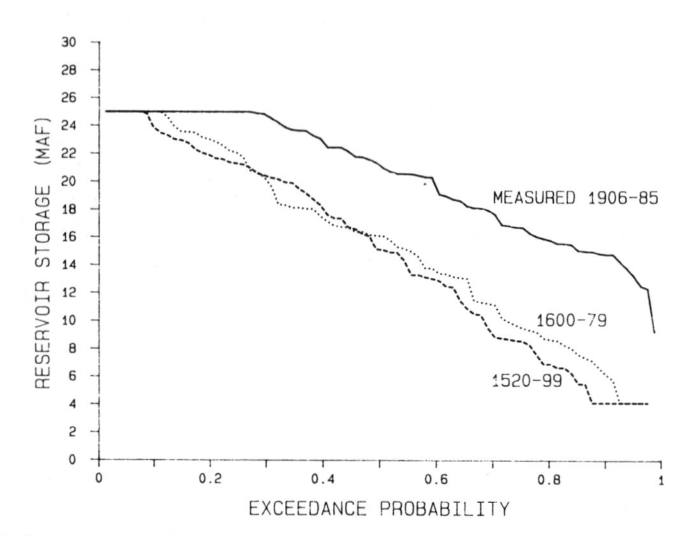

FIGURE 10.9 The amount of storage Lake Powell would contain in the year 2000 under conditions like those that have occurred in three historic hydrologic periods: 1520 to 1599, 1600 to 1679, and 1906 to 1985. The curves show the chance that storage would exceed the indicated level.

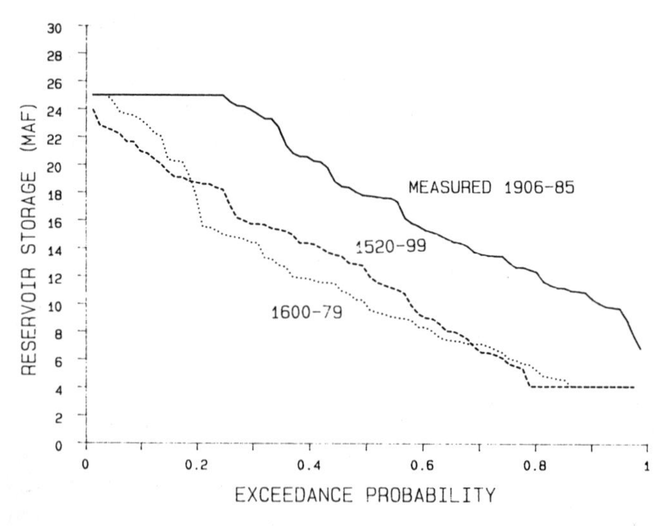

FIGURE 10.10 The amount of storage Lake Powell would contain in the year 2010 under conditions like those that occurred during the periods 1520 to 1599, 1600 to 1679, and 1906 to 1985.

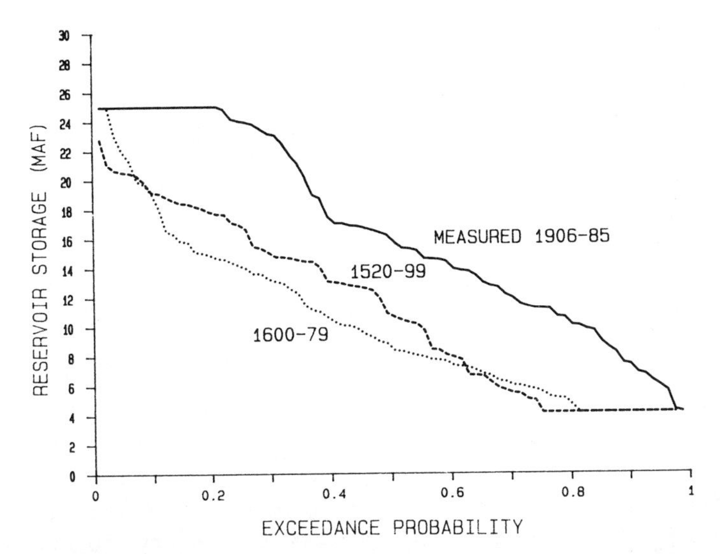

FIGURE 10.11 The amount of storage Lake Powell would contain in the year 2020 under the three historic hydrologic scenarios.

Lake Powell average storage levels were calculated by taking the average from 80 wrapped sequences for each simulation year. Results are shown in Figure 10.12. Note that the average storage level difference under severe sustained drought conditions ranges from approximately 5 to 6 maf (6.2 to 7.4 x 10^9 m^3) from 1998 to 2020.

Lake Mead

Lake Mead simulation results are similar, with the difference between the base case and both drought sequences on the order of 6 maf (7.4 X 10^9 m^3) for the year 2000 at about 90 percent exceedance. Minimum power pool exceedance probabilities are about 80 percent and are shown in Figure 10.13. For the most severe drought scenario (1520 to 1599), minimum power pool exceedance levels decrease to about 50 percent by the year 2010, while those for the sequence 1600 to 1679 decrease to about 70 percent (see

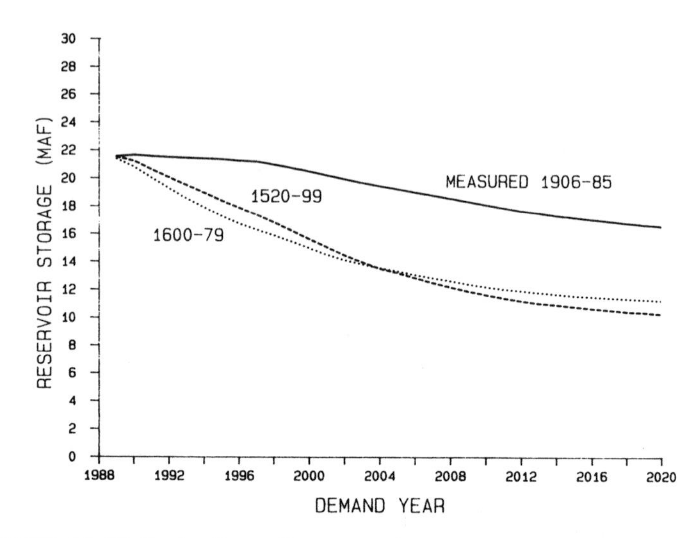

FIGURE 10.12 Lake Powell average storage comparison: the average storage levels the lake would contain if history repeats itself.

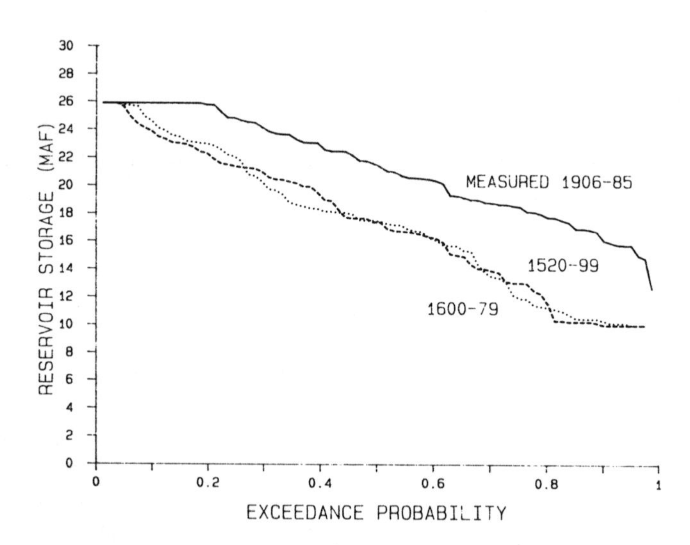

FIGURE 10.13 The amount of storage Lake Mead would contain in the year 2000 under conditions like those from the periods 1520 to 1599, 1600 to 1679, and 1906 to 1985. The curves show the chance that storage would exceed the indicated level.

Figure 10.14). By simulation year 2020, these exceedance levels decrease to about 40 percent and 55 percent respectively (see Figure 10.15). The minimum power pool exceedance probability predicted by the base case is greater than 95 percent.

Average storage levels for Lake Mead are shown in Figure 10.16; the figure indicates a maximum between the base case and severe and sustained drought scenarios.

Upper Basin Shortages

Water shortages in the upper basin were allocated according to the interpretation of the Law of the River and storage provision requirements. Following the same procedure used for the development of exceedance probabilities for reservoir storages, similar cumulative distribution functions were developed for upper basin

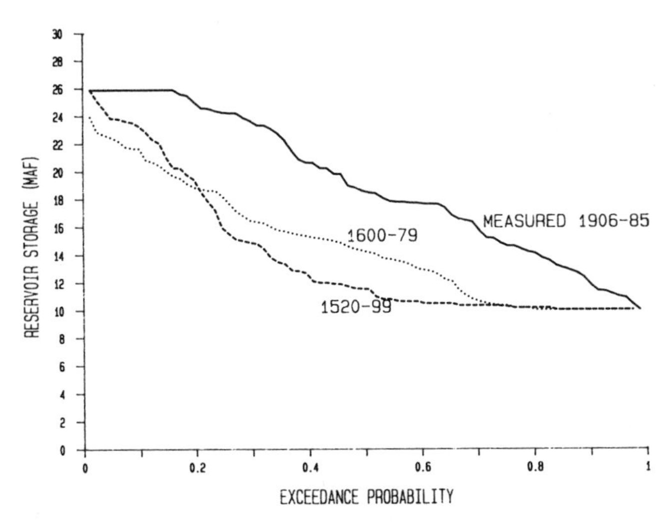

FIGURE 10.14 The amount of storage Lake Mead would contain in 2010 under hydrologic conditions like those that have occurred in the past.

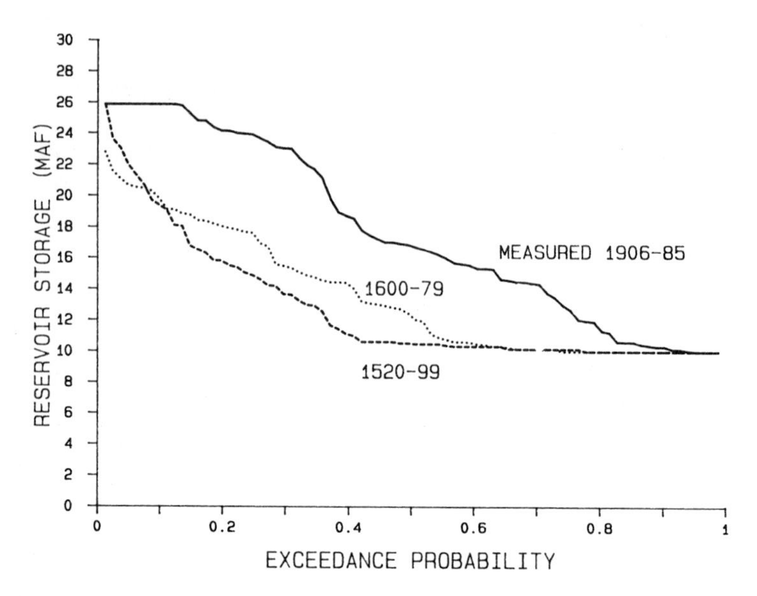

FIGURE 10.15 The amount of storage Lake Mead would contain in 2020 under hydrologic conditions like those from three past periods.

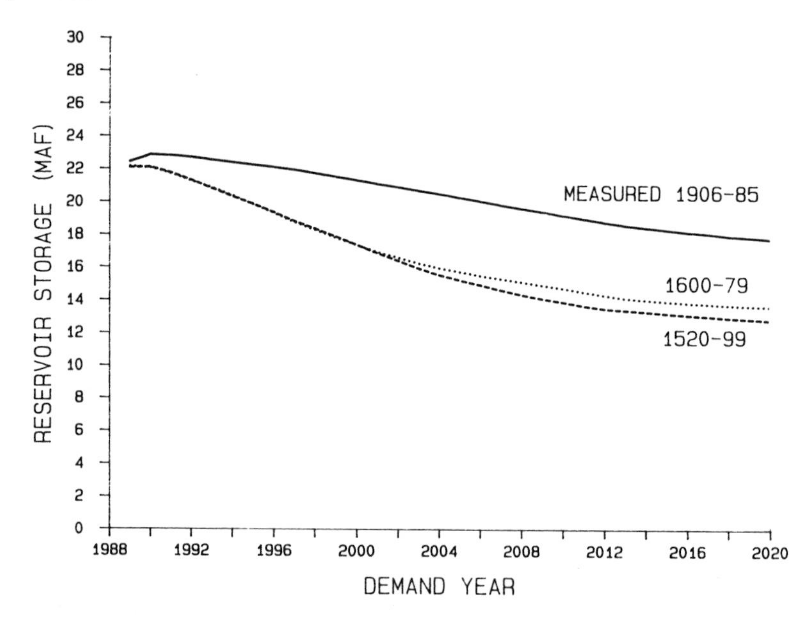

FIGURE 10.16 Lake Mead average storage comparison: the average storage levels the lake would contain if history repeats itself.

shortages. Year 2000 shortage probabilities are shown in Figure 10.17. The base case hydrologic sequence (1906 to 1985) indicates no shortages. The sequence 1520 to 1599 indicates shortages of about 300,000 acre-feet (0.37 x 10^9 m^3) at the 75 percent exceedance level. The 1600 to 1679 sequence indicates upper basin shortages of about 300,000 acre-feet (0.37 x 10^9 m^3) at about the 35 percent exceedance level.

Shortages of about 300,000 acre-feet (0.37 x 10^9 m^3) at the 95 percent and 85 percent exceedance levels are indicated by year 2010 for the two drought sequences (see Figure 10.18). The base case sequence indicates no shortages. Similar results are shown for simulation year 2020, except that exceedance probabilities have increased in all cases, as shown in Figure 10.19. That is, shortages are present at the 98 percent and 88 percent exceedance probability levels for the two drought scenarios, while base case shortages are indicated at 50 percent exceedance probability.

Average shortages for the 32-year simulation period are shown in Figure 10.20. Upper basin shortages are not shown as exceeding 300,000 acre-feet (0.37 x 10^9 m^3), which is consistent with the prevailing Bureau of Reclamation operation policies.

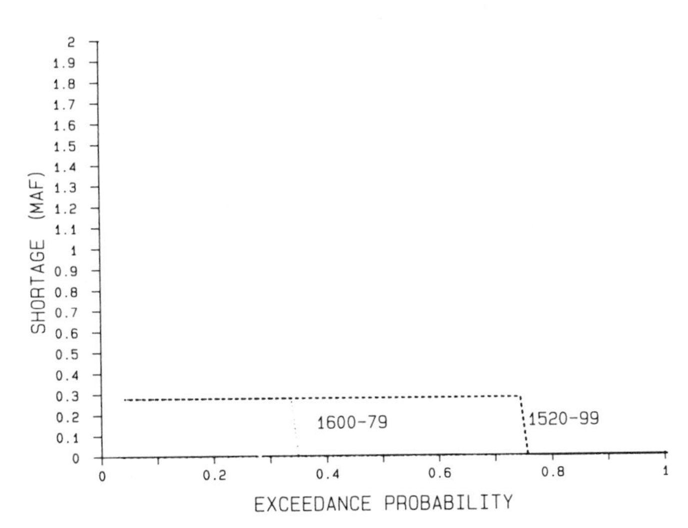

FIGURE 10.17 Possible upper basin water shortages in the year 2000 under hydrologic conditions like those from 1520 to 1599 and 1600 to 1679.

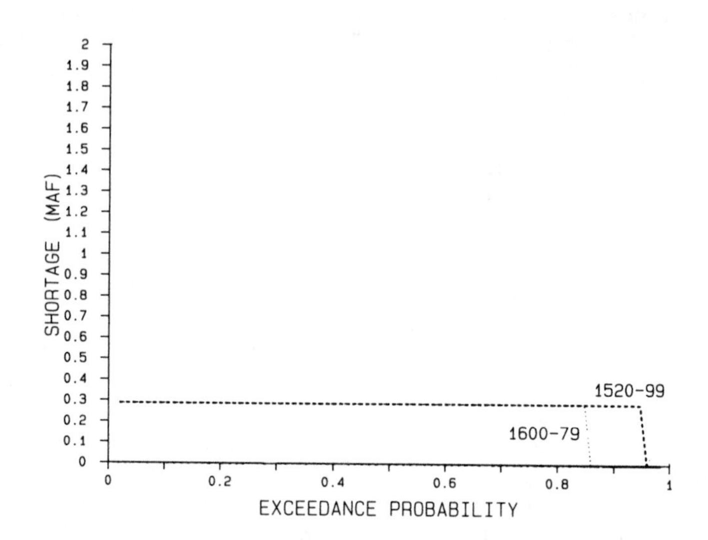

FIGURE 10.18 Possible upper basin water shortages in the year 2010 under two historic hydrologic scenarios.

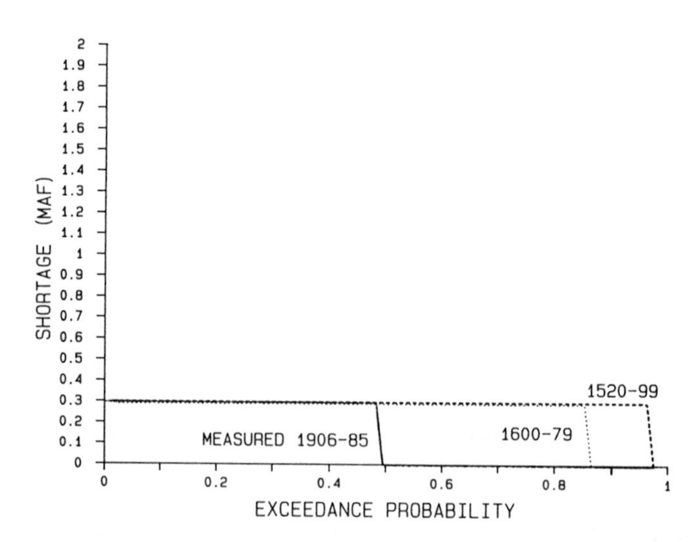

FIGURE 10.19 Possible upper basin water shortages in the year 2020 under three historic hydrologic scenarios.

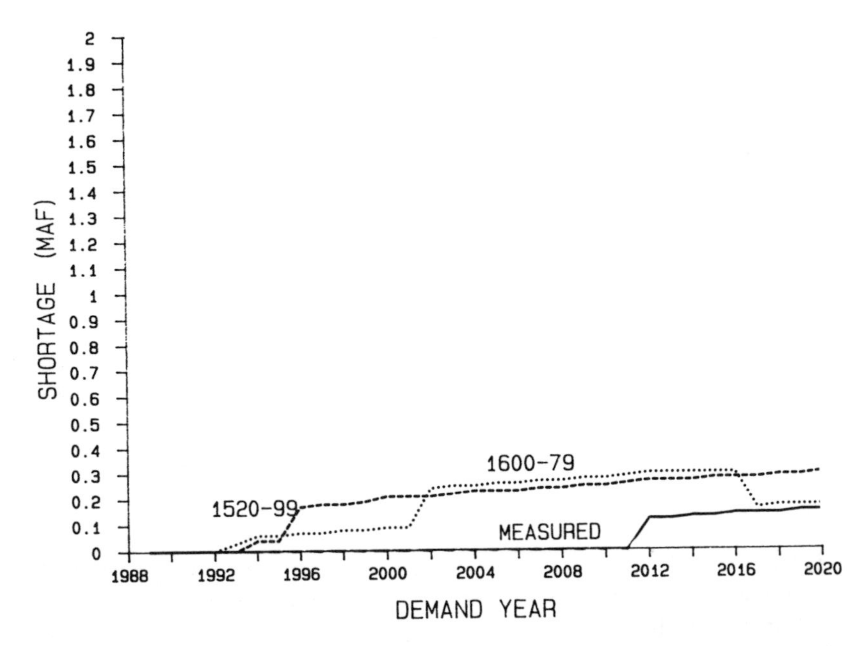

FIGURE 10.20 The average water shortage that would occur in the upper basin if history repeats itself.

Lower Basin Shortages

Calculated lower basin shortages are much more severe than those for the upper basin. For simulation year 2000, no shortages are predicted by the base case hydrology, although 1.45 maf (1.8 x 10^9 m^3) shortages are calculated at the 70 and 30 percent exceedance levels for the drought sequences 1520 to 1599 and 1600 to 1679, respectively, (see Figure 10.21). For the year 2010, the base case hydrology yielded 1.45 maf (1.8 x 10^9 m^3) shortages at about the 50 percent exceedance level, while the 1520 to 1599 drought sequence shows the same shortages at the 95 percent exceedance level. The drought sequence 1600 to 1679 indicates 1.45 maf (1.8 x 10^9 m^3) shortages at 85 percent exceedance, as shown in Figure 10.22. For simulation year 2020, the base case exceedance levels increase to about 60 percent, as shown in Figure 10.23. Note that for the hydrologic periods 1520 to 1599 and 1600 to 1679, the exceedance probabilities stay approximately the same. Lower basin average shortages are shown in Figure 10.24.

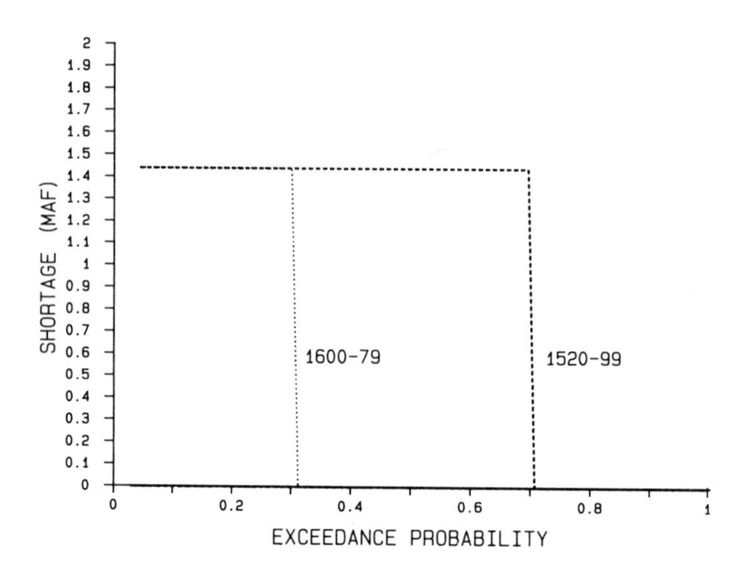

FIGURE 10.21 Lower basin water shortages in the year 2000 under hydrologic conditions like those during the periods 1520 to 1599 and 1600 to 1679.

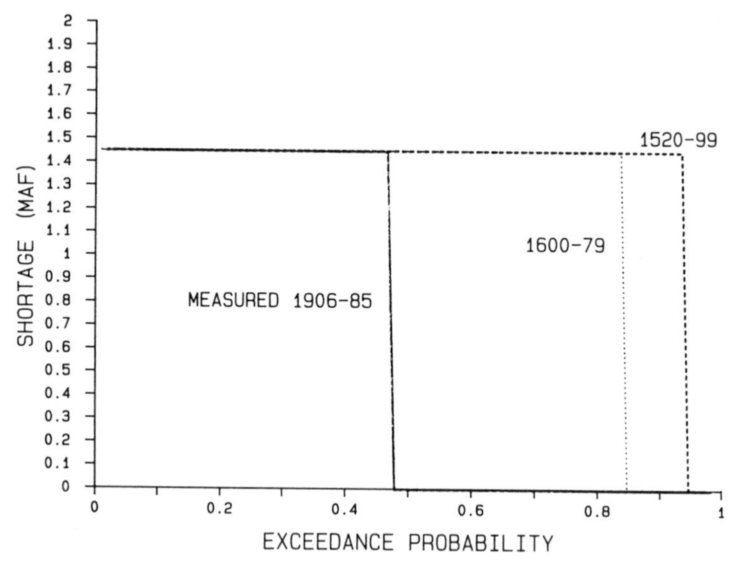

FIGURE 10.22 Lower basin water shortages in the year 2010 under hydrologic conditions like those during the periods 1520 to 1599, 1600 to 1679, and 1906 to 1985.

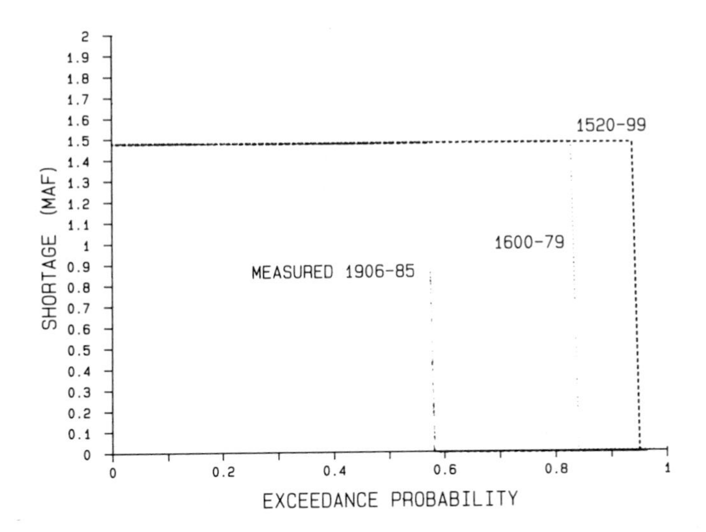

FIGURE 10.23 Lower basin water shortages for the year 2020 under hydrologic conditions like those from three historic periods.

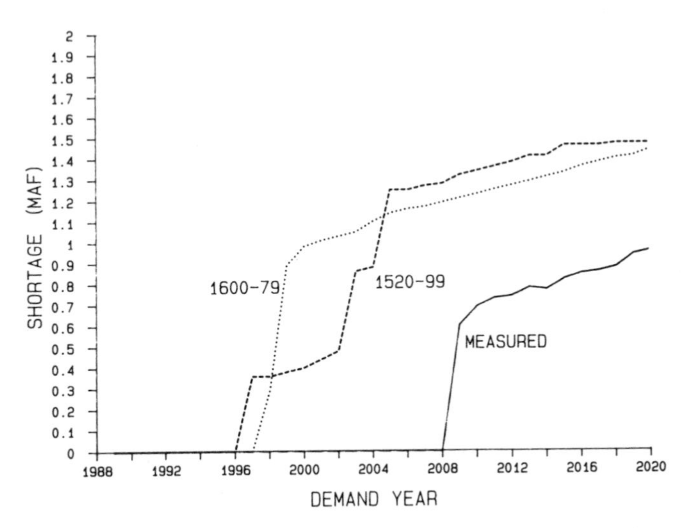

FIGURE 10.24 Lower basin average water shortages that would occur if history repeats itself.

Conclusions of Drought Analysis

A repeat of the hydrologic periods 1520 to 1599 or 1600 to 1679 on the Colorado River would have significant impacts on water availability and shortage allocations in the upper and lower basins. It should be noted that results presented here are not speculative what-if climate scenarios. They represent outcomes of hydrologic sequences that have been reconstructed from the historical past. Observation of Figure 10.6 indicates that the river has had above-average flow of 15.06 maf (18.6 x 10^9 m^3) over the past eighty years as opposed to its previous average of about 13.5 maf (16.7 x 10^9 m^3) per year. The potential for anthropogenic climatic change effects could exacerbate a situation in which the equilibrium level of the Colorado River over the long term (its mean flow) is lower than that measured over the past century. Shortages on the order of 300,000 acre-feet (0.37 x 10^9 m^3) (under current system operation regimes) could occur in the upper basin as a result of hydrologic conditions similar to those revealed in the study of tree-ring data. Lower basin shortages on the order of 1.5 maf (1.9 x 10^9 m^3) have probabilities of occurrence of over 90 percent and about 85 percent, respectively, for the 1520 to 1599 and 1600 to 1679 droughts.

Conclusions and Implications for Climate Change

Flooding of the Colorado River in the lower Colorado River basin caused substantial damage to homes and businesses in the spring and summer of 1983. Abnormal meteorological events—the greater than normal precipitation—contributed to the flooding but were not solely responsible for it. Two other factors that contributed to the flooding were: (1) the practice of maintaining a system of full reservoirs to satisfy demands from consumptive users, hydroelectric generators, and recreational interests; and (2) physical encroachment into the flood plain, made possible by the dams along the river. Although these factors could be managed physically, thereby averting the flood risks seen in 1983, there are many contrary interests that may interfere with steps to mitigate or prevent flood damage in the future. Although priorities for Colorado River management are mandated by U.S. law, such management has historically been the product of political and economic constraints created by the river's many beneficiaries.

If the business-as-usual approach has been an optimistic one, then areas of future study need to focus on activities that can help mitigate negative impacts in the upper and lower Colorado River basins. These include conservation practices, the conjunctive management of surface and ground water, and systems for orderly and mutually agreeable movement of water to higher-valued uses, including temporary water transfers under drought conditions.

The merits of the drought study lie in the use of data drawn from actual climatic events. While there is a degree of bias in the tree-ring reconstructed sequences, estimates of the streamflow mean are considered reliable.

It is doubtful that Lake Powell and Lake Mead would be allowed to drop to minimum power pool levels before other types of shortage procedures were invoked, since in a real-time operation there is no way to forecast whether a drought has ended. This analysis distributes shortages in the upper and lower basins in a manner that would be consistent with existing laws and operating criteria. This analysis is not intended to be a statement of how shortages would actually be allocated. An area for future research is a critical assessment of potential shortage allocations and strategies that might be invoked by the Secretary of the Interior in the event of a severe, sustained drought. It is in the political arena that decisions will be made about ways of responding to, mitigating, and avoiding drought effects.

The extraordinary amount of storage along the Colorado River gives the region robustness and resilience. The lesson of importance to policymakers is to develop flexibility in the operating procedures that matches the variability of the resource being controlled: namely, streamflow. For example, rigid operating procedures may need to be set aside and new procedures developed that allow for changing the current power pool and base flow requirements.

REFERENCES

Arizona Republic. June 22, 1983. The flood facts. P. A2.

Arizona Republic. July 1, 1983. Residents who build near river should have expected trouble, watt says. P. A2.

Arizona Republic. Sept. 3, 1983. $80 million flood was "unavoidable." P. B1.

Benjamin, J. R., and C. A. Cornell. 1970. Probability, Statistics and Decision for Civil Engineers. New York: McGraw Hill.

Broadbent, R. (Commissioner, U.S. Bureau of Reclamation). 1983. Prepared statement for the U.S. House of Representatives, Committee of Interior and Insular Affairs, September 7, Yuma, Arizona.

Brown, B. G. 1988. Climate variability and the Colorado River compact: implications for responding to climate change. Chapter 12 in Glantz, M. H., ed., Societal Responses to Regional Climate Change: Forecasting by Analogy. Boulder, Colo.: Westview Press.

Colorado River Board of California. 1969. California's Stake in the Colorado River. Los Angeles: The Board.

Colorado River Board of California. 1990. Executive Directors Report, October. Los Angeles: The Board.

Debler, ? 1930. Hydrology of the Boulder Canyon Reservoir with Reference Especially to the Height of Dam to Be Adopted. Denver, Colo.: U.S. Bureau of Reclamation.

Dozier, L., and J. Brown. 1983. Management of Colorado River Dams. Presented at the Eighth Annual Workshop of the Rural Communities Institute, Gunnison, Colo., August 1-3.

Dracup, J. A. 1977. Impact of the Colorado River basin and Southwest water supply. In J. Wallis, ed., Climate, Climatic Change, and Water Supply. Washington, D.C.: National Academy Press.

Dracup, J. A., K. S. Lee, and E. G. Paulson. 1980a. On the definition of droughts. Water Resources Research 16(2):297-302.

Dracup, J. A., K. S. Lee, and E. G. Paulson. 1980b. On the statistical characteristics of drought events. Water Resources Research 16(2):289-296.

Dracup, J. A., S. L. Rhodes, and D. Ely. 1985. Conflict between flood and drought preparedness in the Colorado River basin. Pp. 229-244 in J. Lundquist, U. Lohm, and M. Falkenmark, eds., Strategies for River Basin Management. Dordrecht, Netherlands: D. Reidel Publishing, Co.

Freeny, G. B. 1981. Managing conflicts on the lower Colorado River system. In Proceedings of the National Workshop on Reservoir Systems Operations. New York: American Society of Civil Engineers.

Glantz, M. H., ed. 1989. Forecasting by Analogy: Societal Responses to Regional Climate Change. Boulder, Colo.: National Center for Atmospheric Research.

Glantz, M. H., ed. 1988. Societal Responses to Regional Climatic Change: Forecasting by Analogy. Boulder, Colo.: Westview Press.

Hundley, N., Jr. 1986. The west against itself: the Colorado River—an institutional history. Pp. 9-49 in G. D. Weatherford and, F. L. Brown, eds., New Courses for the Colorado River: Major Issues for the Next Century. Albuquerque, N.M.: University of New Mexico Press.

Hundley, N., Jr. 1975. Water and the West. Berkeley and Los Angeles, Calif.: University of California Press.

Jacoby, G. C., Jr. 1975. Lake Powell Effect on the Colorado River Basin Water Supply and Environment. Lake Powell Research Project Interim Report. University of California at Los Angeles: Institute of Geophysics and Planetary Physics.

Kendall, D. R., and J. A. Dracup. 1991. A comparison of index-sequential and AR(1) generated hydrologic sequences. Journal of Hydrology 122:335-352.

Matalas, N. C., M. B. Fiering. 1977. Studies in geophysics. In Climate, Climatic Change, and Water Supply. Washington, D.C.: National Academy Press.

Metropolitan Water District of Southern California, Resources Division. 1989. State Water Project Simulation Model.

Metropolitan Water District of Southern California. 1980. Colorado River Annual System Regulation Model, Report No. 941.

Meyers, C. J. 1966. The Colorado River. Stanford Law Review: pp. 1-75.

Nathanson, M. N. 1978. Operating Criteria. Chapter 7 in Updating the Hoover Dam Documents. Washington, D.C.: U.S. Bureau of Reclamation.

National Research Council (NRC). 1977. Climate, Climatic Change, and Water Supply. Washington, D.C.: National Academy Press.

Rhodes, S. L., D. Ely, and J. A. Dracup. 1984. Climate and the Colorado River: the limits of management. Bulletin of the American Meteorological Society 65:682-691.

Stockton, C. W. 1977. Interpretations of past climatic variability from paleoenvironmental indicators. In J. Wallis, ed., Climate, Climate Change, and Water Supply. Washington, D.C.: National Academy Press.

Upper Colorado River Commission. 1982. Thirty-Fourth Annual Report. Salt Lake City, Utah: The Commission.

U.S. Army Corps of Engineers. 1982. Colorado River Basin-Hoover Dam: Review of Flood Control Regulations, Final Report. Los Angeles, Calif.: The Corps.

U.S. Bureau of Reclamation. 1983. Status of Reservoirs: Colorado River Storage Project. Washington, D.C.: U.S. Department of the Interior.

U.S. Bureau of Reclamation. 1985. Colorado River Simulation System. Washington, D.C.: U.S. Department of the Interior.

Weatherford, G. D., and G. C. Jacoby. 1975. Impact of energy development on the Law of the Colorado River. University of New Mexico School of Law National Resources Journal 15:171-213.

Yevjevich, V. M. 1967. Objective Approach to Definitions and Investigations of Continental Droughts. Hydrology Paper 23. Ft. Collins, Colo.: Colorado State University.

11

Economic Consequences of Climate Variability on Water in the West

Kenneth D. Frederick
Resources for the Future
Washington, D.C.

The economic impacts of hydrologic extremes and variability on specific regions depend on the nature of the economy, the slack in the existing water-supply system, and society's ability to anticipate and adapt to hydrologic change. Demand management and water marketing are potentially important tools for responding to drought and long-term reductions in supply.

A case study of the Missouri River basin illustrates the possible impacts of a general warming on the availability of water within one of the West's principal river basins and indicates how management changes and a reallocation of supplies would help the region adapt to a sizable reduction in streamflow.

Hydrologic extremes have long posed risks to settlements in the western United States. A 5-year drought in the twelfth century may have caused the prehistoric Anasazi people to abandon the Colorado plateau (Kneese and Bonem, 1986). Twice within the last century prolonged drought forced tens of thousands of desperate families to flee the semiarid plains in search of more promising economic opportunities. And currently, a multiyear drought extending from southern California to the Missouri River basin is exacting a toll on a variety of water users.

The temporary transformation of the Trinity River in Texas from a small river to a mile-wide flood in the spring of 1990 provided a recent reminder of what can happen when too much water arrives within too short a time. Even though California has about six million acre-feet of flood control storage and 6,000 miles of levees, floods may pose a bigger problem to the state than earthquakes (Hartshorn, 1986). Floods have consistently been the nation's most deadly atmospheric hazard in recent decades; they accounted for 61 percent of all presidential disaster declarations in the decade starting in April 1974 (Riebsame et al., 1986).

CHANGES IN VULNERABILITY TO HYDROLOGIC VARIABILITY

Factors Tending to Reduce Vulnerability

The susceptibility of the West's economy to hydrologic extremes has changed over time. A decline in the economy's dependence on water and an increase in the control over supplies have tended to make the West less sensitive to changes in water supplies. Water's influence on economic development generally weakened during the last century. Development of steam engines, internal combustion motors, and electricity generation and transmission reduced the significance of on-site water power. Expansion of railroads, highways, and air transport diminished the importance of water-based transport. Water intensive industries such as irrigated agriculture declined in relative importance, and industries in general learned to prosper with less water (National Water Commission, 1973).

The tremendous expansion of the infrastructure to store and transport water and to tap ground water supplies also has tended to reduce the susceptibility of the nation's economy to climate variability. More than 63,000 dams with 869 million acre-feet of storage are included in the 1982 inventory of the nation's dams. More than three-fourths of these dams and two-thirds of the storage were completed since 1945 (U.S. Army Corps of Engineers, 1982).[1] About 47 percent of these dams and 55 percent of the storage are in the 17 western states, giving the region considerable capacity to prevent floods and to supply water during drought. Ground water also provides an important buffer against fluctuations in surface supplies in many areas of the West. Ground water use was essentially limited to areas with low pumping depths or artesian pressure until technological advances in the 1930s made it feasible to pump water from much greater depths. Water stored within deeper aquifers is less susceptible to climate variations, but the economies of some areas have become dependent on the use of nonrenewable ground water supplies.

Factors Tending to Increase Vulnerability

Countering these changes are several trends tending to make the

West more susceptible to hydrologic fluctuations. At least two factors are increasing the costs associated with drought.

First, demands on the resource have increased, making water more valuable and the competition for supplies during drought more intense. Nationally, offstream water use rose from about 40 billion gallons per day (bgd) in 1900, to 180 bgd in 1950, to 440 bgd in 1980 (Picton, 1960; Solley, Merk, and Pierce, 1988). Although natural supplies are much more sparse in the West, nearly half of the nation's fresh water withdrawals are now in the 17 western states.

Second, the rate of construction of reservoirs to assure water supplies has decreased since 1970 and is likely to continue declining. A basic principle of reservoir planning is that the risk of deficiency increases if the storage period (that is, available reservoir storage divided by average daily withdrawals) is not increased as withdrawals increase. The storage period rose from 204 days in 1960 to 216 days in 1970, and it had increased for at least six consecutive decades prior to 1970. By 1980, however, it had fallen to 201 days (USGS, 1984). Moreover, unless withdrawals continue to decline as they did from 1980 to 1985, the storage period is likely to continue to fall for two reasons: (1) the high economic and environmental costs of developing new supplies with additional storage, and (2) the adverse impacts of sedimentation on existing storage. The costs of water supply projects have increased sharply in recent decades, and continued increases are inevitable for three reasons: (1) the best reservoir sites have already been developed; (2) as storage capacity on a stream increases, the quantity of water that can be supplied with a high degree of probability grows only at a diminishing rate; and (3) the opportunity costs of storing and diverting water rise as society places higher values on instream flows. While the data on sedimentation rates for most dams are poor or nonexistent, one estimate suggests annual sediment losses total about 1.4 to 1.5 million acre-feet (Guldin, 1989). Over a decade, this loss is equivalent to about two percent of the nation's aggregate reservoir storage.

Actual flood damages also have been rising over time. Increased development and rising real property values in the flood plains together with upstream developments that increase runoff rates and flood peak frequencies have resulted in greater flood losses despite a growing capacity to regulate intertemporal flows. In the absence of better preventive measures, flood losses are likely to continue rising because urban expansion within the flood plains is increasing at 2 percent per year (Schilling, 1987).

Uncertainties of Climate Change

Past trends may provide a poor guide to the economic implications of future climate variability in a world undergoing anthropogenically induced climate change. As other papers presented at this colloquium indicate, a greenhouse warming would accelerate the global hydrologic cycle and significantly increase the uncertainty as to the water supplies of specific regions. Regional impacts are likely to include changes in precipitation and runoff patterns, evapotranspiration rates, and the frequency and intensity of storms. Even the direction of precipitation and runoff changes are uncertain. Higher temperatures, however, are likely to have particularly large and adverse impacts on annual runoff in arid areas where changes in precipitation and evaporation have amplified effects on runoff. Seasonal streamflow patterns would also be affected, especially in areas where precipitation currently comes largely in the form of winter snowfall and runoff comes largely from spring and summer snowmelt. These conditions characterize much of the West.

ECONOMIC SENSITIVITY

Existing water use patterns, infrastructure, and management practices reflect past climate and water availability. The economic consequences of adjusting to any given climate change would depend on the nature of the economy and its dependence on water supplies, the slack in the supply system, and society's ability to anticipate and adapt to hydrologic change. Effective adaptation to drought involves curbing excessive and low-value uses through demand management and transferring scarce supplies to uses for which the losses from inadequate supplies would be greatest.

Nature of the Economy

Agriculture is one of the most sensitive of human activities to climate conditions and variability. Dryland farming is highly dependent on the timely availability of water. Too much water can make it difficult to plant in the spring or to harvest in the fall. And too little water can reduce or even eliminate yields. Irrigation reduces susceptibility to variations in rainfall unless irrigators depend on fully allocated surface water and must share in any shortfalls. Ground water supplies are less susceptible to drought than surface water, although nonrenewable irrigation supplies might be mined faster under a hotter and drier climate.

Agriculture is certainly not the only sector to be affected by drought. A major drought is likely to affect adversely all instream and offstream water users. The impact on particular sectors would depend in part on the institutional arrangements for allocating scarce supplies. Historically, western water law strongly favored offstream users at the expense of instream users. Thus, reservoir and streamflow levels were drawn down to the detriment of recreation, fish and wildlife, and hydropower. The balance has shifted somewhat in recent years as a result of state and federal legislation and judicial decisions protecting environmental interests such as wild and scenic rivers, unique ecological environments such as Mono Lake, and endangered species. Drought may also adversely impact the forests and the economic interests dependent upon them by increasing the risks of fire, disease, and pest damage.

The overall losses associated with the 1976-1977 drought in California have been estimated at $2,663 million. Agriculture, with losses estimated at $1,475 million over the two years, accounted for more than half of the total; livestock accounted for more than half of these agricultural losses. Energy costs increased by more than $450 million as a result of the drought. And timber interests lost $280 million to fire and $390 million to insect damage (Association of California Water Agencies, 1989). It is too early to know the extent of the economic impacts of the drought that started in 1987 and is now well into its fourth year in California.

Slack in the System

A region's vulnerability to climate variability depends in part on the amount of slack between water supplies and demand and the robustness and resilience of the supply system. When supplies are stretched to meet demand under normal hydrologic conditions, even a mild drought requires adjustments in water use. Water resource systems traditionally have been designed to be robust (able to respond to the range of uncertainties associated with future variability) and resilient (able to operate under a range of conditions and to return to designed performance levels quickly in the event of failure). The recent decline in the storage period noted above suggests a decline in the overall robustness of the nation's water supplies. And the rising costs of and prevailing skepticism toward new water projects suggests that the tradition of building large redundancy into water supply and control projects may be a thing of the past. Moreover, the existing systems were designed and are operated assuming future levels and patterns of precipitation and runoff will be similar to those experienced in the past. The prospect of long-term climate change poses new risks and challenges for managing these systems and raises questions about their vulnerability to climate change.

Gleick (1990) uses five indicators of a region's vulnerability to climate change. These include the ratios of storage capacity and consumptive use to renewable supplies, measures of a region's dependence on hydroelectricity and ground water overdrafts, and a measure of streamflow variability. The critical values that Gleick designates as indicating vulnerability as well as the values of the indicators for the nine principal water resource basins in the western United States are presented in Table 11.1. All nine basins are vulnerable on at least two of the five indicators. The Great Basin exceeds the critical limits on all five criteria and California and the Missouri River basin are vulnerable on four counts. The Missouri, for example, appears to have plentiful storage (equivalent to 112 percent of its mean annual renewable supply). But a relatively high ratio of consumptive use to renewable supplies, a high degree of reliance on hydroelectric power, high rates of ground water overdraft, and high streamflow variability all suggest the region is vulnerable to the hydrologic uncertainties associated with climate change.

Improved Management[2]

Seeking ways to improve management of the existing infrastructure is always prudent; management improvements assume greater importance in view of the vulnerability of water resource systems, the limitations on structural responses, and the prospect of climate change. The economic impacts of future hydrologic change are likely to depend even more than in the past on the ability to anticipate and adapt to these changes.

Joint management of water supply systems that are currently managed independently with separate operating rules and objectives may make it possible to improve significantly the supply capabilities of each system. Integration of the three principal water supply agencies in the Washington, D.C. area illustrates the potential advantages of joint operation of facilities. The combined

TABLE 11.1 Indicators of Vulnerability to Climatic Conditions.

Measure of:	Storage[1]	Demand[2]	Hyrdo[3]	Overdraft[4]	Variability[5]
Water Resource Region					
Missouri	1.12	.29	.25	.25	4.22
Arkansas-White-Red	0.45	.17	.10	.62	5.59
Texas-Gulf	0.61	0.23	0.01	0.77	9.90
Rio Grande	1.89	0.64	0.09	0.28	22.00
Upper Colorado	2.61	0.33	0.04	0.00	4.00
Lower Colorado	4.22	0.96	0.27	0.48	1.42
Great Basin	0.35	0.49	0.25	0.42	3.92
Pacific Northwest	0.19	0.04	0.93	0.08	1.92
California	0.42	0.29	0.30	0.12	4.48
Critical values	0.6	0.2	0.25	0.25	3

[1] Measure of storage. Ratio of maximum basin storage volume to total basin annual mean renewable supply as of 1985. Regions with values below 0.6 have small relative reservoir storage volumes. Large reservoir storage volumes provide protection from floods and act as a buffer against shortages.

[2] Measure of demand. Ratio of basin consumptive depletions (including consumptive use, water transfers, evaporation, and ground water overdraft) to total basin annual mean renewable supply as of 1985. Water is considered a decisive factor for economic development in regions with values above 0.20.

[3] Measure of dependence on hydroelectricity. Ratio of electricity supplied by hydroelectric facilities to total basin electricity production as of 1975. Regions with values 0.25 or above have a high dependence on hydroelectricity.

[4] Measure of groundwater vulnerability. Ratio of annual ground water overdraft to total ground water withdrawals as of 1975. Regions with values of 0.25 or above already have ground water supply problems.

[5] Measure of streamflow variability. Ratio of 5 percent exceedance flow to 95 percent exceedance flow. Values of 3 or above suggest high streamflow variability.

SOURCE: Gleick, 1990.

drought condition water yield of the three systems was increased more than 30 percent at a cost saving of between $200 million to $1 billion compared to the proposed structural alternatives. Although the specific circumstances in this case are unique, Sheer's

(1986) studies suggest major benefits from improved management are also possible in areas with very different characteristics.

The obstacles to integrated management are largely institutional. Separate ownership of water supply systems, multistate jurisdictions, and state laws and administrative practices all hinder reform. Officials in the principal federal water construction agencies have at least begun to talk about the need for change. The Bureau of Reclamation's *Assessment 1987* (DOI, 1987) concludes that "the Bureau's mission must change from one based on federally supported construction to one based on effective and environmentally sensitive resource management." A recent paper by three senior members of the U.S. Army Institute for Water Resources advocates greater emphasis on management measures to meet the problems caused by extreme events and the uncertainties stemming from the prospect of climate change (Hanchey et al., 1987).

Demand Management

Traditionally, water planners adopted a supply-side approach to provide for growing water demands. Offstream water use was projected to grow approximately in step with population and economic growth. These projections were treated as virtual requirements to be supplied with little regard for cost. This approach may have approximated an efficient strategy when the costs of supplies were low and streamflows were sufficient to meet all demands. When large quantities of water can be developed at relatively low cost and when withdrawing water from a stream does not significantly alter its availability for other users, it may be reasonable to assume that the benefits of a water-supply project exceed its costs. These conditions, however, no longer characterize the situation in the West or even in the rest of the nation.

The need to manage the demand for water has gained much wider acceptance within the last decade or so, but there is less agreement as to how it should be done. Regulatory measures such as restrictions on watering lawns and washing cars and sidewalks are common means of reducing water use during drought. Less common and more controversial is the use of regulations such as imposing water conservation standards for toilets, showerheads, and water-using appliances to curb the long-term growth of demand. Some local and state governments have already mandated water conservation measures, and legislation under consideration in the Congress calls for national standards designed to reduce water use.

Water prices also influence use. Planners, however, have traditionally assumed that the demand for water is unresponsive to price (that is, perfectly inelastic with respect to price). Prices rarely reflect the full cost of water use. Indeed, water has been treated as a free resource for which there has been no charge for withdrawing water from or for discharging pollutants into a lake or stream. Water prices are set to cover the costs of delivery and treatment, but even these costs are sometimes subsidized. Wahl (1989) estimates that federally-supplied irrigation water receives a subsidy equivalent to 80 percent of the economic costs of developing supplies and delivering them to an irrigation district. The urban water supply industry usually sets rates just high enough to cover average costs including a return to capital. Average cost pricing in a rising cost industry such as water results in prices below marginal costs. These low prices encourage consumption in excess of socially efficient levels. Efficient pricing would set price equal to marginal social cost to limit use to the point where the benefits derived from use of the last unit are equal to the costs of producing that unit.

Water Marketing

Water is a scarce resource in the United States, and it is almost certain to become scarcer as the supply and demand for the resource continue to change over time. Making the best use of the West's water requires an efficient way to reallocate scarce supplies in response to changing supply and demand conditions. In the United States, markets are the usual mechanism for allocating scarce resources. Well functioning markets allocate scarce resources to their highest-value uses and they provide incentives to conserve and develop new supplies. Water markets, however, are generally crude and are relatively uncommon.

The nature of the resource as well as government regulations pose problems for developing efficient water markets. Efficient markets must satisfy two conditions, both of which may be difficult to meet for water resources. There must be well-defined, transferable property rights and the buyer and seller must bear the full costs of the transaction. It can be difficult to establish property rights over ground and surface waters that are fugitive in time and space. Supplies may be common property resources that belong to no one until they are extracted for use. When ownership is only established by extraction, the individual does not pay the full costs of that use and there is an incentive to overuse the resource. Transferring water from one use or location to another is likely to affect third parties by altering the quantity, quality, timing, or location of water available to others. Another obstacle to the development of efficient water markets is that some of the services provided by water, such as the amenities of a free-flowing stream, are public goods that are usually not marketed. Furthermore, water utilities tend to be natural monopolies that have their prices set by regulatory agencies and utility managers rather than by the interaction of supply and demand (Frederick and Kneese, 1990).

In spite of these difficulties, water marketing does occur in the West, and with more appropriate state and federal policies marketing could play a much greater role in allocating supplies and encouraging conservation. Transaction costs are often unnecessarily high because of long delays, uncertainties, and legal fees. And the introduction of marginal cost pricing by utilities would curb use and provide investment funds to repair inefficient supply systems.

IMPLICATIONS OF A HOTTER AND DRIER CLIMATE: A CASE STUDY OF THE MISSOURI RIVER BASIN

The following case study of the Missouri River basin illustrates the possible impacts of a general warming on the availability of water within one of the West's principal river basins and indicates how management changes and a reallocation of supplies among alternative uses would help the region adapt to such hydrologic changes. Although the impacts of a global warming on the Missouri River basin are unknown, global climate model results suggest that the basin might become hotter and drier. The decade starting in 1931 was such a period within the basin. Superimposing the climate of that decade on the basin as it exists today provides some idea of the water issues that might arise under such a climate.

If the climate of the 1931-1940 analog period became the norm, runoff and evaporation rates would differ from those of the current climate. Estimates of the impact of these differences on the basin's renewable water supplies are presented in Table 11.2.[3] The mean assessed total streamflow represents the renewable supply available before consumptive use. Renewable supplies under the analog climate are only 69 percent of the long-term mean at the outflow point of the basin (subregion 1011), and they range from 64 to 99 percent measured at the outflow points of the various water resource subregions.

The analog climate would also affect water demand. The demand for irrigation and domestic water, especially for lawn watering, would probably rise as a result of the hotter and drier conditions. Quantification of the changes in water demand would be highly speculative, and the subsequent analysis assumes that consumptive uses are unchanged by the analog climate.[4] To the extent that this omission understates the competition for water, the analysis may understate the water problems likely to emerge under the analog climate.

Water is a scarce resource in the Missouri basin even in the absence of any anthropogenically-induced climate change. Society has placed increasing values on instream water uses such as recreation and protection of fish and wildlife habitats in recent decades. The rising demand for these very water-intensive uses in combination with the recent drought within the region have contributed to growing conflicts over water use in the basin. Streamflows during the 1988-1989 drought, which has aggravated water conflicts in the region, exceeded the mean flows during the analog period by 16 to 23 percent at the gauging stations used to reconstruct the analog flows. These conflicts have been particularly evident in the management of the main stem of the river and in the opposition to several water projects proposed for the basin. The U.S. Army Corps of Engineers is under pressure from the upper basin states of North and South Dakota, Wyoming, and Montana to give greater weight to the economic, recreational, and environmental values within the upper basin that are affected by management of the main stem reservoirs. Several proposed water projects including Two Forks Dam in Colorado, Deer Creek Dam in Wyoming, and Catherland irrigation project in Nebraska have encountered strong opposition from environmentalists.

Table 11.3 provides an indication of the adequacy of mean assessed total streamflows to supply 1985 consumptive uses and "desired" instream flows under the current and analog climates.[5] Total use, defined as the sum of cumulative consumptive use and "desired" instream flows, exceeds the mean assessed total streamflow under the current climate in two of the eleven subregions. Under the analog climate, total use exceeds these streamflows by at least seven percent in eight of the subregions; even in the other three subregions, total use is 94 percent or higher of mean assessed total streamflow. When total use exceeds mean assessed total streamflow, then ground water supplies are being mined and/or desired instream flows are not being met.

The amount of water actually available to meet instream uses is derived by subtracting consumptive use and adding ground water

TABLE 11.3 Water use as a percent of mean assessed total streamflow in the Missouri River basin under the current and analog climate.

	Current Climate			Analog Climate		
Subregion	Cumulative Consumptive Use	Desired Instream Flow	Total Use	Cumulative Consumptive Use	Desired Instream Flow	Total Use
1001	13	60	73	18	79	96
02	12	60	72	16	78	94
03	12	60	72	16	80	96
04	18	75	93	23	97	119
05	18	61	78	24	83	107
06	16	61	77	23	85	109
07	76	55	130	77	55	132
08	52	34	86	74	48	123
09	25	60	85	37	88	125
10	50	61	111	79	95	174
11	21	60	81	31	87	117

SOURCE: Frederick, 1990.

overdrafts to assessed total streamflow. This quantity is the current streamflow. Table 11.4 shows desired instream flows as a percentage of current streamflow under the current and the analog climates. Under the analog climate, instream flows are generally well below desired levels as estimated in the Second National Water Assessment. The desired flows would exceed the actual flows by 25 percent or more for sustained periods in five subregions under the analog climate. The values that would be lost are not easily estimated, but they are likely to be high. Moreover, conflicts as to the preferred timing of flows for aquatic habitat, navigation, hydropower, and other uses increase as the resource becomes scarcer.

Preliminary results from an ongoing study by the U.S. Army Corps of Engineers (1990) indicate opportunities for mitigating the overall costs of low-flow conditions within the Missouri basin. A repeat of the 1931-1940 climate would have major impacts on water users in the Missouri. Under current operating criteria for the 6 main stem reservoirs operated by the U.S. Army Corps of Engineers, the navigation season would be reduced from its normal of eight months to about five months during six of the ten years despite the relatively high priority navigation receives in the management of the river. Hydroelectric power production would decline to about half

TABLE 11.4 Desired instream flow as a percent of current streamflow in the Missouri River basin under the current and control climates.

	Current Climate	Analog Climate
1001	69	
02	68	92
03	68	95
04	91	125
05	74	109
06	72	111
07	156	160
08	59	125
09	77	131
10	80	153
11	72	114

SOURCE: Frederick, 1990.

its normal level and the reservoirs would contain less than half their normal quantities of water during these years. These outcomes implyprofound negative implications for the recreational services as well as the fish and wildlife habitat provided by these reservoirs.

Table 11.5 presents preliminary estimates of the average annual benefits by principal use categories derived from the operations of the Missouri main stem system under existing operating criteria and historical streamflows. The large water flows required to support navigation and the high priority navigation receives in the current operating scheme are in striking contrast to the relatively small contribution navigation makes to the overall benefits from the system. Missouri River navigation accounts for less than 2 percent of total system benefits. Even when the contribution to traffic on the Mississippi River is included, navigation accounts for less than 3 percent of total annual benefits. In contrast, hydropower provides 58 percent; flood control a total of 16 percent; water supplies within the lower basin states of Iowa, Kansas, Missouri, and Nebraska 11 percent; and upper basin reservoir recreation 8 percent of the overall benefits.

Even during periods of average or above-average flow, conflicts may emerge among alternative operating criteria. For instance, maintaining navigation flows may conflict with the interests of the upper basin in maintaining high and relatively stable lake levels and with power production at Gavins Point, where releases in support of navigation may exceed the capacity of the power plants. As more space

TABLE 11.5 Estimated Annual Benefits of Missouri Main Stem System Operations.

	Millions of dollars	Percent of totals
Hydropower	470	58
Flood control (Missouri River)	95	12
Flood control (Mississippi River)	36	4
Water supply (downstream)	93	11
Water supply (reservoir)	N/E[2]	—
Recreation (reservoir)	67	8
Recreation (downstream)	3	a[3]
Navigation (Missouri River)	14	2
Navigation (Mississippi River)	6	1
Other[1]	30	4
Total	814	100

[1] The U.S. Army Corps of Engineers lists total benefits of system operations at $814 million but they itemize only $784 million of these.

[2] N/E indicates no estimate available.

[3] a indicates less than 0.5 percent.

SOURCE: U.S. Army Corps of Engineers, 1990.

is devoted to flood control, reservoir levels may be subjected to wider seasonal fluctuations and less storage is available for protection against drought. Nevertheless, these conflicts pale in comparison to those that emerge under drought conditions.

The guidelines for managing the main stem reservoirs have come under attack during the recent drought and are currently under review. Table 11.6 compares the differences between the preliminary average annual benefits for the various users (power, reservoir recreation, downstream recreation, water supply, navigation on the Missouri, navigation on the Mississippi, and flood control on the Mississippi) when selected operating criteria are varied from the base case reflecting current policy. These preliminary results provide strong

TABLE 11.6 Comparative analysis of annual incremental benefits for alternative operating criteria for the main stem of the Missouri River[1] (millions of dollars).

		Recreation		Water	Navigation		Flood Control	
Alternative	Power	Reservoir	Downstream	Supply	Missouri	Mississippi	Mississippi	Total
Base Case	0	0	0	0	0	0	0	0
A	2.35	0.20	-0.03	70.31	-0.18	2.91	0.08	75.65
B	4.80	3.41	-0.03	70.10	-0.44	3.55	0.11	81.51
C	14.57	3.40	-0.26	70.31	-0.51	1.07	0.09	88.67
D	23.01	3.45	-0.23	70.28	-0.54	-0.63	0.13	95.47
E	34.56	3.91	-0.27	57.31	-1.17	-3.80	0.21	90.75
F	42.79	3.91	-0.32	54.48	-1.70	-2.51	0.39	97.04

[1] The dollar values represent the differences in the average annual benefits or costs when each alternative is compared to the base case. The annual averages are derived by simulating the hydro-logic record from 1898 to 1989 with the existing infrastructure and water demands.

SOURCE: U.S. Army Corps of Engineers, 1990.

NOTES TO TABLE 11.6 :

Explanation of Alternative Operating Criteria Base Case:

(1) System storage is divided such that:

- the first 18.3 maf is for the permanent pool
- the next 39.3 maf is for carry over multiple use
- the next 11.6 maf is for annual flood control and multiple use
- the top 4.7 maf is for exclusive flood control.

(2) Length of navigation season: the current rules for determining the length of the navigation season are in effect. The navigation season is curtailed if system storage on July 1 is less than 41 maf.

(3) Minimum winter season release rate is 6,000 cfs.

(4) Minimum summer release rate is 6,000 cfs.

Variations from the Base Case incorporated in the alternatives.

Alternative A

(3) Minimum winter releases 12,000 cfs.

(4) Minimum summer releases 18,000 cfs.

Alternative B

(1) Permanent pool storage increased to 31.0 maf.

(2) Changes in navigation rule curve with the season curtailed if storage is less than 41 maf on July 1.

(3) Same as A.

(4) Minimum summer releases 12,000 cfs.

NOTES TO TABLE 11.6 (continued)

Alternative C

(1) Same as B.
(2) Changes in navigation rule curve with the season curtailed if storage is less than 54 maf on July 1.
(3) Same as A.
(4) Same as A.

Alternative D

(1) Same as B.
(2) Changes in navigation rule curve from Alternative C with the season curtailed if storage is less than 54 maf on July 1.
(3) Same as A.
(4) Same as A.

Alternative E

(1) Permanent pool storage is increased to 44 maf.
(2) Changes in navigation rule curve with the season curtailed if storage is less than 58 maf on July 1.
(3) Minimum summer releases of 9,000 cfs.
(4) Same as B.

Alternative F

(1) Same as E.
(2) Changes in navigation rule curve from Alternative E with the season curtailed if storage is less than 58 maf on July 1.
(3) Same as E.
(4) Same as A.

signals as to the types of changes in operating criteria that would increase the benefits the nation derives from the Missouri River main stem reservoirs. Some of the more promising changes are:

- Increasing minimum winter and summer releases (Alternative A) adds nearly $76 million to the average annual system benefits. Most of these benefits accrue to lower basin communities that would have greater security in their water supplies.
- Increasing the size of the permanent pool (the minimum level of water that is maintained in the reservoirs for fish and wildlife, recreation, operation of hydropower generating units, and municipal, industrial, and agricultural water-supply intakes) and curtailing navigation flows sooner when reservoir levels are low (Alternatives B through F) increase the benefits from hydropower production and reservoir recreation in the upper basin. The dollar value of the adverse effects on navigation of increasing the permanent pool and altering other operating criteria that currently favor navigation are overwhelmed by the positive impacts on power, water supplies, and reservoir recreation.

The results summarized in Table 11.6, which are based on simulations for the entire hydrologic record from 1898 to 1989, suggest the potential benefits of alternative management criteria even in the absence of climate change. The alternative operating criteria are designed to deal with conflicts that emerge during relatively low-flow periods, and the largest gains in annual benefits occur during such periods. Consequently, under a scenario in which the climate of the 1931-1940 decade becomes the norm, the annual incremental benefits of the alternative operating criteria would be much higher than the values presented in Table 11.6.

NOTES

1. The inventory only includes dams that were at least 6 feet in height with a storage capacity of at least 25 acre-feet or at least 25 feet in height with a capacity of 15 acre-feet.

2. The next two sections draw on Frederick and Gleick, 1989.

3. The methodology underlying these estimates is described in Frederick (1990). In brief, the estimates start from data in the Second National Water Assessment (U.S. Water Resources Council,

1978) for mean natural streamflows (flows in the absence of any diversions, man-made reservoirs, and consumptive use) and total assessed streamflows as of 1985 (equal to natural flow minus net evaporation from manmade reservoirs and net exports) for the eleven water resource subregions within the Missouri basin. The adjustments to runoff are based on the differences between observed flows in the 1951-1980 control period and the 1931-1940 analog period at specified gaging stations. Streamflow data going back to 1931 are available for hundreds of gaging stations in the Missouri basin. At most of these stations, however, measured flows are not a reflection of natural streamflows; they have been altered by diversions, dams, or other human impacts. Consequently, flows at seven gaging stations within the basin that were unaffected by human impacts are used as proxies to estimate the changes in natural streamflows attributable to the analog climate. The differences between the reconstructed flows during the 1931-1940 analog and the 1951-1980 control periods capture the effects of changes in precipitation and evaporation from land surfaces. The impacts of temperature and other climatic changes on reservoir evaporation are based on estimates of net evaporation from the six large reservoirs in the main stem of the Missouri during the two periods. The estimated average change in net evaporation rates from these reservoirs is used to estimate evaporation from reservoirs throughout the basin.

4. The impacts of the analog climate on water withdrawals and consumptive use are examined in Frederick (1990).

5. The estimates of "desired" instream flows are from the Second National Water Assessment. The assessment suggests that desired instream flows are "that amount of water flowing through a natural stream channel needed to sustain the instream values at an acceptable level. Values of instream flows relate to uses made of water in the stream channel that include fish and wildlife population maintenance, outdoor recreation activities, navigation, hydroelectric generation, waste assimilation (sometimes termed water quality), conveyance to downstream points of diversion, and ecosystem maintenance that includes freshwater recruitment to the estuaries and riparian vegetation and flood-plain wetlands" (U.S. Water Resources Council, 1987). Fish and wildlife were determined to be the dominant use because the flows that would ensure full fish and wildlife benefits would also provide for all other instream values. The desired instream flows in the assess-

ment are (conservative on the side of identifying more water for instream uses than further study might reveal to be justified" (Bayha, 1978).

REFERENCES

Association of California Water Agencies. 1989. Coping With Future Water Shortages: Lessons From California's Drought. Sacramento Association of California Water Agencies.

Bayha, K. 1978. Instream flow methodologies for Regional and National Assessments. Instream Flow Information Paper No. 7. FWS/OBS-78/61. Washington, D.C.: U.S. Fish and Wildlife Service. P. 4.

Frederick, K. D. 1990. Working Paper IV - Water Resources. Report prepared for the project Processes for Identifying Regional Influences of and Responses to Increasing CO_2 and Climate Change—the MINK Project. Washington, D.C.: Resources for the Future.

Frederick, K. D., and P. H. Gleick. 1989. Water resources and climate change. In N. J. Rosenberg, W. E. Easterling, III, P. R. Crosson, and J. Darmstadter, eds., Greenhouse Warming: Abatement and Adaptation. Washington, D.C.: Resources for the Future.

Frederick, K. D., and A. V. Kneese. 1990. Reallocation by markets and prices. In P. E. Waggoner, ed., Climate Change and U.S. Water Resources. New York: John Wiley & Sons.

Gleick, P. H. 1990. Vulnerability of Water Systems. In P. E. Waggoner, ed., Climate Change and U.S. Water Resources. New York: John Wiley & Sons.

Guldin, R. W. 1989. An Analysis of the Water Situation in the United States: 1989-2040. USDA Forest Service General Technical Report RM-177. Fort Collins, Colo.: U.S. Forest Service.

Hanchey, J. R., K. E. Schilling, and E. Z. Stakhiv. 1987. Water resources planning under climate uncertainty. In Proceedings of the First North American Conference on Preparing for Climate Change: A Cooperative Approach, Washington, D.C., October 27-29.

Hartshorn, J. K. 1986. Drought . . . for flood? Western Water (January/February). The Water Education Foundation, Sacramento, California. Pp. 4-10.

Kneese, A. V., and G. Bonem. 1986. Hypothetical shocks to water allocation institutions in the Colorado basin. In G. D. Weatherford and F. L. Brown, eds., New Courses for the Colorado

River: Major Issues for the Next Century. Albuquerque: University of New Mexico Press.

National Water Commission. 1973. Water Policies for the Future. Final Report to the President and to the Congress of the United States. Washington, D.C.: U.S. Government Printing Office.

Picton, W. L. 1960. Water Use in the United States 1900-1980. Report for Business and Defense Services Administration, U.S. Department of Commerce. Washington, D.C.: U.S. Government Printing Office.

Riebsame, W. E., H. F. Diaz, T. Moses, and M. Price. 1986. The social burden of weather and climate hazards. Bulletin of the American Meteorological Society 67(11):1378-1388.

Schilling, K. E. 1987. Water Resources: The State of the Infrastructure. Report to the National Council on Public Works Improvement. Washington, D.C.: The Council.

Sheer, D. P. 1986. Managing water supplies to increase water availability. In U.S. Geological Survey, National Water Summary 1985: Hydrologic Events and Surface-Water Resources. Water Supply Paper 2300. Washington, D.C.: U.S. Government Printing Office.

Solley, W. B., C. F. Merk, and R. R. Pierce. 1988. Estimated Use of Water in the United States in 1985. U.S. Geological Survey Circular 1004. Washington, D.C.: U.S. Government Printing Office.

U.S. Army Corps of Engineers. 1982. National Program of Inspection of Non-Federal Dams: Final Report to Congress. Washington, D.C.: U.S. Department of the Army.

U.S. Army Corps of Engineers. 1990. Draft Phase I Report for the Review and Update of the Missouri River Main Stem Master Water Control Manual. Omaha, Nebraska: Missouri River Division. P. 11; 128.

U.S. Department of the Interior (DOI), Bureau of Reclamation. 1987. Assessment 1987 . . . A New Direction for the Bureau of Reclamation. Washington, D.C.: The Bureau.

U.S. Geological Survey (USGS). 1984. National Water Summary 1983: Hydrologic Events and Issues. Water Supply Paper 2250. Washington, D.C.: U.S. Government Printing Office.

U.S. Water Resources Council. 1978. The Nation's Water Resources 1975-2000, Second National Water Assessment, Vol. 1. P. 42. Washington, D.C.: U.S. Government Printing Office.

Wahl, R. W. 1989. Markets for Federal Water: Subsidies, Property Rights, and the Bureau of Reclamation. Washington, D.C.: Resources for the Future.

12

Western Water Law, Global Climate Change, and Risk Allocation

A. Dan Tarlock
Chicago Kent College of Law
Chicago, Illinois

For behold, The Lord, the Lord of Hosts doth take away from Jerusalem and from Judah the stay and the staff and the whole stay of health, and the whole stay of water . . . I will command the clouds that they rain no rain upon it.

—Isaiah, Books 3 and 5

INTRODUCTION: WHAT THE DOOMSAYERS SAY

Both the urban and rural West are extremely vulnerable to the predicted adverse consequences of global warming (EPA, 1988). Coastal dwellers along the Pacific Ocean face rising sea levels and the loss of littoral land. Estuarine areas face the risk of destruction from sea level rises as the vital balance between fresh water and salt water will be destroyed. Regional fresh water supplies will be adversely affected in difficult-to-predict ways. The head of the Advanced Study Program at the National Center for Atmospheric Research advises that the change will not be a simple shift to a warmer but stable climate. Instead, we must plan for a new climate each decade. The new climate will bring both surface temperature increases and large year-to-year weather variations (Firor, 1990b). Urban areas that survive sea-level rises will probably face severe water shortages as spring runoffs decline and forests die.

Both farmers and wildlife will suffer from the heightened competition for diminished supplies. In California, for example, there may be less snowpack, higher winter runoff, and lower spring and summer runoff. Annual deliveries to the State Water Project could decline by 7 to 15 percent (EPA, 1988). At the same time, the demand for electricity will increase by 4 to 6 percent over the increase that would occur without global warming. The growing competition between municipal and industrial water users and agricultural water users will exacerbate existing supply shortfalls in populous arid areas, and cities may use their political power at both the state and federal level to bar all but the most essential crops from being irrigated. Simultaneously, the efforts to allocate more water to in-situ uses that began in the 1970s may literally evaporate. A recent global warming disaster scenario includes the prediction that "[i]n northern California, low water levels and high temperatures deoxygenated Tule Lake, inducing epidemics of botulism that eventually killed off immense flocks of ducks and geese that had made Tule the greatest single gathering around the world for migratory waterfowl" (Oppenheimer and Boyle, 1990).

GLOBAL WARMING RESPONSE STRATEGIES: WHAT SHOULD WE BE DOING?

Three interrelated responses to global climate change have been identified: (1) further research, (2) adaptation to temperature rises, and (3) the reduction of the root causes of the warming (resource demand). The merits of the first option are a given (Guruswamy, 1990). The current debate centers on the comparative merits of the second two options. Carbon dioxide and other greenhouse gases must be reduced to slow the warming, but implementing the third option will require a radical change in energy generation and consumption and thus a radical shift in the economic and social organization of all countries.

Climates have historically varied throughout the world, and civilizations accepted variations more or less as fate. In the past, the causes of climate change were unknown natural phenomena rather than human activities; man did not try to manage climate change. However, the legacy of the enlightenment is that climate can be adapted to man through technological progress. The entire settlement of the West can be understood as a living example of this faith. We have refused to accommodate to the limitations of aridity and have sought to turn deserts into gardens for all who would cultivate them.

Global warming is forcing a modest reexamination of this practice. Most moderate alarmists counsel adoption of the second strategy: the decade-by-decade adoption of flexible response strategies to prepare us to live with long-term change. Water shortages are one very important category of the full range of possible adverse effects that can be addressed through adaptation. If we can adapt to water shortages, the adaptation will represent the first major reversal of faith in technological progress as the solution to the limitations of nature.

Ultimately, adaption to a changing climate is at best a temporary strategy and is not a substitute for more fundamental shifts in resource use. Globally, the answer lies in shifting from nonrenewable to renewable energy sources and in curbing explosive population growth (Firor, 1990a). In the West, the answer lies in confronting the relationship between water demand and urban growth. In all the major arid states, unlimited population growth is taken as an article of faith and the function of water policy is to supply all the water necessary to accommodate this growth. Many serious observers of the West think that the question is backwards. We should first set growth limits and use them to temper water demands to the more realistic use of available, possibly diminishing supplies. The Bureau of Reclamation cannot do this alone; nevertheless, the global warming debate may place the Bureau at the center of debate about the future of land use in the West. What happens in Fresno is related to the anti-growth debate in Los Angeles.

As the major federal water manager in the West, the Bureau of Reclamation will be affected by global warming-induced water shortages. Flexible adaptation strategies will require the ability to capture and store decreased rainfall and snowfall and to move available, reduced supplies to the areas of greatest demand with speed. However, existing technical and institutional barriers may make this adaptation difficult. This paper addresses the capacity of state water law and federal reclamation law to adapt to the possibilities of shortages as normal rather than abnormal events. It does not address the strategies needed to achieve a new energy balance. Rather, it assumes that the West faces a substantially increased risk of water shortages and speculates about how the existing law of prior appropriation will respond to these shortages, when and if they occur, as well as the likely effect of recent trends in western water law on global warming adaptation. The basic conclusion is that the law of prior appropriation is not well suited in practice to achieve an optimum allocation in times of shortages because of the gap between priority rights holders and demand, but that reallocation trends currently underway can form the basis for a western global warming adaptation strategy.

PRIOR APPROPRIATION: IS IT A RISK MANAGEMENT SYSTEM?

In theory, state and federal reclamation law have a great capacity to respond to global warming-induced water shortages because the function of western water law has been to allocate a scarce resource among competing users in times of shortage. The law of prior appropriation was developed to allocate water among California miners and to distribute water throughout the West. The law has endured in the face of sharp criticisms about its efficiency (Reisner and Bates, 1989) and equity (Freyfogle, 1986) because it has been able to accommodate changing use demands—by adding indefinitely to the classes of claimants eligible to acquire water rights and by allowing water to be shifted among uses. Irrigators, hydroelectric generators, cities, recreationists, and spokespersons for fish and wildlife have all been accommodated. Thus, the law of prior appropriation, supplemented by federal and state reservoir management, is a potential complete risk allocation strategy.

There are two major problems with the use of prior appropriation for risk allocation. First, the law has never been used for this function. As a result, there are major political, institutional, and legal barriers to its use to declare winners and losers, which must be done if water is to be allocated in times of severe shortages. Second, the risk allocation schedules produced by the strict application of prior appropriation will be widely perceived as perverse. The highest priorities are often the lowest-valued uses. For example, the highest priority on the Colorado River remains irrigation, although the highest values of water are for municipal and industrial supplies and the enhancement of environmental values. Perverse priorities are not an absolute barrier because water can be voluntarily reallocated. However, we are just starting to market water on a large scale, and the jury is out on the success of this method of reallocation.

Prior appropriation allocates the risks of shortages by a simple principle: priority of use. The question is whether the magnitude of the global climate change risks can be allocated within the framework of prior appropriation. Western water law is premised on shortages and priority schedules that provide clear risk allocation schemes. But we do not expect the risks to occur with any regularity. The whole thrust of federal and state water policy has been to reduce the risk of shortages to as close to zero as possible by the construction of large carryover storage facilities. In some places, such as California, ground water pumping serves the same back-up function. Thus, we expect that reservoirs and ground water will avoid all but the mildest forms of rationing during droughts. States have tried to accommodate unlimited growth on a limited water budget by providing ample margins of safety against shortages. When water deliveries have been reduced or stopped according to a strict priority schedule, the losers have generally been small farmers, Indian tribes, and fish and wildlife. Most irrigators have been buffered by the harshness of prior appropriation by both carryover storage and formal and informal mechanisms that share the burdens of shortages by pro rata rather than pro tanto delivery reductions. Thus, although the law of prior appropriation is a risk allocation mechanism, the expectation that it will be used for this purpose is low.

The strong expectations of user security will impede the Bureau of Reclamation should it seek to introduce flexibility (e.g., reallocation) into its mission. Historically, that mission has been to support local users by reducing the risks of shortages to as close to zero as possible by providing sufficient carryover storage to keep water flowing downstream from its reservoirs during dry years and to deliver water to the beneficiaries of the original project at subsidized rates. Our model of natural disaster is the seven-year cycle of plenty and famine experienced by Egypt in the book of Genesis rather than Anasazi long-term drought scenarios. Just as the Pharaoh heeded Joseph's advice and stored the harvests of plenty, so too has the Bureau of Reclamation heeded the vision of scientists and western promoters and stored spring runoffs in wet years to provide reserves for dry years. The faith in our ability to reduce the risks of shortages has powerful and insufficiently noted influence on the western water law. Fear of shortage has been used as the rationale for large projects and has crowded other adaptation strategies off of the political agenda (Stegner, 1986).

The issue that prior appropriation poses for global warming adjustment strategies is how flexible the system will be in shifting water to areas of greatest need and in promoting maximum access to a scarce resource. Global warming adaptations will place a premium on both technical and allocative efficiency. Users in water-short areas will have to conserve existing supplies by using less, and they will face increased pressures for reallocation. Economists have long criticized western water law because it ignores higher, alternative values of water. Many western water observers argue that the historic allocation pattern is grossly inefficient. Too much water is used to grow surplus or low-valued crops, and too much water is wasted (Reisner and Bates, 1990). In almost all western areas, agriculture preceded urbanization. Thus, agricultural users hold the most senior water rights. For most of this century, water allocation has been relatively static because the three major uses—agriculture, hydroelectric power generation, and municipal and industrial consumption—were able to share the available water budget without unduly disrupting each other.

Until recently, there was a widespread perception that the allocation of western water was eternal, but the system was never completely static. It contained reallocation mechanisms that allowed minor adjustments, though, in general, prior appropriation remained watershed-based in practice. Transfers were the exception rather than the norm (although marginal agricultural areas did shift to urban uses). Today, the exception may become the norm. There is a growing consensus in the water community that water needs to be reallocated from irrigated agriculture to municipal, industrial, and instream uses to protect a broad range of environmental and recreational values. Water marketing has been endorsed by the national environmental community as well as by urban suppliers. Transfers can be used to meet both urban and environmental demands with minimum disruption for existing users.

Prior appropriation contains two principles that could become the basis for global climate adaptation. Appropriative rights are usufructuary property rights. The original Edenic vision of the West as a land of small irrigators assumed that water rights should be tied to the soil. However, most states have rejected the appurtenancy principle, and have made water rights transferable property rights. In addition, water has a social value; it can only be used for a beneficial purpose. In this century, beneficial use has been defined only as nonwasteful use. Waste has long been defined by local custom, with the result that few irrigation practices are found to be nonbeneficial. A redefined concept of beneficial use could play a larger role in the future. For example, beneficial use could be defined as efficient use; the beneficial use doctrine would then form the basis for requiring substantial water conservation measures. The operating criteria imposed on the Newlands Project in the Truckee-Carson basin of western Nevada is a possible model of how beneficial use can form the theoretical basis for increased farm and urban water conservation requirements (DOI, 1966). The beneficial use doctrine can be complemented by the public trust doctrine. In California, this doctrine has been used to reallocate vested rights to trust purposes, which include environmental protection. For example, some commentators have argued that the public trust requires reductions in water use and the reallocation of water to dilution flows to redress the adverse effects of agricultural runoff. Water marketing has been endorsed by the national environmental community as well as by urban suppliers. Transfers can be used for both urban and environmental purposes with minimum disruption for existing users. Water marketing could be the cornerstone of an adaptive strategy because water can be shifted to areas of highest demand regardless of its original priority and use. The agreement between the Imperial Irrigation District (IID) and the Metropolitan Water District (MWD) of Southern California could be a model of future transfers. The MWD has paid the IID $120 million to save 100,000 acre-feet of water, which will be added to Los Angeles's lower priority on the Colorado River for the next 30 years, and this is only IID's first trip to the fat farm. Overall, however, we now have more water market theory than we have water markets, largely because proponents have underestimated the complexity of water transfers.

The transfer debate centers on two related questions. The first question is, what are the barriers to transfers? Although most western water rights are transferable, the transaction costs of a transfer can be high. The vested rights of third parties must be protected under state law, and in an increasing number of states transfers are subject to public interest review. These barriers are not insurmountable, however. A comprehensive survey of water transfers in six western states (University of Colorado School of Law, 1990) illustrates that a variety of transfers occur both among similar users and from existing to new users. The transaction costs vary from minimal to very high in Colorado, but in all states transfers are generally supported by state law and are taking place. The second question in the transfer debate is, what is the relevant range of third party interests with a stake in the transfer? In past decades, states have begun to include a variety of previously excluded interests in the allocation and transfer process. Environmental representatives, Indian tribes, ethnic communities, and areas of origin now have a greater stake in water allocation processes than they have had in the past. The net result of these developments is to complicate water transfers. As Professor Joseph L. Sax of the University of California, Berkeley, has observed, water transfers are more like diplomatic negotiations than commercial transactions. The expanded compass of protected interests is legitimate; however, it poses new challenges to the water community to distinguish between good and bad transfer barriers (National Research Council, 1992).

BUREAU OF RECLAMATION RESOURCES

Transfers

The 27 million acre-feet of water that the Bureau of Reclamation supplies to farmers throughout the West have been targeted for a starring role in water marketing. Federal reclamation projects have been identified as a major source of water for municipal, industrial, and environmental uses. Reclamation projects use large amounts of subsidized water, often at low technical efficiencies. However, the Bureau faces two major institutional barriers to reallocating the supplies that it controls to adapt to global warming. First, reclamation law creates strong expectations that the original project beneficiaries will be the eternal beneficiaries of project water; every proposed transfer or conservation requirement will be met with substantial, although not insurmountable, opposition. Second, the Bureau's attempt to recast its mission as that of a multiple-purpose manager to deflect criticism that it has helped pollute western waters and degrade or destroy prime fish and wildlife habitats may be inconsistent with global climate change adaptation.

Efforts to promote efficiency through cost increases and conservation plans have not been aggressively pursued, although a court has held that there is no constitutional right to federally subsidized water.[1] Water marketing advocates argue that voluntary transfers may overcome resistance to transfers and "can be as effective as appropriate pricing in leading to efficient use of water" (Wahl, 1989). However, there are many legal and political barriers to the movement of Bureau of Reclamation water away from the original projects. Federal reclamation law was designed to promote family farms. The legacy of this largely unsuccessful experiment is that the law provides no incentives for transfers. As a leading expert has concluded, "Reclamation law is devoid of any explicit Bureau [of Reclamation] water transfer policy" (Driver, 1987).

Transfers of project water may take place both under federal and state law, but the prevailing assumption is that they will be the exception rather than the rule. Section 8 of the Reclamation Act of 1902 provides "[t]hat nothing in this Act shall be construed as affecting or intended to affect or to in any way interfere with the laws of any State or Territory relating to the control, appropriation, use or distribution of water in irrigation, or any vested right acquired thereunder, and the Secretary of the Interior, in carrying out the provisions of this Act, shall proceed in conformity with such laws."[2] Section 8 was initially construed to mean that the Bureau of Reclamation is "simply a carrier and distributor of water . . . with the right to receive the sums stipulated in the contracts as reimbursement for the cost of construction and annual charges for operation and maintenance of the works."[3] However, in the wake of the New Deal expansion of federal powers, the Supreme Court held that Congress may preempt state law.[4] The Court adhered to these cases in *California v. United States*,[5] but in 1983 the Court again described federal ownership of rights as "at most nominal" because the beneficial interest was held by owners of project land.[6]

Section 8 of the Reclamation Act of 1902 further provides that "[t]he right to the use of water acquired under the provision of this Act shall be appurtenant to the land irrigated, and beneficial use shall be the basis, the measure, and the limit of the right."[7] Section 8 also requires that the Secretary of the Interior proceed in conformity with state law in "the control, appropriation, use, or distribution of water used in irrigation, or any vested right acquired thereunder." The Reclamation Projects Act of 1939 allows the U.S. Army Corps of Engineers and the Bureau of Reclamation to impound water for municipal and industrial use.

The net effect of Section 8 is that project water based on state water rights cannot be reallocated by the federal government alone unless Congress has preempted state law. Individual Bureau of Reclamation project contracts may present additional problems. For example, all projects generate return flows, but control of these flows varies. Some contracts give the United States control over the flows for project use; other contracts give the district power to use them within the district or, in the case of the Central Arizona Project, to sell the flows.

Recent changes in Bureau of Reclamation policy indicate a greater receptivity to transfers, but the new policies do not eliminate the long-standing bias toward appurtenancy in federal reclamation law. Late in 1988, the Bureau, in its new management mode (as opposed to its engineering mode), announced a seven-principle transfer policy. The policy does not amount to a radical switch to water marketing. It reaffirms traditional Bureau deference to state law and generally announces a reactive, rather than a proactive, position on transfers. For example, the Bureau will become involved only where there is a potential effect on federal projects and services and the transfer has been requested by an appropriate nonfederal political authority. Transfer agreements that are part of an Indian water rights settlement (of which there are many, either negotiated or being negotiated) is the major exception to this passive stance. The policy reaffirms the protection of third party interests and the mitigation of adverse environmental effects. Water will not be transferred unless third party effects can be avoided or mitigated. The policy only touches on the volatile issue of subsidy recapture. The issue is whether project beneficiaries can receive the current market value of subsidized water or whether the government should recapture some or all of the increment of value added by decades of underpriced water. The policy states only that transfers will not be burdened with costs beyond those actually incurred. This response seems inadequate, although most analysts argue that subsidy recapture should be subordinated to the removal of transfer restrictions. Still, subsidy recapture will be an issue in both contract renewals and transfers.

Reservoir Operation

The Bureau of Reclamation operates carryover storage reservoirs throughout the West. Bureau operations are subject to varying levels of discretion that may change as the result of severe shortages caused by global climate change. Ironically, global climate change may constrain the Bureau's operating discretion more than it is now. Projects may actually have to be operated to meet legally binding allocations rather than to maximize power revenues. Bureau projects are subject to a complex state and federal scheme of priorities and preferences, and these priorities and preferences vary from reservoir to reservoir. The legal position of the Bureau is not clear. The orthodox analysis is that the Bureau is only a carrier for water allocated by state law. Section 8 of the Reclamation Act requires that the Bureau perfect project water rights under state law. This analysis was developed at a time when the Bureau's constitutional powers were not as broad as they are now and the Bureau operated smaller-scale projects. Congress may preempt state law and delegate to the Bureau the power to allocate water as it chooses.[8] Thus, the modern rule is that the Bureau must presumptively follow state law unless Congress has chosen to preempt it.

The carrier analysis works to structure the operation of small Bureau of Reclamation reservoirs; the Bureau stores the maximum amount of water possible during the spring runoff and answers calls during the irrigation season, refusing to honor a call only if there is not enough to satisfy senior water rights holders. The carrier analogy may also be legally correct for large, multipurpose reservoirs, too, but it is often irrelevant. The amount of water available gives the Bureau considerable discretion to operate the reservoir. As long as supplies are relatively abundant over a three- or 4-year period, there is usually a difference between the de jure operating rules and the de facto operating procedures.

Glen Canyon Dam's operation provides an example of the changes in operating procedures that severe shortages must produce.[9] Glen Canyon Dam is the linchpin of Colorado River management because it enables the upper Colorado River basin states to store sufficient water to meet their 10-year delivery obligation to the lower basin states. Paradoxically, the Law of the River controls the yearly operation of the dam but does not constrain daily operations for power generation. The reason is that the Law of the River only affects power generation in the case of long-term, extreme water shortages while reservoir law specifies annual fill targets. The dam was constructed as part of the Colorado River Storage Project Act to provide a large reserve to enable the upper basin to withstand prolonged periods of drought and meet its obligations under the 1922 Colorado River Compact to deliver 7.5 million acre-feet to the lower basin every ten years. Because the irrigation and other projects along the upper Colorado River could never pass a clean benefit-cost analysis and could not be subsidized by the beneficiaries, the upper basin states used power revenues from the storage projects to cover a large percentage of the repayment obligations. Glen Canyon Dam is presently operated to maximize power revenues, although hydroelectric generation is a low priority use on the river. In theory, Glen Canyon Dam is controlled by the Law of the River, a complex mass of compacts, international agreements, statutes, judicial decrees, and informal operating procedures. In practice, however, it is controlled by the Bureau of Reclamation and Western Area Power Administration operators, who manage the dam like an automatic bank teller machine for the southwest power grid. There is concern about the adverse effects of pulsating flows on the riverine environment of the Grand Canyon, but global warming could affect the dam's virtually unrestricted use for power generation in other ways. The Law of the River is largely irrelevant to day-to-day operations because it is only a law of mass allocations between regions and among states. The compacts that form the core of the law reflect the prevailing water use values at the time of their negotiation: irrigation and municipal and industrial use, as well as conservation storage to meet the demands for these uses in times of shortage.

The negotiators of the 1922 Colorado River Compact assumed that they were allocating an average annual flow at Lee's Ferry, Arizona, of 16 million acre-feet. Article III of the compact apportioned "in perpetuity" the "exclusive beneficial consumptive use" of 7.5 million acre-feet to each basin. The lower basin was given the additional right to the assumed one million acre-feet surplus. In anticipation of the assertion of claims by Mexico, the Mexican burden was divided equally between the two basins. The power of the lower basin was augmented by two provisions of Article III. The first provides that the upper basin states will not cause the flow at Lee's Ferry to be depleted by 75 million acre-feet for any consecutive 10-year period, the second provides that the upper basin states cannot withhold the delivery of water "which cannot reasonably be applied to domestic and agricultural use." A reciprocal duty on the lower basin not to demand deliveries on the same condition is meaningless, because the lower basin puts all its entitlement to domestic and agricultural use. The only mention of power is in Article IV, which subordinates the use of the river for navigation to "domestic, agricultural and power purposes."

The relationship between power generation and other uses of the river has been the subject of some speculation among commentators, but there has not yet been a conflict that tests the relationship. The upper basin's 10-year delivery obligation is absolute, thus, the upper basin states are precluded from objecting to the use of this water for power generation before it is consumed by the lower basin states. To establish the relationship between power generation and other river uses, there would have to be a prolonged drought making it impossible for the upper basin to meet its 10-year delivery obligations, and the lower basin states would have to demand the release of water for power generation. The late Dean Meyers suggested that if lower basin consumptive demands are met, Article III (e) prohibits the lower basin from demanding water solely for power generation, because the compact expresses a clear preference for domestic and agricultural uses over power generation (Meyers, 1966).

Prolonged warming may create the Anasazi scenarios that river watchers fear. If downstream priority right holders—the several large irrigation districts along the Colorado River in Arizona and California—make a call that triggers the 10-year obligation, however it is defined, the Bureau would have to let water flow through the dam to serve these priority uses, regardless of power contracts. Similarly, the 10-year obligation may be a basis for the upper basin states to require that water not needed for immediate downstream priority uses be stored in Glen Canyon Dam as a reserve against their 75-million-per-decade delivery obligation (Getches, 1985).

New Interests

Management of the Colorado under drought conditions is further complicated by the recent efforts to accommodate new interests. The post-Colorado River Compact experience with the accommodation of new interests contains mixed lessons for global warming scenarios. Three major classes of uses were traditionally excluded by reclamation-era allocations: recreation, environmental quality preservation (including public health and fish and wildlife habitat enhancement), and Indian claims. Most of the major developments in water law have revolved around the incorporation of these values at the federal and state levels.

In 1922, the full range of relevant interests was not represented in the negotiations. The two most obvious exclusions were Indian tribes and the government of Mexico. Both of these claimants have been accommodated by superimposing new rights onto the original mass allocation. Under the Supreme Court's *Winters* doctrine, Indian tribes have been given priority to large amounts of water.[10] Federal reserved rights may also be claimed by federal land management agencies, and there is an argument that the Grand Canyon National Park enabling legislation allows the National Park Service to assert a federal reserved water right to protect the ecological integrity of the canyon. These rights are assumed by the state in which the reservation lies. Likewise, the government of Mexico obtained quality rights by treaty and quality rights by subsequent international agreements.

The accommodation of newer resource values—the use of the river for habitat maintenance, recreation, and the stabilization of riparian corridors such as the Grand Canyon—has proved more difficult. These interests can be recognized through the creation of new rights, but their protection requires management of the river. The web of interconnected statutes, international agreements, and cases that make up the Law of the River is not designed to manage the river for the full range of resource values. All states have resisted management because of a fear that it will dilute their mass allocation consumptive use rights. This defect in the Law of the River is becoming more acute as new values assert themselves.

New values are incorporated in federal statutes passed after the 1922 compact and the Boulder Canyon Project Act, but these new statutes are not well integrated into the Law of the River. In the 1960s, Congress began to enact a number of national environmental statutes. The two most relevant for the Colorado River are the Clean Water Act and the Endangered Species Act. These statutes superimpose environmental protection mandates onto existing water allocation regimes without specifying the extent to which prior allocations are modified, but the few court cases involving these statutes suggest that existing allocations must be accommodated to accomplish the federal objectives.

Indian water rights are protected under the *Winters* doctrine. Indians assert large claims to both surface and ground water. The issue with respect to Indian water rights is the range of purposes for which the water can be used. Western water users take the pastoral people analysis of *Winters* literally, arguing that these rights are restricted for agricultural use on reservations. The Indian tribes generally claim the right to the full range of modern beneficial water uses and the right to lease water off the reservation. Off-reservation use has been allowed on an ad hoc basis in recent Indian water rights settlements. Thus, Indian water rights can be added to the pool of water available to respond to global warming-induced demands.

Recreational and environmental uses are at risk because they are junior in fact and in law. These uses are just beginning to get rights status under state law. Generally, under federal law they are protected as regulatory property rights. When they are recognized, they have a low de facto or de jure priority. Thus, they would probably be the first to be curtailed in a global warming management scenario. The recent incorporation of environmental values on the Colorado River illustrates their fragile legal status (Goldenman, 1990) and the challenges ahead for the Bureau of Reclamation to assure that global warming does not destroy the gains made as a result of the recognition of the value of protecting, to the maximum extent possible, natural environments.

CONCLUSION

Global warming may be the most serious environmental threat facing the West. If, as many believe, global warming is occurring, there is increased urgency to begin the necessary modification of our historic water allocation policies, which are premised on an unlimited ability to outwit nature to accommodate all people attracted to the West. Federal and state western water managers are presented with a unique opportunity to begin the task of designing a set of water allocation institutions that will allow the modern West to continue as a viable region, even as aridity becomes an operational fact of daily life.

NOTES

1. *Peterson* v. *United States Department of the Interior*, 899 F.2d 799, U.S. Court of Appeals (1990).

2. 33 U.S.C. Section 383 (1986).

3. Ickes v. *Fox*, 300 U.S. 82 (1937).

4. Ivanhoe Irrigation Dist. v. *McCracken*, 257 U.S. 275 (1958) (*acreage limitation*); *City of Fresno* v. *California*, 372 U.S. 672 (1963) (*federal preference for irrigation*); and *Arizona* v. *California*, 373 U.S. 546 (1963) (*federal allocation in times of shortage*).

5. 438 U.S. 645 (1978).

6. *Nevada* v. *United States*, 463 U.S. 110, 126 (1983).

7. 43 U.S.C. 383 (1986).

8. *Arizona* v. *California*, 363 U.S. 546 (1963).

9. This portion of the paper is drawn from Ingram, Tarlock, and Oggins, 1991.

10. *Winters* v. *United States*, 107 U.S. 564 (1908).

REFERENCES

Driver, B. 1987. The effect of reclamation law on voluntary transfers of water. Rocky Mountain Mineral Law Institute 33(1):26-27.

Firor, J. 1990a. The Changing Atmosphere: A Global Challenge. New Haven: Yale University Press.

Firor, J. 1990b. The heating of the climate. Colorado Journal of International Environmental Law and Policy 1:29.

Freyfogle, E. 1986. Water justice. University of Illinois Law Review 1986:481-519.

Getches, D. 1985. Competing demands for the Colorado River. University of Colorado Law Review 56:413.

Goldenman, G. 1990. Adapting to climate change: a study of international rivers and their legal arrangements. Ecology Law Quarterly 17:741.

Guruswamy, L. 1990. Global warming: integrating United States and international law. Arizona Law Review 32:221.

Ingram, H., D. Tarlock, and C. Oggins. 1991. The law and politics of the operation of Glen Canyon Dam. In Colorado River Ecology and Dam Management. Washington, D.C.: National Academy Press.

Meyers, C. 1966. The Colorado River. Stanford Law Review 17:1.

National Research Council. 1992. Water Transfers in the West: Efficiency, Equity, and the Environment. Washington, D.C.: National Academy Press.

Oppenheimer, M., and R. Boyle. 1990. Dead Heat II. New York: Basic Books.

Reisner, M., and S. Bates. 1989. Overtapped Oasis: Reform or Revolution for Western Water. Covelo, Calif.: Island Press.

Stegner, W. 1986. The American West as Living Space. Ann Arbor: University of Michigan Press.

U.S. Department of the Interior (DOI). 1966. Draft Environmental Impact Statement for the Newlands Project Proposal Operating Criteria and Procedures, May. Washington, D.C.: Department of the Interior.

U.S. Environmental Protection Agency (EPA). 1988. The Potential Effects of Global Climate Change in the United States. Washington, D.C.: EPA Office of Policy, Planning and Evaluation.

University of Colorado School of Law. 1990. The Water Transfer Process as a Management Option for Meeting Changing Demands. Boulder, Colo.: Natural Resources Law Center.

Wahl, R. 1989. Markets for Federal Water: Subsidies, Property Rights and the Bureau of Reclamation. P. 130. Washington, D.C.: Resources for the Future.

13

Water Resources Forecasting

John C. Schaake, Jr.
National Weather Service
Silver Spring, Maryland

Thomas Carlyle, Scots essayist and historian, called economics the "dismal science." A problem with economics is that ultimately all accounts must balance—there's no free lunch. This rule holds in physics, too. Perhaps if Carlyle had known the second law of thermodynamics he may have had a more depressing view of that science. Balancing accounts of all kinds—economic, environmental, social, physical and international—seems to become more challenging each year. Prices, both economic and environmental, seem to be rising. Capital is hard to find. Interest rates seem high and unsteady. The federal budget has not been balanced for more than a decade. The trade deficit continues to grow, as does the number of homeless. In the three decades since Rachel Carson wrote *Silent Spring*, the nation has learned to expect environmental consequences of our activities. To see that this awareness has led to an improved environment, one only has to travel to other parts of the world (developed and developing) where the environment has not been valued so highly. But environmental quality is not free; awareness of environmental value limits choices and raises costs.

Within the context of this complex social and environmental background, the nation manages its water resources. The twentieth century has been a time of water development—both physical (as evidenced by the building of reservoirs, levees, and irrigation projects) and institutional (as evidenced by a proliferation of legislation and agencies to regulate water). Hundreds of billions of dollars were invested by government at all levels. Now, the best reservoir sites are taken, the best irrigation projects developed. New stresses on water resources and the environment continue to come from the nation's growing population and economy. Unforeseen environmental consequences of existing water projects require attention. And in the international arena, water (both its quantity and quality) may become far more important than oil in the next century as a source of conflict and, potentially, war.

Water must also be managed in a climatic as well as a social context. Water resources decisions are made under the assumption that the climate of the recent past is representative of the future. The possibility of global change suggests that this assumption may be invalid and adds uncertainty to the water management agenda. While climate change may impact water resources significantly, it is interesting that a major source of uncertainty in climate change also involves water. Water is important in the energy balance of the earth: as an agent for energy transport and as a factor in the optical properties of clouds and the atmosphere. A major goal of the World Climate Research Program (a joint program of the World Meteorological Organization and the International Council of Scientific Unions) is to improve understanding of the climate system through a Global Energy and Water Cycle Experiment (GEWEX). Much of the work that must be done by GEWEX to improve our understanding of the hydrologic cycle of the earth is also needed to improve water resources management. Conversely, much of the work needed to improve water resources management should help (in part through GEWEX) reduce uncertainty about climate change and the effects of climate change on water resources. This is a very important issue that will be developed below.

Water management has two parts: one long range, involving the planning, design, and construction of new facilities; the other short range, involving the operation of existing facilities. Both long-range and short-range decisions are sensitive to the hydrological effects of climate variability and change. Water management decisions depend on water resources forecasts, and these forecasts must take into account the effects of climate variability and change.

The drought of 1988 focused national attention on the role of water resources in our country. Print and electronic media brought clear images of the consequences of drought: barges stranded on sandbars in the Mississippi River, empty reservoirs, and withered corn fields. The drought dramatically illustrated that industries tied to the availability of water are inherently risky. Risky decisions in water-based industries are a daily fact of life: barge companies decide how heavily loaded each barge should be; electric and water utilities decide how much effluent can be safely discharged into an estuary; dam managers determine hydropower and irrigation release schedules and operate reservoirs to balance flood control against water supply. In many cases these water management decisions are based on ad-hoc information systems that force inefficient and wasteful utilization of the nation's water resources.

Water resources forecasts for short-range operational decisions are important for several reasons. First, the incremental benefits to the nation of improving the operation of existing water facilities may be at least as great as the present value of all future water facilities; one way to achieve some of these benefits is through improved decisionmaking using water resources forecasts. Second, water resources forecasts must consider climate variability and the current state of the climate system, which requires forecasts of or assumptions about the short-term local climate regimen. The procedures used to translate the current climate regimen into a runoff regimen can also be used to understand the vulnerability of water resource systems to climate variability or change. Third, the hydrologic methods required for water resources forecasting can contribute to improved understanding of the global climate system. This will require research into how we can apply hydrologic models to the many parts of the world for which data are inadequate to calibrate conventional models. Research is also needed to improve models for the significant fraction of the United States that lies between gaging stations and for which available data are inadequate to support model calibration.

SCIENTIFIC BASIS FOR WATER RESOURCES FORECASTING

The science of real-time hydrologic forecasting has reached the point where significant advances can be made to provide improved information for water managers. A water resources forecast may range from the estimation of the stage or discharge of a river for the next one or two days to the prediction weeks or months into the future of quantities such as volume, maximum flow, minimum flow, and time until an event occurs. As the duration of the forecast period increases, the level of uncertainty in the forecast also increases. Information about uncertainty is an important part of a water resources forecast. Thus, water resources forecasts are probabilistic statements about the future. They are particularly useful in decisionmaking when uncertainty is considered explicitly.

Hydrologic Forecast Models

A wide range of hydrologic forecast models have been developed, and many of these could be used as part of a forecast system. On one hand, these models are good enough to produce useful results; a few have been included in the National Weather Service River Forecast System (NWSRFS) that will be described next. But there are important limitations of these models that require improvement, as explained below under Scientific Opportunities.

The NWSRFS is a software system (containing more than 350, 000 lines of computer code) consisting of many programs that are used to perform all of the steps necessary to generate streamflow forecasts. The main components of the NWSRFS (Figure 13.1) are the Calibration System (CS), the Operational Forecast System (OFS), and the Extended Streamflow Prediction System (ESP). The NWSRFS is a modular system that allows the hydrologist to select from a variety of models and to configure them in a way that is appropriate to the application. The models simulate snow accumulation and ablation, calculate runoff, time distribute runoff from the basin to the basin outlet, and channel route streamflow. All of the models are available to the CS, OFS, and ESP systems.

The Calibration System performs tasks needed to process historical hydrometeorological data and to estimate model parameters for a specific basin. In model calibration, simulated streamflow is statistically and visually compared to the observed streamflow. Model parameters are adjusted until the model simulated streamflow best matches the observed streamflow. Brazil and Hudlow (1980) discuss calibration procedures in more detail.

The Operational Forecast System is used to process real-time hydrometeorological data and to make forecasts. The OFS contains three main components: data entry, preprocessor, and forecast. The data entry component is a set of programs that transfer hydrometeorological data from a variety of sources to the observed data base. The Preprocessor component reads raw station data, estimates missing data as required, and then uses these data to calculate mean areal time series of precipitation, temperature, and potential evapotranspiration for a particular basin. These processed time series are stored in a data base for the forecast component.

Forecast component operation is directed by a forecast parameter data base that includes model parameter values as well as information that describes both the basin connectivity of the river system and the sequence of models to operate for a given subbasin.

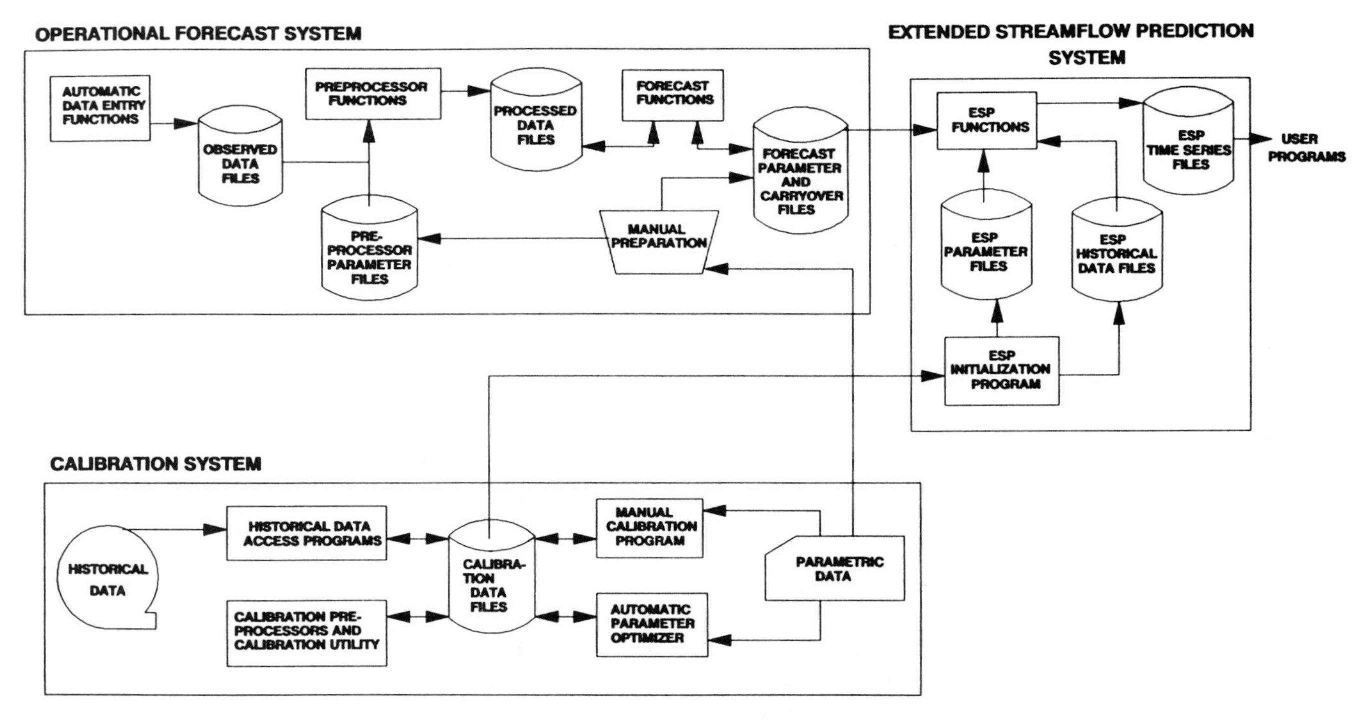

FIGURE 13.1 National Weather Service River Forecast System.

The initial state of the river system at the beginning of the forecast run is held in a carryover data base. Carryover data are usually updated daily based on observed data. This is done to limit how far into the past the OFS must go to make a forecast and to assure that the model initial conditions produce model simulations that are reasonably consistent with current observations.

Extended Streamflow Prediction

ESP is the portion of NWSRFS that permits hydrologists to make probabilistic forecasts of streamflow and other hydrologic variables (such as soil moisture) for extended future periods (Day, 1985). A schematic of ESP is shown in Figure 13.2. ESP assumes that historical meteorological data are representative of possible future conditions; it uses historical data as input to hydrologic models, obtaining the current states of these models from the forecast component of the OFS. A separate streamflow time series is simulated for each year of historical data using the current conditions as the starting point for each simulation. The streamflow time series can be analyzed for peak flows, minimum flows, flow volumes, and so forth, for any period in the future. A statistical analysis is performed using the values obtained from each year's simulation to produce a probabilistic forecast for a particular streamflow variable. This analysis can be repeated for different forecast periods and different streamflow variables. Short-term quantitative forecasts of precipitation and temperature can be blended with the historical time series to take advantage of any skill in short-term meteorological forecasting. In addition, knowledge of the current climatology can be used to weight the years of simulated streamflow based on the similarity between the climatological conditions of each historical year and the current year.

WATER RESOURCES FORECASTING SERVICES

The National Weather Service (NWS) has begun a program for improving its Water Resources Forecasting Services (WARFS). The NWS provides weather and river warning services that have some utility for water management planning and operations; however, the NWS's existing services are structured primarily for flood warning and have limited ability to provide longer-term water supply forecasts. WARFS is a natural extension of the NWS's existing services, because flood forecasting and forecasting for water management interests are based on the same science and technology.

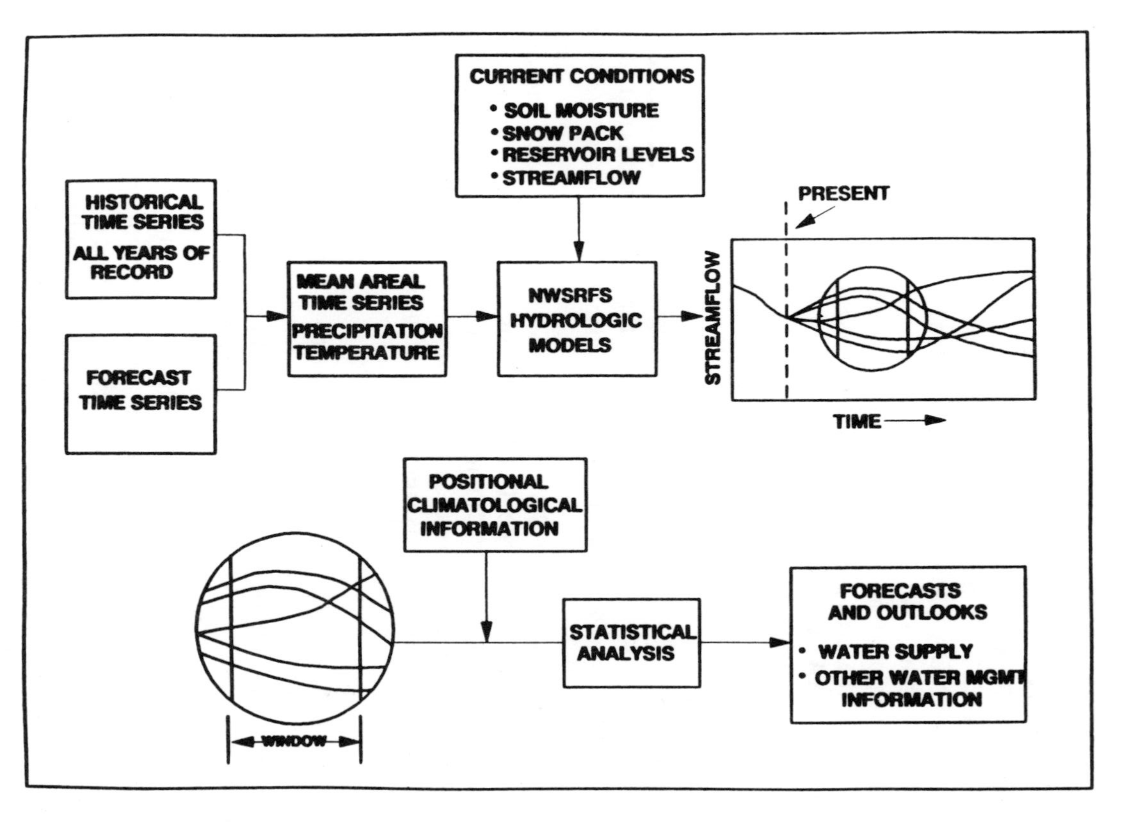

FIGURE 13.2 The ESP procedure.

The NWS WARFS program (Figure 13.3) is built on the following components:

1. Existing National Oceanic and Atmospheric Administration (NOAA) river flood warning programs, weather and climate forecasting services, and private sector value-added service links, with the existing NWSRFS as the heart of WARFS;
2. NOAA's NWS modernization program;
3. Other NOAA scientific support capabilities; and
4. Data about river, snow, and precipitation conditions that cost local, state, regional, and other federal agencies more than $60 million annually but that are provided to NOAA essentially free or in return for other NOAA services.

Existing Services

The existing NWS river and flood forecast system produces forecasts of river levels or discharges at more than three thousand locations. Some of these forecasts are produced daily, others only when there is a risk of flooding.

In the western United States, the NWS cooperates with the Soil Conservation Service of the Department of Agriculture to provide seasonal volume forecasts during the snowmelt season. These are issued monthly and are based on regression relationships between the water content of snow measurements and future runoff volumes.

NWS Modernization

Three major NOAA technological improvement programs are underway as part of a strategy to modernize and restructure the National Weather Service. Two of these programs, the Next Generation Weather Radar (NEXRAD) and the Automated Surface Observing System (ASOS) will provide much (but not all) of the technology needed to observe precipitation nationwide. Achieving the required accuracy of precipitation estimates, however, will require data management, integration, and analysis procedures that incorporate a large variety of precipitation data sources from other federal, state, and local gage networks. The third component of the NWS modernization program is the Advanced Weather Interactive Processing System (AWIPS). The function of AWIPS is to provide a modern interactive processing environment that will be the center of all forecast operations in a particular office. Data analysis and quality

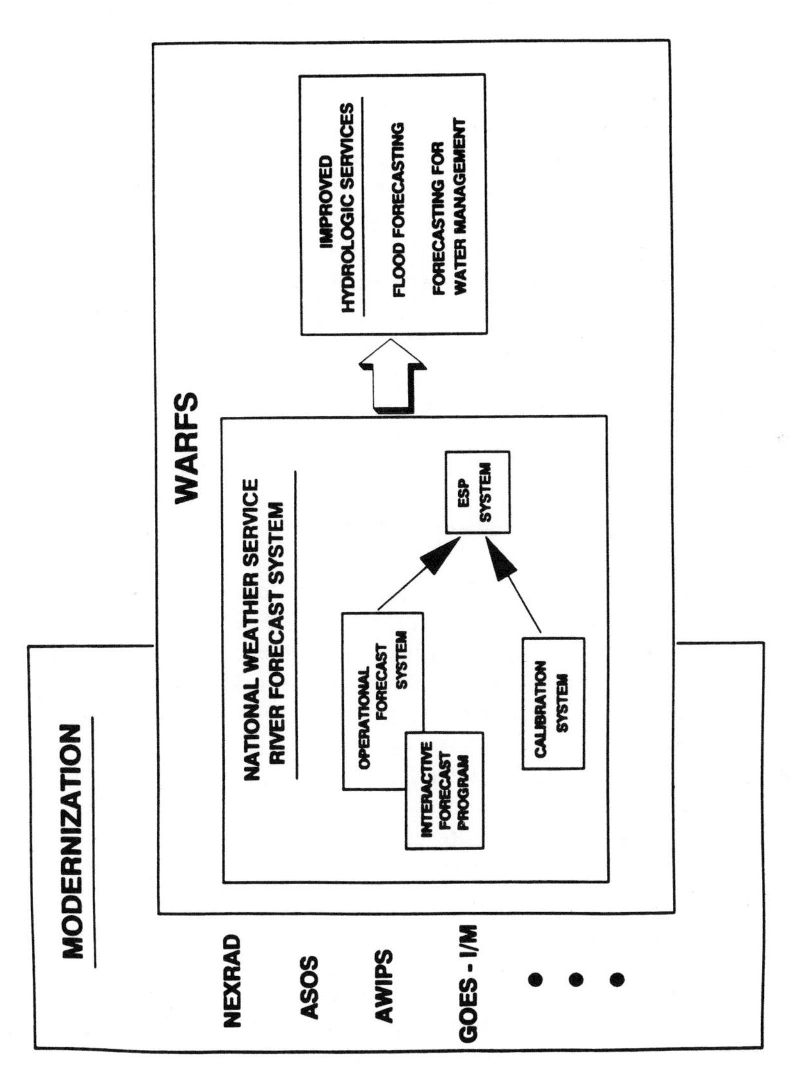

FIGURE 13.3 Water resources forecasting services.

control, forecast modeling, forecast interpretation, and product formulation will be carried out interactively at modern workstations. The data and computer programs provided by the NWS modernization programs, along with the technology of advanced hydrologic and climate forecast models will be used to:

1. support forecast service requirements of government and quasi-government water managers;
2. provide basic water resources forecasts to private sector intermediaries who will tailor forecasts to serve specific industries;
3. satisfy the needs for forecast services at a variety of time scales and for a variety of water use situations nationwide;
4. provide critical information on hydrometeorological forecast reliability that can be used for risk-based water management decisionmaking;
5. incorporate improved weather and climate forecast information into hydrological models; and,
6. improve other forecast capabilities.

Example Applications of WARFS Technology

In a few locations, the NWS has already implemented ESP procedures, and these are being used to make water resources management decisions on a limited basis. In 1977, the NWS began developing water resources forecasting technology for the Potomac River basin. This activity came in response to requests for assistance in alleviating an impending water supply crisis in the Washington, D.C., metropolitan area. During drought periods, water demands from a growing population exceeded the capacity of existing water facilities, which relied on the unregulated Potomac River and two small regulated streams in adjacent basins. Proposals to solve the water supply crisis for the politically divided region had emphasized construction projects (such as building as many as 16 major dams and reservoirs) with high costs and little public support. By the early 1980s, it was clear that local water managers were sold on the soundness of WARFS technology and its cost-effectiveness. In 1983, a series of contracts were signed by the federal government, the state of Maryland, and the Commonwealth of Virginia. These contracts provided for a limited construction program of two reservoirs and joint operation of the water resources of the region using prototype WARFS technology. The project was nominated for the Outstanding Civil Engineering Achievement Award of 1983.

More recently, the Southeast has experienced extended drought conditions. Beginning in 1986, ESP forecasts were generated for the Lake Lanier basin on the Chatahoochee River. Lake Lanier is the primary water supply source for the Atlanta, Georgia, metropolitan area. By 1988, soil moisture conditions were extremely dry. Figure 13.4 compares soil moisture conditions on June 1, 1988 with those on June 1 for the years 1958 through 1986. These values were generated using the NWSRFS. ESP was used to produce probabilistic streamflow forecast information, such as that shown in Figure 13.5. This plot shows the reservoir volume for June 1988 as a function of probability. The line labeled "historical" is based on an analysis of the historical streamflow for June. The line labeled "WARFS forecast" is based on ESP analysis that takes into account the extremely dry soil moisture conditions that existed on June 1, 1988. The historical streamflow analysis indicates that the probability of an inflow volume of 100,000 acre-feet or larger is 76 percent (for any year taken randomly), while the ESP analysis indicates that the corresponding probability is only 3 percent for 1988. This difference is particularly large because of the extreme drought conditions of 1988. It illustrates the potential importance of state of the art forecast information to water supply operations.

It is possible to generate many additional forecast products for water resources using WARFS technology. Figure 13.6 shows spring flood potential conditions across the nation that the NWS estimated for 1990. This map is based on the qualitative judgment of hydrologists using information from a variety of sources (including models, the Palmer Drought Index, and observed precipitation). Figure 13.7 shows the type of information that could have been made available with WARFS technology. Using ESP, it would be possible to estimate quantitatively the probabilities of flooding. This type of product would provide the forecast information needed to make complex water management decisions using a risk-based analysis. Although the example is based on probabilities of flooding, the same type of information could have been generated for droughts. Evidence that valid information is present in such long-range forecasts is suggested in Figure 13.8, which shows flood conditions as of May 1990.

SCIENTIFIC OPPORTUNITIES

Although water resources forecasting methods are well enough advanced to produce useful results for water management, they have important limitations and deficiencies that need to be corrected. Some of these are discussed below.

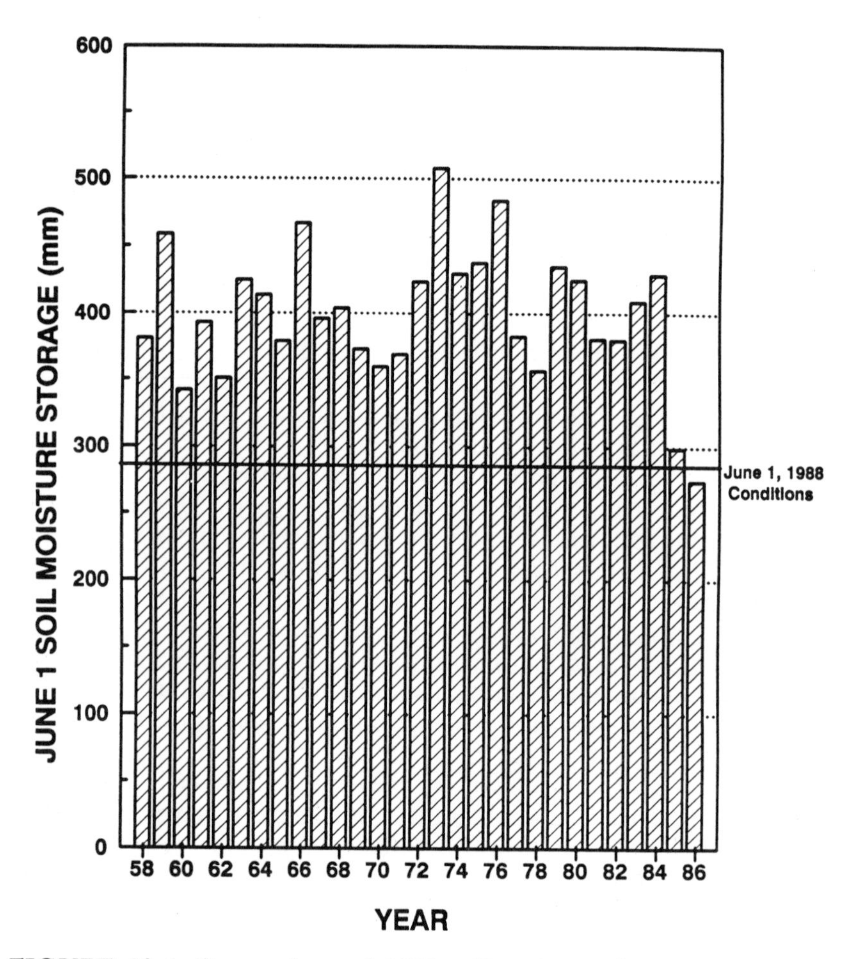

FIGURE 13.4 Comparison of 1988 soil moisture in the Lake Lanier (Georgia) basin with historical values.

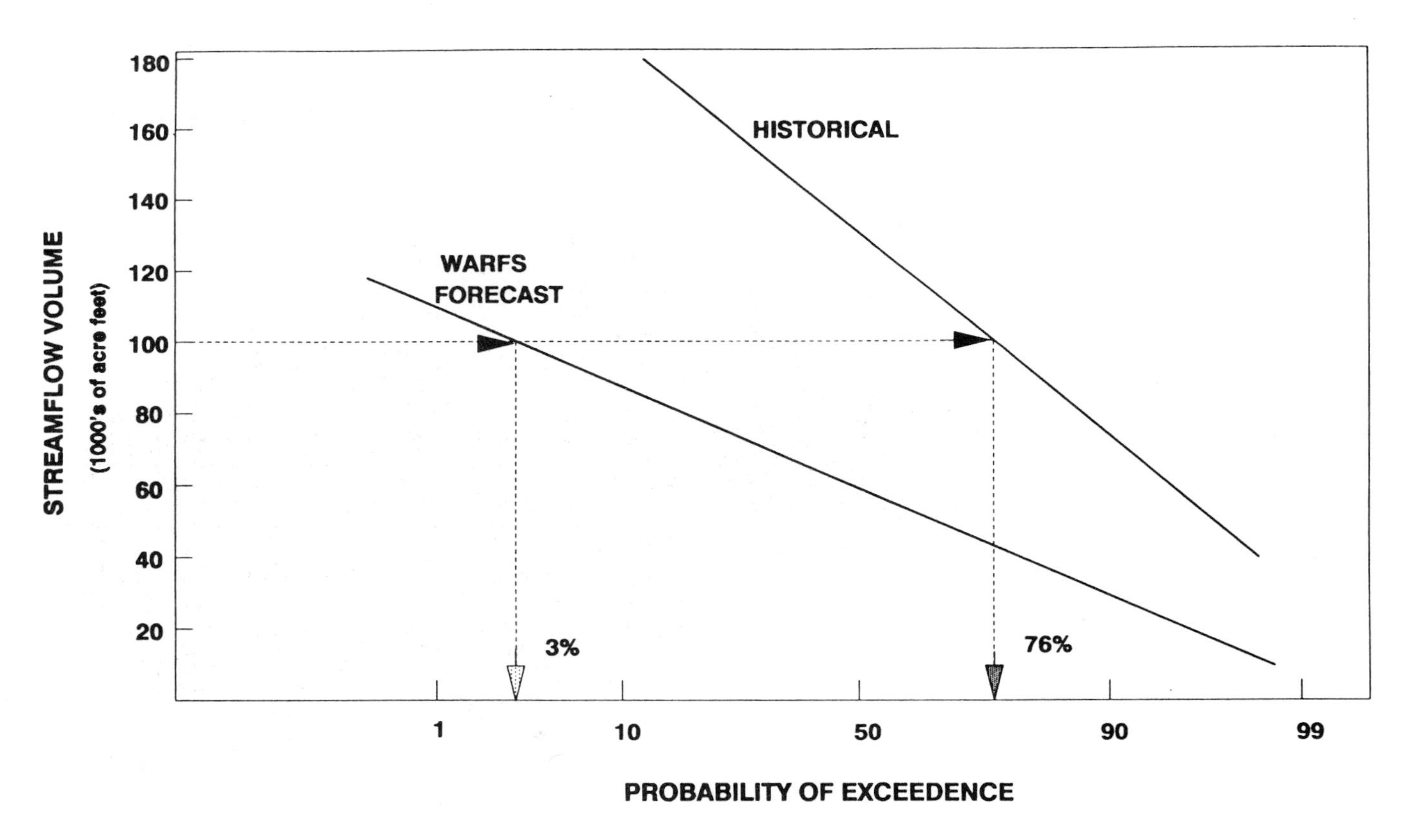

FIGURE 13.5 Reconstructed Lake Lanier inflow forecast for June 1988.

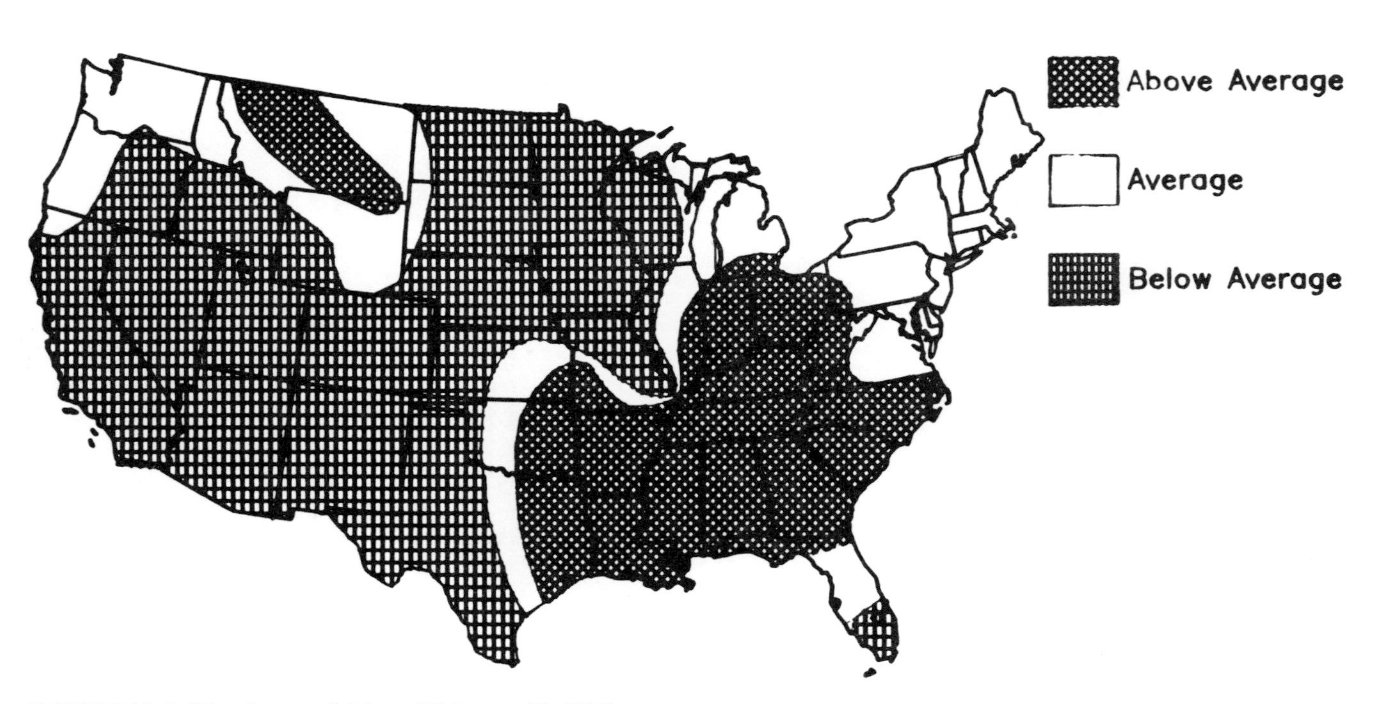

FIGURE 13.6 Flood potential (as of February 28, 1990).

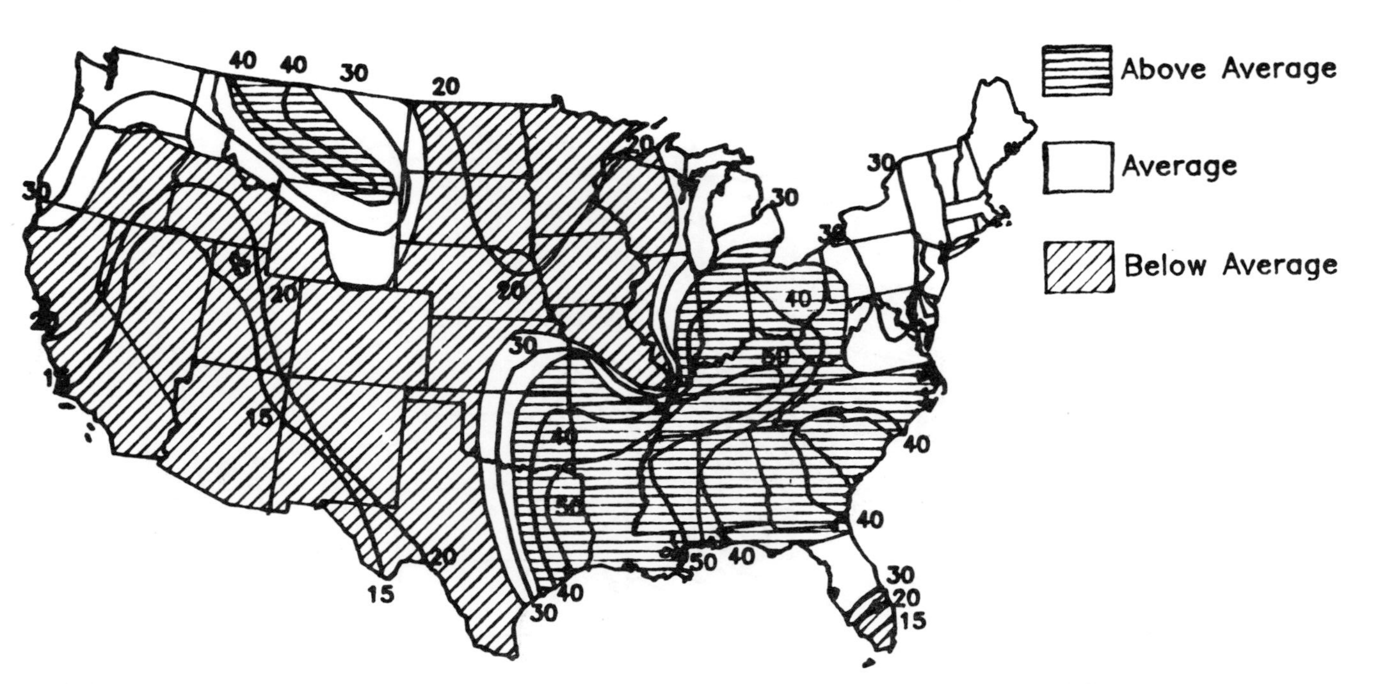

FIGURE 13.7 Hypothetical probabilities of flooding based on WARFS-type technology (as of February 28, 1990).

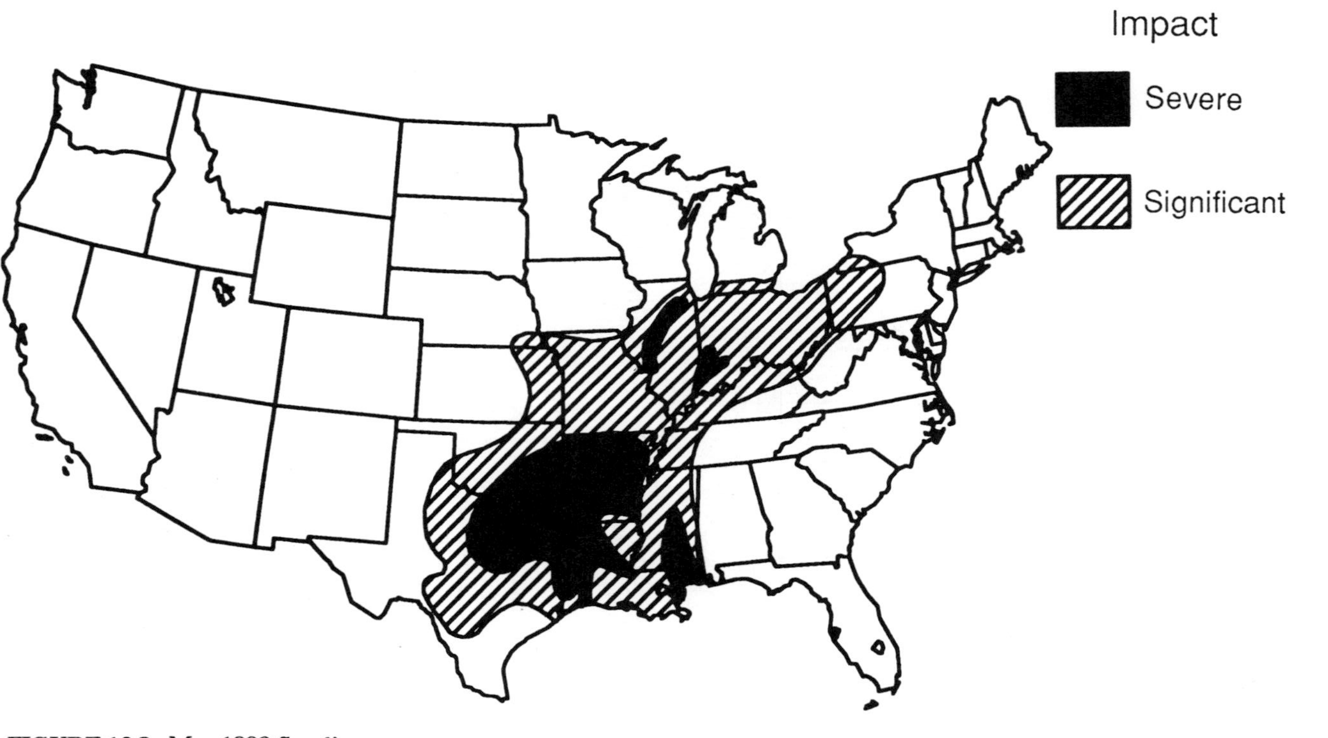

FIGURE 13.8 May 1990 flooding.

Models and Parameter Estimation

Existing hydrologic models do not have a very strong physical/mathematical basis for their structure. The models' parameters depend on the physical characteristics of basins and on the climate. The lack of physical/mathematical foundations for the models means that their parameter values cannot be estimated on the basis of physical properties or climate information.

Currently used model calibration procedures do not adequately consider the potential biases in precipitation and runoff measurements or potential evaporation estimates. In any large river system, model calibration is possible only over a small part of the total area—the headwater area where streamflow is unregulated and represents the natural runoff. Parameters estimated for these areas are then assumed to apply to other areas. If model parameters could be related to physical characteristics and climate, multiple basins could be calibrated simultaneously, and the effects of local measurement biases might be reduced.

GEWEX Continental-Scale International Project

Because water is an important factor in the redistribution of energy from the equator toward the poles and in the optical properties of the atmosphere that control the local energy balance, it is necessary to understand better the hydrologic cycle to improve the estimates of how increasing greenhouse gas levels will affect the climate. Accordingly, a major international project is being planned to acquire data (both historical and new) and to conduct research on the water and energy balance of the land and atmosphere over a large continental area. The area selected is the Mississippi River basin and as much of the rest of North America as it may be feasible or necessary to include in the study. Certain process studies may occur in other parts of the world as well.

There are some important relationships between the GEWEX Continental-Scale International Project (GCIP) and WARFS that need to be developed. For example, one of the specific goals of GCIP will be to account for the water balance of ungaged river basins. This is necessary to improve the way atmospheric models represent hydrologic processes globally. In most parts of the world there are insufficient data to calibrate the parameters of hydrologic models. The studies that GCIP will carry out would be useful to WARFS and may produce a new generation of hydrologic models and parameter estimation techniques that would be easier to use than the present ones. On the other hand, many of the studies and data preparation steps that will be important to GCIP will be needed by water management and forecast agencies. Therefore, GCIP must rely on these activities for partial support.

Long-Range Weather Forecasting Improvements

Although there may be better ways to use existing long-range weather forecast information for hydrologic forecasting, there may be still better ways to develop new long-range weather forecasts specifically for use by WARFS. Recent analysis of the El Niño and its relation to the climate system suggest that new climate forecast products could be developed. In the next few years, these are likely to be mostly statistically based forecasts. In the long run, general circulation models of the global atmosphere and oceans may add new information to the statistical analysis. Because there will always remain substantial uncertainty in long-range forecasts, careful attention must be given to both the statistical and the physical bases for such forecasts.

Orographic Precipitation Analysis and Precipitation Forecasting

In the western United States, most water comes from winter precipitation. This precipitation is spatially highly variable because of the mountains. Using relatively simple water and energy balance models of the atmosphere, the orographic component of precipitation can be estimated. The output from orographic precipitation models can be combined with limited surface measurements of precipitation at low elevations and with snow water content at higher elevations to make improved estimates of the spatial distribution of precipitation and snow water content. The procedures for doing this are analogous to the procedures that will be used in AWIPS to merge NEXRAD precipitation estimates with rain gage data to make optimal combined rainfall estimates.

The same orographic precipitation models useful for analyzing precipitation that has already occurred can potentially be used in conjunction with regional meteorological forecast models to make quantitative precipitation forecasts. Also, it may be possible to use observations of precipitation from the atmosphere to update the moisture content variables of atmospheric forecast models. The method for doing this would be analogous to the methods used in hydrologic forecasting to update soil moisture variables using observed streamflow data.

Data Base Improvements

Improvements are needed in both historical and operational data systems to make hydrologic analyses more efficient and more accurate. Historical data bases for all of the hydrometeorological information needed for WARFS are not economically accessible, and some critical information is available only in printed publications (such as historical observation times of daily precipitation data, which change from time to time at many stations). Fortunately, these same data are needed for the GCIP, and there is a NOAA initiative to improve access to NOAA's environmental data.

Operational data systems are being improved by many different agencies. The Geostationary Operational Environmental Satellite (GOES) data collection system is being used to improve real-time availability of data. Nevertheless, the random reporting capacity of existing data collection platforms and of the GOES satellite are not being used efficiently. This results in time delays of many hours before data are received by forecasters and water managers. This means that ground truth data that are vital to the NEXRAD precipitation estimation algorithms may not be available when they are needed most. Also, about 500 more automated precipitation gages are needed nationwide to support NEXRAD precipitation estimation.

Improvements in ESP

Although ESP software allows short-term quantitative forecasts of precipitation and temperature to be used, it has generally not been used in practice. Procedures for using short-term probabilistic precipitation forecasts in ESP are still being developed. These procedures must account for the temporal and spatial covariance of future precipitation patterns, as well as for the general magnitude or intensity of precipitation. In addition to designing procedures to incorporate short-term forecasts in ESP, guidelines for deciding historical weights of various years need to be developed. The procedure of assigning weights to historical year simulations has been used only in experimental applications of ESP.

One of the implications of climate change for ESP is that climate events far into the past may not be representative of the present. Therefore, the number of historical years of data that should be used in ESP simulations is limited, perhaps to only 20 or 30, depending on how fast climate changes. Also, it is essential to keep existing hydrometeorological data systems operating, even if there already is an extensive historical record. Continuing to operate existing data systems is necessary so that recent data may be used in ESP and so that model parameters (which are also a function of climate) may be updated. Yet, because of the current federal budget pressures, proposals to reduce data networks are being made to meet budget targets.

An alternative to using historical hydrometeorological data to drive ESP is to use stochastically generated data. This involves making assumptions about the hydrometeorological processes that are not necessary if historical data are used. On the other hand, if climate is changing and if data networks are reduced, the stochastic alternative may be needed. One advantage of the stochastic approach is that forecasters would have more control over the current climatic regimen and could consider not only the effects of long-term change but the current state of the atmosphere, oceans, and continents.

ACKNOWLEDGMENT

Most of the figures and examples and many of the ideas expressed in this paper are based on the work of Gerald N. Day, who is responsible for technical oversight of NWS ESP procedures and who is serving as the program manager for the NWS WARFS initiative. I gratefully acknowledge his contribution and advice.

REFERENCES

Brazil, L. E., and M. D. Hudlow. 1980. Calibration procedures used with the National Weather Service River Forecast System. Pp. 371-380 in Preprint Volume, IFAC Symposium on Water and Related Land Resources Systems, Case Institute of Technology, Case Western Reserve University, Cleveland, Ohio, May 28-31.

Day, G. N. 1985. Extended streamflow prediction using NWSRFS. American Society of Civil Engineers, Journal of Water Resources Planning and Management 111(2):157-170.

14

Some Aspects of Hydrologic Variability

Stephen J. Burges
University of Washington
Seattle, Washington

Much of the discussion at the colloquium reminded several members in the audience of the debate in the research community about the "Hurst phenomenon" which occupied some extremely thoughtful scientific hydrologists and water resource engineers for a 15-year period after Hurst's book summarizing his findings was published (Hurst et al., 1965). Possible explanations of the Hurst phenomenon based on threshold conditions are given in Klemes (1974) and Potter (1976).

The essence of Hurst's observations were that after examining numerous geophysical time series throughout the world (annual streamflow volumes, rainfall, lake varves, etc.), he determined that the degree of apparent persistence (long intervals of well below or well above "normal" trends) could be indexed to a coefficient "H" which we now know as the Hurst coefficient. A random process has H = 0.5 while one that can describe prolonged multiyear duration droughts is in the range of 0.7 to perhaps as large as 0.85. The time series associated with such "Hurst like" measure resemble remarkably the sequences now considered in "alternative scenarios" resulting from "climate change." This means that the tools of stochastic hydrology that were developed over a 20-year period starting in the mid-1960s might be useful for assessing water resource system reliability under variable climates. If the annual scale is considered, both the range of what could be experiences in any year (the marginal probability distribution), and the connectedness of low flow years, high flow years, etc. (the serial persistence or autocorrelation) can be represented by these tools. Shorter time increments can be examined by disaggregating annual quantities (e.g., Lane and Frevert, 1990).

Three figures are included to illustrate variability and operational consequences. Figure 14.1 shows a portion of a tree ring growth index time series from Dell, Montana (Lettenmaier and Burges, 1978). The time between A and B is 50 years, a typically data rich record length when determining system reliability. One form of variability is an apparent increasing trend within this 50-year period. When viewed in a larger time frame, it is clear that there is no trend but the series indicates different types of variability—extreme swings from low to high as well as an apparent increasing trend superposed on extreme swings. Tree-ring series are being used increasingly in regression equations as surrogates of streamflow that occurred before stream gauges were installed with the objective of providing an equivalent long historical record of "streamflow that might have occurred."

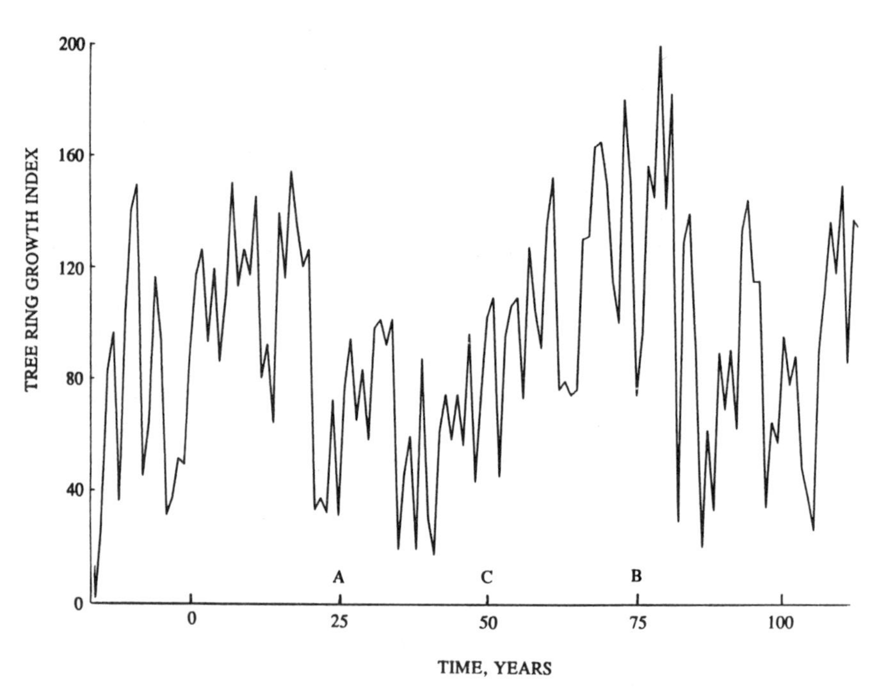

FIGURE 14.1 Annual tree ring index for Limber pine at Dell, Montana showing the range of variability that can be observed in a geophysical time series. The time period between A and B corresponds to 50 years. C corresponds approximately to the year 1810.

Figure 14.2 shows estimated natural annual flow for the Logan River, Utah (USGS Gauge 10-1090) for a 70-year period. The measure of long-term persistence, the Hurst coefficient, is approximately 0.8 and the lag-one correlation coefficient, a measure of shorter-term persistence, is 0.4. This is a highly persistent river flow situation with prolonged excursions from the mean level of both high and low flow with associated swings in flow volume from year to year. The variability about the mean level is related to the standard deviation of the annual flow volume.

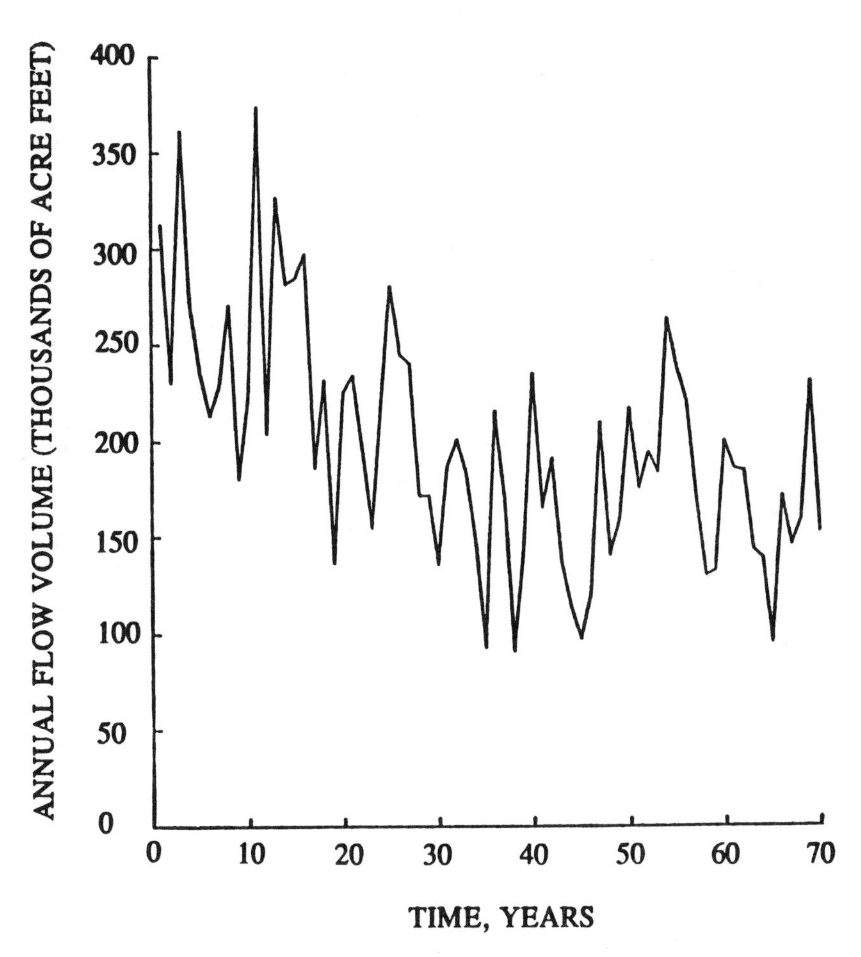

FIGURE 14.2 Annual flow time series, Logan River, Utah (USGS Gauge 10-1090), Hurst coefficient H = 0.80, lag-one correlation coefficient = 0.4, record length 70 years.

Figure 14.3 shows the estimated theoretical empirical storage-reliability diagram for a single reservoir having historical input annual streamflow persistence almost identical to the natural case shown in Figure 14.2. The annual coefficient of variation (standard deviation divided by the mean) for this hypothetical situation is 0.5 which is not overly variable. (Hydrologically benign rivers in Western Washington State have annual coefficients of variation between about 0.18 and about 0.25, depending on catchment size; Arroyo Seco, California has a coefficient of variation of approximately 0.75; many Australian rivers have annual coefficients of variation in excess of 1.0). Three contracted supply levels (physical as opposed to economic demand) are shown: 0.5, 0.7, and 0.9 of the mean annual flow volume. The cumulative probability distribution of needed storage size for a particular physical demand was determined by routing 1000 (independent) stochastically generated sequences, each having length 40 years through a mass curve analysis (sequent peak algorithm, Fiering, 1967) to yield 1000 values of storage. These derived quantities were ranked from high to low order and plotted in Figure 14.3 using an extreme value Type I distribution page. (An extreme value Type I cumulative probability distribution plots as a straight line on such a scale). Reliability is interpreted as follows. For the 98 percent level and a demand of 0.5 of the mean annual flow volume, if the reservoir capacity was approximately three times the annual flow volume, then in 98 percent of all 40-year long stochastic sequences that are all statistically equivalent to the indicated streamflow parameters (mean, variance, and persistence), demand would be satisfied fully. In the remaining 2 percent of sequences, there would be some short fall. Relevant methods for generating annual scale single site and multiple site stochastic or "synthetic" streamflow volumes are given by Salas et al. (1980), and Stedinger et al. (1985) among others.

To place Figure 14.3 in context, consider the capacities of Glen Canyon and Hoover Dams. Their combined capacities are approximately four times the mean annual natural flow of the river or approximately eight times the present flow volume at Lee's Ferry. (Actual Hurst coefficient estimates and annual coefficient of variation for the natural river flow volume are not readily available). Figure 14.3, which is not meant to represent conditions for the Colorado River system, shows clearly that as the physical demand level increases for the hypothetical threshold hydrologic condition modeled, the reliability for a given storage decreases markedly. The stochastic hydrology tools (the formal schemes for generating synthetic flow sequences) for making such assessments

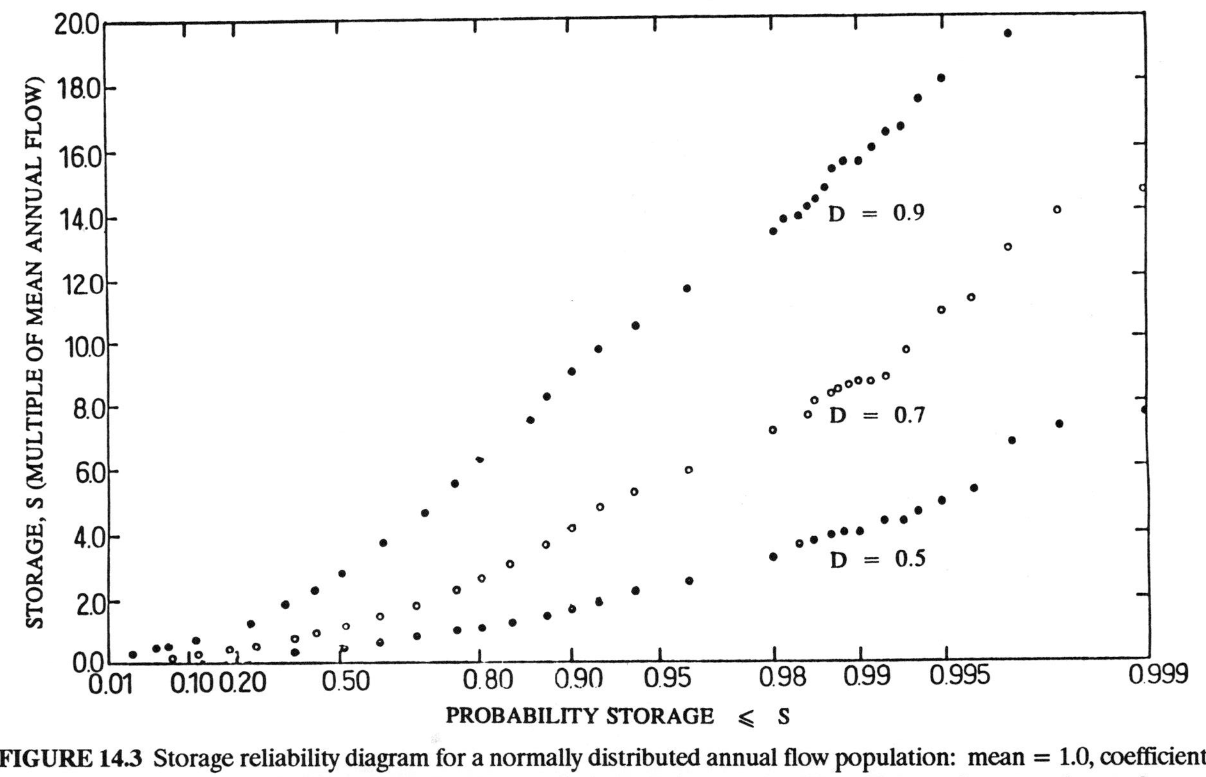

FIGURE 14.3 Storage reliability diagram for a normally distributed annual flow population: mean = 1.0, coefficient of variation = 0.5, Hurst coefficient = 0.8, lag-one correlation = 0.4; operation life = 40 years. Storage volumes, S, were determined using the sequent peak algorithm (Fiering, 1967) for three fixed annual physical "contracted supply" levels, D = 0.5, 0.7, and 0.9 of the mean annual flow volume.

are well tested and have been available for some time. Those tools could be extremely useful for considering both present and future variability.

REFERENCES

Burges, S. J., and D. P. Lettenmaier. 1975. Operational comparison of stochastic streamflow generation procedures. Technical Report 45, Harris Hydraulics Laboratory. Seattle: Department of Civil Engineering, University of Washington.

Fiering, M. B. 1967. Streamflow Synthesis. Cambridge, Mass.: Harvard University Press.

Hurst, H. E., R. P. Black, and Y. M. Simaika. 1965. Long-Term Storage—An Experimental Study. London: Constable.

Klemes, V. 1974. The Hurst phenomenon: a puzzle. Water Resources Res. 20(4):675-688.

Lane, W. L., and D. K. Frevert. 1990. Applied Stochastic Techniques (LAST Personal Computer Package Version 5.2), Users Manual. Denver: U.S. Bureau of Reclamation.

Lettenmaier, D. P., and S. J. Burges. 1978. Climate change: detection and its impact on hydrologic design. Water Resources Res. 14(4):679-687.

Potter, K. W. 1976. Evidence of nonstationarity as a physical explanation of the Hurst phenomenon. Water Resources Res. 12(5):1047-1052.

Salas, J. D., J. W. Delleur, V. Yevjevich, and W. L. Lane. 1980. Applied Modeling of Hydrologic Time Series. Littleton, Colorado: Water Resources Publications.

Stedinger, J. R., D. P. Lettenmaier, and R. M. Vogel. 1985. Multisite ARMA (1,1) and disaggregation models for annual streamflow generation. Water Resources Res. 13(2):497-509.

15

Management Responses to Climatic Variability

Gilbert F. White
University of Colorado
Boulder, Colorado

Yesterday, we reviewed the problem of certainty and uncertainty with respect to climate change from the standpoints of the scientific framework, the recent record, and the paleo record. We explored, in a tentative fashion, some of the implications of change for crop production, hydrologic characteristics of the water cycle, and management of large water systems. We moved on from that to consider economic and legal implications, and a number of other possible effects of climate change.

Today, we shift from the question of possible implications to the question of what might be done about it. As a kind of a bridge between these two considerations, we have the background given yesterday by the Commissioner of Reclamation, who spoke of how climate uncertainty looks to a major water agency in course of transition. We also had John Dracup's suggestion of a new kind of warning service now being prepared for the water community.

We might recall some of what is known from behavioral science about the nature of effective warnings. We know that to get across new information that will affect the manner in which people respond to a hazard warning, there are certain desirable characteristics. One is that the source of the warning be credible. I can look here and say, it's hard to think of a more credible organization to transmit information than the Water Science and Technology Board of the National Research Council. Certainly, it wouldn't be viewed as lacking in solidity or probity in its activities. One wouldn't expect that Bob Dickinson or Roger Revelle would lead us beyond the boundaries of solid scientific understanding. But in addition to the credible channel for the warning, we need a lucid message. That's what we were trying to elucidate yesterday—a lucid message of just what we know and what we don't know. Beyond that, we need some indication of the evidence on which people can make their own risk calculation—make their own judgment on the probability of a specified magnitude of effect. I think we had begun to get at that yesterday.

We know that people are much more likely to act intelligently if they are provided with information, not only about the costs and benefits or what the impact might be, but also about what the feasibility is of their doing something in response. If they are presented, for example, with a course of action that they don't think is practicable or that isn't within their reach, they are less likely to perceive it accurately, let alone to act on it, than if they feel there is something they can pick up and do.

This morning we consider what sorts of actions people here, or people we can influence, could exercise. Of course, the kind of action we exercise depends upon the social setting in which we are placed—upon the organization or society. I'm reminded of this by a joke that's making the rounds in Moscow these days. It came out of discussions of the catastrophe at the Aral Sea. The formulation runs as follows: In a new Soviet society, when you're confronted with a looming catastrophe you turn to the market and let the market decide what's an effective course of action. In the United States, in a capitalistic society, when you're confronted with a catastrophe the attitude is, "Don't worry, the government will bail you out." So, one's perception of the kind of framework in which we are operating has a considerable influence on how accurately we may perceive the problems we are confronting. Thus, I would suggest that there's another consideration in this whole matter of viewing climate change from a global standpoint.

As many of you know, there has been going on recently through the Intergovernmental Panel on Climate Change (IPCC) an analysis of the whole range of topics that we've been considering here—going far beyond water. The IPCC has now reported. (Bob Dickinson referred to some of the findings.) The IPCC's report was considered at the World Climate Conference in Geneva last month. There now have evolved several suggestions about steps that should be taken. The Western European nations are strong on moving to the acceptance in an international convention of limits on production of greenhouse gases in various forms. They are proposing a treaty that would be executed by the time of the United Nations (U.N.) Conference on Environment and Development scheduled to take place in Brazil in June 1992. The United States has been opposed to certain features of that proposed treaty on the grounds that they don't want to specify the maximum or minimum figures for limitation of effluents.

There's been a good deal of distress in the environmental community about the reluctance of the United States to enter into such agreements. But there's still another group—best perhaps represented by the "group of 77" (the developing countries)—that is taking a different position. I haven't heard how that's coming out in the U.N. General Assembly this week in New York. But the other position—as I understand it (and we'll all know better in the course of a few weeks when the debates are over)—is that if one looks at the environmental situation around the world in relation to development, it's all very well to think about limiting chlorofluorocarbons or carbon dioxide or methane, but if we are concerned about the welfare of the planet, much more important these days is the wise management of what appears to be a severely deteriorating set of resources of soil, water, and vegetation. The Western European nations are talking about limiting carbon dioxide, which will have an effect over the next 20 or 30 years. The argument of this group is that we need only look around to see that the planet is under severe pressure and its society is decaying. Take action now, and don't satisfy ourselves in thinking we're coping with it by simply limiting effluents. This question is being raised in connection with preparations for the U.N. conference.

You may ask, which of the actions that they may be proposing or appraising ought to be taken regardless of climate change? To what extent are they truly urgent requirements for our world society? I hope we shall have this in our minds as we consider the management alternatives. And then, what is it that has still further urgency because of climate change? It is with those questions that I introduce our four speakers.

16

Improvements in Agricultural Water Management

Dale Bucks
U.S. Department of Agriculture, Agricultural Research Service
Beltsville, Maryland

The agricultural community was always been at the mercy of water supplies. Now, with the prospect of larger variations in temperature and precipitation brought about by climate change, farmers are faced with even greater uncertainties about water supplies for crops, forage, and livestock. In the next 30 to 50 years, we could witness large-scale changes in agriculture resulting from the continued depletion of our natural resources—land and energy as well as water—and our heightened awareness that some agricultural practices are driving environmental change. This paper discusses the role of agriculture in improving water management under conditions of climate uncertainty.

THE PRESSURES ON AGRICULTURE

We have known for some time that there is not enough land, water, and fossil fuel to go around. Despite this knowledge, our adoption of conservation practices has been woefully inadequate. As the population continues to multiply (predictions are that the world population will reach 12 billion people in the next 30 to 35 years), the pressures on agriculture will increase.

Until now, inexpensive energy and high labor costs have driven the mechanization of agriculture. Thanks to the low cost of petroleum, food and fiber production more than doubled between 1950 and 1970. But the days of cheap fossil fuels are numbered and prices will skyrocket as supplies dwindle. The increasing scarcity of fossil fuels could have grave consequences for U.S. agriculture, which has become increasingly reliant on inexpensive petroleum products for fuel, pesticides, and fertilizers. The quality of life that we've come to expect here in the United States—the inexpensive food supplies—may not continue.

THE CONTRIBUTION OF AGRICULTURE TO ENVIRONMENTAL CHANGE

Since the beginning of the industrial revolution, the abundance of carbon dioxide, methane, chlorofluorocarbons, nitrous oxide, and other gases in the atmosphere has increased. The buildup of these gases may alter global temperatures, the worldwide distribution of precipitation, and the quantity and quality of our water resources—all of which impact the productivity of our crop land, range land, and forest land. Yet, agriculture is not only impacted by environmental change, it is also a cause of that change. A recent EPA report shows, for example, that agriculture contributed an estimated 26 percent to the increase in atmospheric gases in the 1980s; agriculture accounted for 13 of the 18 percent increase in atmospheric methane. Ruminant animals, rice paddies, fertilizers, cultivated natural soils, biomass burning, and land use conversion are all sources of agriculture's contribution to environmental change.

Though we know agriculture is contributing to the increase in atmospheric gases, our understanding of the links between agriculture and climate variability is still limited. Understanding how agriculture affects and is affected by environmental change requires improved understanding of basic hydrologic processes and improved water supply forecasting techniques.

THE NEED FOR NEW KNOWLEDGE

We need to improve our quantitative understanding of basic hydrologic processes: ground water recharge, snow accumulation, rainfall variations with elevation, evaporation, and how all of these processes influence streamflow. We also need to develop improved fundamental process models that link appropriate hydrologic processes to relative factors in environmental change.

We need improved predictions of future water supplies. We need to improve the accuracy of water supply forecast models through new developments in hydrologic process models and new technologies, such as remote sensing, geographic information systems, digital elevation techniques, and real-time data. We have some of the pieces of this puzzle, but we need to start using forecasting models in agriculture.

Atmospheric science has made great progress in the last 25 years in understanding atmospheric conditions and developing an improved capability to make predictions for very short time periods. But we need to improve our predictions of precipitation, temperature, and evaporation for a month or more in advance. Existing meteorological information could be used more often to provide statistical probabilities that certain events will occur—for example, to provide the probable beginning and duration of rainy seasons or the statistical likelihood that a drought will continue for a specific period. Such knowledge is useful in planning the introduction of drought-resistant cultivars, selecting the crop to be produced, introducing new crop rotation systems, and selecting improved agricultural practices. We also need to improve predictions of interannual or decadal variations in temperature, precipitation, and evaporation. Though such predictions have been made, accurate predictions of the onset, duration, cessation, or likelihood of recurrence of drought for site-specific locations are not available.

AGRICULTURAL STRATEGIES FOR COPING WITH ENVIRONMENTAL CHANGE

Crop production under environmental change is affected by two competing tendencies. As carbon dioxide levels go up, yield will more than likely increase. But the likely increase will be counterbalanced if precipitation declines, in which case we could find that production decreases or remains the same. The wise course for dealing with this uncertain future is to increase water use efficiency. More efficient water management technology is evolving at an accelerated pace, but implementing new technologies may require decades and will usually require additional investments in physical and supportive infrastructure. National water policies may in some cases hinder the implementation of new technologies.

Dryland Agriculture

The key to successful dryland farming is using systems and practices that can take better advantage of favorable years. Three components of water conservation for dryland production are retaining precipitation on the land, reducing evaporation loss, and using crops that best fit rainfall patterns. Retaining precipitation on the land starts with preventing runoff. For example, furrow diking, conservation bench terraces, and precision land leveling have proven effective. We need to increase the use of these methods. For reducing evaporation, the most promising practice is application of a mulch or residue cover. Crop residues are the only practical source of mulching, which makes crop rotations and predictions of precipitation and snow on a monthly or annual bases extremely important. Matching cropping systems with rainfall patterns—essential in dryland farming for ensuring a good probability of producing a harvestable crop—requires predictions of climate variability on an interannual or decadal basis. There is a potential to change our cropping patterns, but it is very difficult to make adjustments if we do not know the direction or magnitude of precipitation change.

Irrigated Agriculture

Irrigated agriculture consumes 80 percent of our fresh water supply. Although it accounts for only 11 percent of our total crop land (16 percent of total harvested crop land), it provides 51 percent of the nation's total crop value (though that value declines to 38 percent if you look at total marketable products sold). The point is that irrigated agriculture not only consumes a significant portion of our water supplies, it provides a major contribution to the nation's agricultural economy.

All irrigation systems could be operated more efficiently to help solve the problem of limited water supplies. Many individuals have the misconception that irrigation efficiency depends only on irrigation system design. However, an important aspect of improving water use efficiency is improving irrigation system management, operation, and maintenance. Automated irrigation scheduling can have a major impact on improving on-farm irrigation efficiency. Excessive seepage of irrigation water from canals constructed in permeable soil—a major cause of high water tables and saline soils in many irrigation areas—can be reduced by lining the canals. Finally, automation of delivery systems can sometimes provide a higher level of water control: many delivery systems encourage over irrigation because water is supplied for fixed periods or at fixed amounts, irrespective of seasonal variations in on-farm water needs.

In addition to increasing irrigation efficiency, another method for coping with limited water supplies is to intercept drainage and sewage effluent waters. The drainage or reclaimed water can be blended or used separately for irrigation or other purposes after it has been fully used for crop production.

Whatever the strategy chosen for adapting irrigated agriculture to limited water supplies, the strategy must be matched to the crop, management, and labor conditions.

Mixed Range Land and Crop Land

The interaction between rand land and crop land is regulated primarily by climate, available irrigation water, and management. Weather modification, water harvesting, snow harvesting, livestock management, water spreading, and other techniques are available for improving water-use efficiency of range and crop plants. However, environmental impacts of altering meteorological parameters such as precipitation have not been determined scientifically, and intensive, well-designed experimentation is still needed to evaluate properly the potential usefulness of weather modification techniques.

Many range lands are now invaded by brush and need vegetative modification to control runoff and improve infiltration. Successful brush control, reseeding, reforestation, and other range land improvement actions must be integrated into the total water supply management system.

THE NEED FOR CHANGES IN AGRICULTURAL INSTITUTIONS

The institutions involved in allocating and delivering water can influence the effectiveness of water use in irrigated agriculture. Some economists suggest that a change in the pricing scheme for water is needed so that water will be allocated more efficiently. Others argue that jumping immediately to increased water pricing would ignore the need for an institutional structure that will allow improved decisionmaking with respect to water use. Water policy and law, when properly employed and enforced, become not the single solution to improving water use for increased production but an essential ingredient of efficient and effective management.

Many authorities are recommending that greater attention should be focused on improving existing irrigation systems. Others suggest that irrigation systems should not be changed unless the economic benefits have been considered and increased education and training have been provided along with advanced technology. Sound education and technical assistance programs can reduce the gap between research and actual improved irrigation practices.

Some authorities suggest that increased public financial assistance is required before any significant shift toward water conservation and environmental protection can occur. On the other hand, technical assistance, interest-free or low-interest loans, or direct cost sharing can only be provided where benefits to the public sector are documented.

Generally, new and improved irrigation systems that integrate automated irrigation water management systems and water measurement need to be developed. Such systems can simplify or reduce the number of management decisions to be made by the water managers and irrigators. The result is improved efficiency of water application, reduced energy use, and environmental protection.

CONCLUSIONS

The next 30 to 50 years could bring large-scale changes in agriculture. Even without environmental changes, agriculture will be faced with the seemingly impossible challenge of providing for twice as many people—each person with increasing demands—as exist now. With a relatively fixed land base and water supply and with diminishing reserves of petroleum and mineral resources, our options for increasing agricultural production are severely limited. Uncertainties in future water supplies, precipitation, and other climatic variables could be the straw that breaks the agricultural camel's back.

No single technology can solve all of the water quantity and quality problems confronting agriculture. However, water conservation technologies, when properly selected and implemented, can improve water use and management of crop land, range land, and forest land. Improvements in water management are required at all levels of water use. These management improvements will require bold changes in institutions and organizations, water policy and law, farming systems, education and training programs, and research and development. We must all recognize the critical role of water resources in human life. Improvements in agricultural water management are required both to cope with environmental change and to ensure environmental protection.

17

Creative Water Management

Daniel Sheer
Water Resources Management, Inc.
Columbia, Maryland

Let me begin by saying how pleased I am to be here and to share the same podium with Gilbert White. I believe you'll find that most of what I am about to present has already been stated by Gilbert White in one form or another.

At the outset, I want to make clear my support for global climate change research and my belief in the importance of this topic. However, I think that for water management—and only water management, not the broader issues—global climate change is not an issue. This is for three reasons:

1. Uncertainty. In my opinion, there is simply not enough certainty in any of the climate change information we have at this time to make any quantitative use of it in water management.
2. Rate of Change. Global climate changes are occurring much more slowly than weather changes due to normal climate variability. Therefore, if we are able to respond adequately to changes produced by normal climate variability, we should be able to cope with global climate change.
3. Other Variables. Other variables that affect the need for water management are changing much more rapidly than the climate. These variables will impact water demands regardless of how global climate change affects water supply. Therefore, let's address these other variables. Why worry about the tail—let's worry about the dog first.

The topic of my talk is creative water management as a response to climate variability. I'll begin by answering the question "What is water management?"

WHAT IS WATER MANAGEMENT?

In its simplest terms, managing water is no more than exercising whatever control measures are available to direct the utilization of water. Management implies an objective. Typically, the objective of water management is to achieve "the best possible mix of benefits" from the resource, which sounds quite reasonable and simple. So if defining water management is so simple, why is managing water so difficult?

The initial difficulty actually involves the end of the process—defining the objective "best possible mix of benefits." I know of no place in the water resources field where there exists general agreement as to what constitutes the best possible mix of benefits. In fact, it is extremely rare to find a comprehensive listing of all the benefits that might be achieved in a particular water management program. At Water Resources Management, whenever we begin a new project, the first item of business is to provide the client with a comprehensive list of exactly what we are trying to achieve. We find this list extremely valuable. Developing the list of objectives is not easy—it may take several months—but it is very worthwhile. An attempt to make a comprehensive list of benefits here would be impractical. Instead, I will list a few that are critical.

OBJECTIVES OF WATER MANAGEMENT

First of all, it should be recognized that most existing water law is aimed at preventing individuals from resorting to violence to resolve water disputes. Water law has nothing to do with economics or preserving people's property rights. It is my opinion that in relatively humid regions of England and the eastern United States, where water is plentiful, the riparian doctrine survives because water disputes are few. Further, disputes that do occur are rarely life-and-death situations; society can afford to allow the courts to take their time resolving fairly broad social issues on a case-by-case basis. On the other hand, in the water-short West the appropriative doctrine was developed because it could provide the administrative simplicity required to resolve quickly a relatively large number of life-and-death disputes over water.

A second kind of objective is those that are deeply rooted in our culture. As an example, a few years ago I was in Calcutta speaking to water suppliers. I was listening to a water manager who was complaining bitterly of the destruction of standpipe valves that had been installed to eliminate waste caused by freely running water. Apparently the people in the poorer sections of Calcutta were deliberately destroying the valves. When I asked him why, as this made no sense to me, he responded that freely running water had an important religious significance for Hindus. I pointed out to him that most of the homes in Calcutta had open water reservoirs in their basements, and that these were fed by pipes without valves. The overflow went into the streets and sewers. There were many more of these pipe connections than there were free-flowing standpipes in the poorer sections. It seemed to me that water metering, or a regulation requiring float valves in toilets, would save a good deal more water and energy costs (his actual objective) than would installing valves on a few standpipes. He responded that meters were too expensive and that some of the valves would break. And besides, the people living in these homes had a deep conviction that they had the right to all the water they wanted. How interesting—another religious objective, perhaps an expansion of the one he had already mentioned.

The only other religious objectives I have ever run across involved instream water rights for Native Americans. (You might also add to the religious objectives category the goals of a curious band of religious zealots who hold that Adam Smith is the only true prophet!) Nonetheless, I have noticed that the legacy of Frederick Law Olmstead has survived right here in Phoenix—a whole lot of lawn. The point is that many management objectives for water can have cultural or societal roots.

A third objective is an often forgotten one. Ask the typical water user, "What's the worst thing that can happen when there's no water at the tap?" and almost invariably the answer is, "I won't be able to flush my toilet, I'll get sick, and it's going to smell." While that may be accurate, there's a more important consequence that I worry about: if there's no pressure at the tap, there's no water at the fire plug, and if there's no water at the fire plug what happens when there's a fire? That question has some serious social consequences. We forget that fire protection was the real reason for developing of many of the early water systems in this country, and we particularly forget to include this factor when we look at the impact of climatic variation and drought. I get very upset when people talk about having shortages in urban water supply systems that they do not address fire protection in their future planning. When these shortages do occur, there will be no fire protection, and that's serious business, even over the short term.

Perhaps the greatest difficulty in developing objectives for water management lies not in the lack of consensus, but in the that there exists no effective forum for resolving our differences. Most resource allocations take place in the market, while allocations of publicly held resources such as public lands or taxes take place in the political arena. Although it may be difficult to establish a water market, it is not impossible and in fact is highly desirable. Nonetheless, I don't believe that water marketing has been tried in anything but very limited circumstances. And, with the exception of "water quality" legislation (which does, in some ways, address the allocation of water resources), political bodies have usually stayed way from the individual allocation of water rights, leaving these issues to the courts and the existing structure for resolving disputes I mentioned earlier. But enough about objectives.

DIFFICULTIES IN CONTROLLING WATER

Difficulties in control also make water management particularly problematic for the following reasons:

1. Because of its weight and volume, the large quantity of water required by our society is both difficult and expensive to control.
2. Although water is a renewable resource, unlike wood or grain it arrives in extremely variable quantities with little or no notice by processes that effectively cannot be controlled or even influenced by man.
3. Water is an elusive substance—it won't stay in one place. It evaporates, runs downhill by itself, runs uphill to money (but that's management). This "elusiveness" creates all kinds of what economists call "externalities." Any water management action directly, and often tangibly, affects either "downhill" users or those lacking money. I believe that the appropriative doctrine has succeeded admirably in preventing people in the West from killing each other over water. This has led to a tremendous increase in the number of people downhill of proposed water management changes in the West. The increase in population, coupled with the technical difficulty of quantifying the economic externalities, has destroyed the administrative simplicity that was the driving force behind the appropriative doctrine in the first place. And that, in my opinion, is why western water lawyers do so well.
4. Finally, because it is so bulky, heavy, unpredictable and elusive, physical control of water is often best accomplished by

coordinated, cooperative actions of separate and distinct groups of people. It is my observation that coordinated actions by separate and distinct groups of people run counter to human nature. Therefore, effective water management is not only difficult, it is by its very definition unnatural. I find further support for this concept that effective water management is unnatural in Newton's Third Law of Thermodynamics, which says that entropy is increasing—but that is another matter altogether.

MANAGEMENT MEASURES

Now that I've shown that water management is unnatural and difficult, what can be done to manage water and how does this relate to climate variability and climate change? Actions to manage water can be placed into four categories: structural measures, operational measures, allocations measures, and measures that increase the efficiency of use.

Structural measures are the easiest to define and discuss. Dams, canals, pipes, wells, water treatment plants, desalting plants, and hydropower facilities are all examples of structural measures used to manage water. They all control the movement and the elusiveness of water. Because we have no control without structures, these structures—limited though they may be—become very important. We've built a lot of structures. We need to build more.

But structures themselves do not respond to climate variability or climate change—operational measures do. If you want to know what you need to build in the water business, you should know exactly how you intend to operate it. Yet, most of the time little attention is paid to operations relative to the cost of construction.

The basic operating questions in water management today are, "How does one best defy gravity?" and, "Where and how much water should be left in storage at any given time?" Excellent operating policies for individual facilities are usually self-evident. But good joint operating policies for multiple facilities must be based on estimates of climate variability, and these are often quite uncertain. Moreover, implementation of joint operating policies requires coordination among groups. Still, the benefits of good coordinated operations of municipal supplies can be enormous. For example, joint operations of four large reservoirs along the Kansas River increased the reliable supplies by as much as an additional large reservoir. This additional reservoir would have cost a billion dollars. Clearly, operational measures can be critical.

Allocation policies and laws are enormously important in coping with climate variability. We need to develop allocation policies that address explicitly the temporal variability of water supply. I disagree strongly with the remark made yesterday that "We need to learn how to waste water wisely." Rather, we need to think about how we can best allocate whatever water is available at any given time—very different from wasting water wisely. To allocate rationally, we must expand our objectives beyond those that prevent people from killing each other over water disputes. There's one way of rationalizing water allocation without addressing the problem directly and that is to develop a water market. For many objectives a market is an effective means of allocation, although for other objectives it is not.

To have a water market, two conditions must be satisfied. First, the seller must be able to transfer the goods to the buyer. Second, the cost of the transaction cannot be prohibitive. Under eastern water law, establishing a water market is impossible, because you cannot market the goods—you don't own them. Under western water law, establishing water markets is not quite—but almost—impossible. Where no externalities (such as transfers between similar users in nearby locations) are involved, transfers can occur. But where externalities are involved, as in the majority of cases, western water law makes it very difficult to determine just what the seller can sell. Appeals to the court make the transfer costs enormous. To develop a water market in the West, water law must be modified to provide for an administrative procedure, not a court procedure, for defining which externalities are allowable and which are not. Systems analysis tools could be used for making decisions about externalities on a real-time administrative basis. In some cases, provisions for the sale also must be made.

Of course, many allocations shouldn't be left to the market. Examples of nonmarket types of allocations include allocations measures for water quality, instream flow, social welfare, and fire fighting.

Finally, creative water management involves devising ways to promote efficiency of use. Perhaps creative is not the right word—effective and nontraditional might be better terms. The answer to the question of how to be creative is short and sweet. If you want to be creative, provide a method to review and obtain the benefits of joint operations, with demands dependent on water availability. My experience is that, in the long run, people will thank you for it.

18

Weather Modification as a Response to Variations in Weather and Climate

Wayne N. Marchant and Arnett S. Dennis
U.S. Bureau of Reclamation
Denver, Colorado

Global warming due to increased concentrations of greenhouse gases, if it occurs, will be the most significant change in weather and climate ever produced by human beings. One can speculate about the possibility of offsetting this inadvertent, long-term weather modification by deliberately modifying the weather through cloud seeding. Because some clouds warm the earth's surface and some cool it (depending upon cloud composition and the temperature at the top of the cloud), it is theoretically possible to modify clouds to offset global warming, at least partially. However, the present knowledge base is not complete enough to provide a working hypothesis.

The use of weather modification to offset some of the undesirable changes in precipitation, temperature, and streamflow that may arise from global warming or other climatic variations will be limited for the foreseeable future. At present, the primary application of weather modification is the augmentation of precipitation and runoff. The Bureau of Reclamation has been conducting research on precipitation augmentation for the past 30 years, with the efforts in recent years concentrated on the seeding of winter orographic clouds to increase snowpack and runoff.

THE SEEDING OF WINTER OROGRAPHIC CLOUDS

In the face of water demands that often outstrip supplies, western water managers have been using weather modification to augment water supplies each year since 1948. The most commonly used method is silver iodide (AgI) seeding of winter orographic clouds to increase precipitation, snowpack, and runoff.

Orographic clouds containing supercooled liquid water (SLW) (liquid water at temperatures below 0°C) offer a unique opportunity for increasing precipitation. They may persist for hours at a time over a mountain range producing little or no precipitation, apparently because they do not contain enough natural ice crystals to form snow efficiently. The introduction of AgI crystals or other artificial ice nuclei into such clouds results in the formation of ice crystals, which grow into snowflakes at the expense of the SLW. The lifting of the air on the upwind side of the mountain barrier suggests that orographic clouds can be seeded cheaply and efficiently by AgI generators on the ground.

A typical winter orographic cloud seeding project involves the deployment of 10 to 20 manually operated AgI generators on the ground upwind to seed a target area of several thousand square kilometers. Some operators use radio-controlled generators to improve coverage of remote areas. Aircraft seeding is conducted on some projects to supplement the effects of the ground-based generators or to reach otherwise inaccessible areas.

FACTORS INFLUENCING THE ACCEPTANCE OF CLOUD SEEDING

The initial acceptance of weather modification technology was due to a perception that large economic benefits would be obtained from it. Subsequent analyses have confirmed that the economic impact of a successful weather modification program could be very large. A Bureau of Reclamation analysis of a hypothetical 10 percent increase in Colorado River streamflow showed average annual benefits of $48 million from increased water supplies for irrigation and for municipal and industrial use; $34 million from increased hydropower generation; and $62 million from improved water quality (chiefly from a reduction in salinity). As a program to seed all the major runoff-producing subbasins of the Colorado could be mounted for some $10 million per year, the potential benefit-cost ratio is large.

One point of concern in considering economic impacts of weather modification is that a weather modification program may harm people receiving few or none of the perceived benefits. The proposed program to increase snowpack in the Colorado River basin is one such case. The program, if implemented, would involve inconvenience and additional costs for snow removal for residents of the high-elevation regions within the basin, notably in southwestern Colorado, while the perceived benefits would accrue principally to the lower basin states of California, Arizona, and Nevada.

The possibility of undesirable ecological effects is another retarding factor for the adoption of weather modification technology. People have expressed concerns about the seeding agents themselves and about the expected effects of weather modification programs. However, extensive studies over the last 30 years have shown that the ecological effects are very small and would be difficult to detect.

STATISTICAL EVALUATIONS

In view of the large stakes involved, it was inevitable that, after several years of operational cloud seeding, the sponsors would want an evaluation of the effects. One of the most comprehensive evaluations was conducted by the Advisory Committee on Weather Control, which was established by the United States government in 1953. The advisory committee's statistical analysis of orographic cloud seeding projects was based on comparisons of precipitation in target and control areas during seeded storms and during storms that had occurred before any seeding was done. On the basis of a study of 299 seeded storms, which occurred over several seasons in different projects in the western United States, in 1957 the advisory committee reported apparent increases in precipitation of 9 to 17 percent above the amounts that would have fallen without seeding. The advisory committee also stated that there was only a very small probability that the apparent increases were due to chance.

Although the advisory committee's report was criticized on various grounds, including the possibility that natural changes may have accounted for some of the precipitation increases, its findings have been supported by later review panels. The latest policy statement of the American Meteorological Society on weather modification states: "Precipitation amounts from certain cold orographic cloud systems apparently can be increased with existing technology in the western United States. Increases of the order of 10 percent in seasonal precipitation are indicated in some areas."

It is difficult to translate indications of precipitation increases at individual gauges into estimates of additional runoff from an entire watershed. However, target-control analyses can be performed with runoff data as well as precipitation data. Several authors have found evidence of increases—generally ranging from 5 to 15 percent—in seasonal runoff from mountainous basins in the western United States.

THE INFLUENCE OF PRECIPITATION VARIABILITY

Commercial cloud seeders try to tailor their operations to the needs of their sponsors. Projects to enhance precipitation are most likely to be undertaken following dry periods, when water supplies are low. The most favorable seeding opportunities in orographic clouds are the shallow clouds, rather than the deep clouds associated with major storms. Nevertheless, suspension criteria are often used in an attempt to avoid contributing to floods, avalanches, or other hazards associated with excessive precipitation. The general tendency, then, is for cloud seeding as commonly practiced to increase precipitation slightly during dry years and normal years, but to have little or no impact during wet years. Therefore, it tends to reduce the variability of precipitation and runoff.

THE SEARCH FOR LARGE-SCALE EFFECTS

The operational projects mentioned above were concerned only with increasing precipitation from relatively small portions of extensive cloud systems. Many people have postulated that such projects may affect precipitation over larger areas. One commonly heard argument is that any additional precipitation in a target area must result in a decrease in precipitation downwind. This argument is flawed, principally because precipitation rates are controlled primarily by vertical motions in the atmosphere, with the moisture content of the affected air mass being of secondary importance. Furthermore, the condensate (the moisture present as clouds at any time) is only a small fraction of the total atmospheric water supply. Only a fraction of the total condensate is precipitated over a mountain barrier, and the increases associated with cloud seeding are much smaller than the natural precipitation. Therefore, drying of the atmosphere downwind of a target area must be slight. Searches for downwind effects using actual precipitation data have not turned up any statistical evidence for them.

RESEARCH ON WINTER OROGRAPHIC CLOUDS

A number of field experiments have been carried out to refine estimates of the effects of cloud seeding on orographic precipitation and to determine which clouds are seedable (that is, which clouds can be made to yield additional precipitation by seeding). Experts agree that seedability is associated with the presence of SLW. In recent years, investigators have searched for SLW with new instruments, including dual-channel microwave radiometers and aircraft equipped with laser-based optical probes for counting, sizing, and classifying cloud and precipitation particles. The research on winter orographic cloud seeding has also involved randomized seeding trials and the use of computer programs to compute the trajectories of snowflakes produced by both natural and artificial ice nuclei.

Present indications are that SLW is most likely to be associated either with shallow stratiform clouds with top temperatures in the -5 to -12°C range or with moderate updrafts within deeper cloud layers. Surprisingly high ice crystal concentrations are observed on occasion—for example, 50 per liter at the -5°C level, which greatly exceeds the concentration of natural ice nuclei active at that temperature. Unusually high ice crystal concentrations are often associated with large cloud droplets and are attributed to a variety of secondary ice crystal production mechanisms. High ice crystal concentrations are important because they suppress the concentration of SLW and thereby reduce the seedability of a cloud.

Results of the various experiments and operational programs suggest that a variety of outcomes is possible and that tailoring a seeding operation to a particular project area requires careful consideration of wind fields, generator siting, seeding agent diffusion and transport, and cloud microphysical characteristics (including the presence or absence of secondary ice formation).

THE AVAILABILITY OF SEEDABLE CLOUDS IN DRY PERIODS

If climate variability were to lead to drier conditions for many years in a given watershed, would there be enough seedable clouds left to offer a chance for practical results?

Research results are of some value in answering this question. As noted above, the most seedable clouds are shallow clouds associated with weak storms rather than the deep clouds associated with the central pans of intense storms. Therefore, one can expect to find seedable clouds in some areas that are normally dry. Experimental confirmation for this expectation is provided by apparently successful cloud seeding projects in such arid areas as San Diego County, California and northern Israel.

Radiometer data from Arizona show that seedable conditions occur several times each winter and sometimes last for 20 or 30 hours at a stretch, with only intermittent, light snow falling during that period. Data from the Bureau of Reclamation's Sierra Cooperative Pilot Project (SCPP), which was conducted in the Sierra Nevada from 1976 to 1987, showed that seedable clouds existed for several hundred hours each winter and, if anything, were more frequent during dry years than during wet years. This finding is further evidence that a drier climate would not necessarily lead to a decrease in the number of cloud seeding opportunities.

CONCLUSION

Unavoidably, we will continue to live with climate variability and the uncertainty it produces. Furthermore, that uncertainty is compounded by some predictions of profound, long-term global climate change.

It would be naive to suggest that cloud seeding can eliminate climate variability or overcome uncertainty about climate. Indeed, uncertainty about its effectiveness continues to hinder cloud seeding itself. Nevertheless, precipitation augmentation through cloud seeding has enormous potential to help us moderate the adverse societal and economic costs of wide fluctuations in weather and climate. Given the great potential that our models have shown may be realized through cloud seeding in at least one major basin, it is a tool too attractive to ignore.

REFERENCES

Advisory Committee on Weather Control. 1957. Final Report of the Advisory Committee on Weather Control. Washington, D.C.: U.S. Government Printing Office.

Braham, R. R., Jr., ed. 1986. Precipitation Enhancement: A Scientific Challenge. Meteor. Monographs, Vol 21, No. 43. Boston: American Meteorological Society.

Dennis, A. S. 1980. Weather Modification by Cloud Seeding. International Geophysics Series, Vol. 24. New York: Academic Press.

Hess, W. N., ed. 1974. Weather and Climate Modification. New York: John Wiley & Sons.

19

Managing Water Supply Variability: The Salt River Project

John Keane
Surface Water Resources
Phoenix, Arizona

In a state generally cursed with a lack of surface water supplies, the Salt and Verde river watersheds play a pivotal role. These rivers provide much of the water supply for the greater Phoenix area. Their watersheds stretch in a dividing band across the state (Figure 19.1). To the north lies the drier, relatively flat Colorado Plateau. To the south and west lies the Sonoran Desert, much of which is a series of very broad alluvial valleys separated by metamorphic mountain ranges. The Salt and the Verde watersheds have played an enormously important role in defining Arizona and shaping its growth. These highlands are an orographic barrier to winter storms from the Pacific, forcing cool, moist air to rise and to leave a substantial amount of moisture behind. The highlands also serve as the point of origin for many of the summer thunderstorms that develop in July, August, and September, when moisture flows up from Baja, California and the Gulf of Mexico. The slightly cooler and wetter conditions of the highlands are responsible for much of Arizona's forest land, streamflow, and riparian habitat and provide a place to escape the summer heat of the desert below.

BACKGROUND

The Salt and Verde watersheds have very dry spring and fall periods, separated by a winter season (lasting from December through mid-March) of occasional storms and a summer of scattered thunderstorms. The summer rains do not provide much of the year's streamflow volume because the rains are absorbed by soils left extremely dry by the endless sunshine and very low

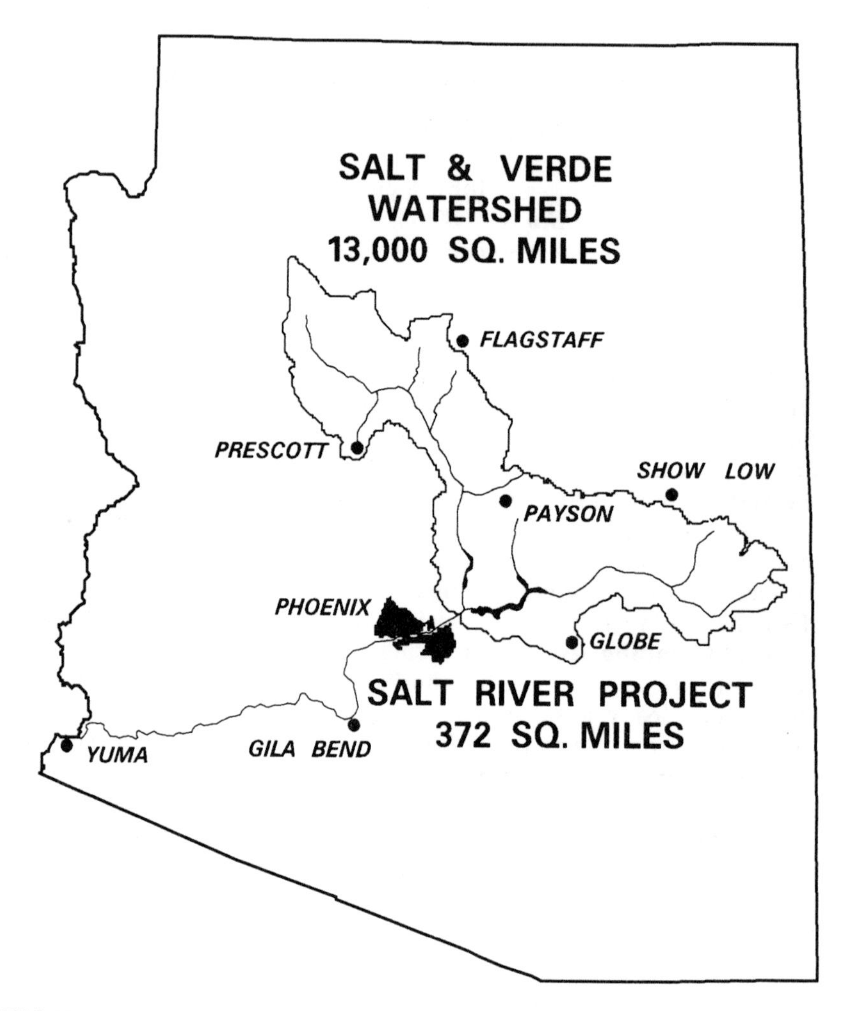

FIGURE 19.1 The watersheds of the Salt and Verde rivers stretch in a dividing band across Arizona.

relative humidity of the preceding May and June. River flows do increase slightly during the summer monsoon season, but all save the largest tributaries go dry again during the fall. Winter precipitation is the dominant force in determining the volume of annual streamflow (Figure 19.2). Some of the increased river flow during the winter and spring comes from rainfall runoff in the area below about 6,500 feet. An even larger portion comes from late winter and early spring snowmelt from the areas above this elevation. In either case, enough rain or snowmelt must be absorbed by the soil to satisfy the considerable soil moisture deficit built up during late summer, fall, and early winter before river flows begin to increase noticeably. The first several storms of winter disappear into the soil without affecting streams at all.

The Salt River valley has a remarkable combination of traits: flat, well-drained soil, available surface water and a long growing season. These characteristics attracted Native American agriculturalists for hundreds of years before Anglo settlement (Figure 19.3), but the Native Americans abandoned the land along the Salt River long before Anglo settlers arrived (although they were still occupying land along the nearby Gila River). There is evidence that water supply variability in the form of both floods and droughts played a role in the Native American abandonment of the Phoenix area. Nonnative settlers began rejuvenating the old native canals in the late 1860s, and then expanded them and built new ones. Over the next 40 years, the same extreme water supply variability that is the focus of this conference plagued the farmers in the Salt River valley. These farmers, however, showed the determination and the tireless skill as promoters that have been the hallmark of the area up until very recently. They, like so much of the West before and after, promoted a federal solution to this supply variability. They were instrumental in the passage of the Reclamation Act of 1902; they succeeded in having Roosevelt Dam, begun in 1904 and completed in 1912, approved as the Bureau of Reclamation's first big project. Other dams along the Salt and the Verde followed over the years and, unlike many other Bureau of Reclamation projects around the West, these were paid off in full in 1955. There is no outstanding debt on the extensive system of dams and canals.

The Salt River Project (SRP) dams store water for the lands within the project boundaries—lands originally judged by a 1910 court decree to have established rights to the Salt and Verde rivers. Over the years, however, growth in the Phoenix area has expanded far beyond the lands having these rights (Figure 19.4). The lands outside the project area could be settled only by using the limited supply of the Agua Fria River in the West Valley or by using deep well pumps to draw water from the enormous alluvial aquifer underneath the valley.

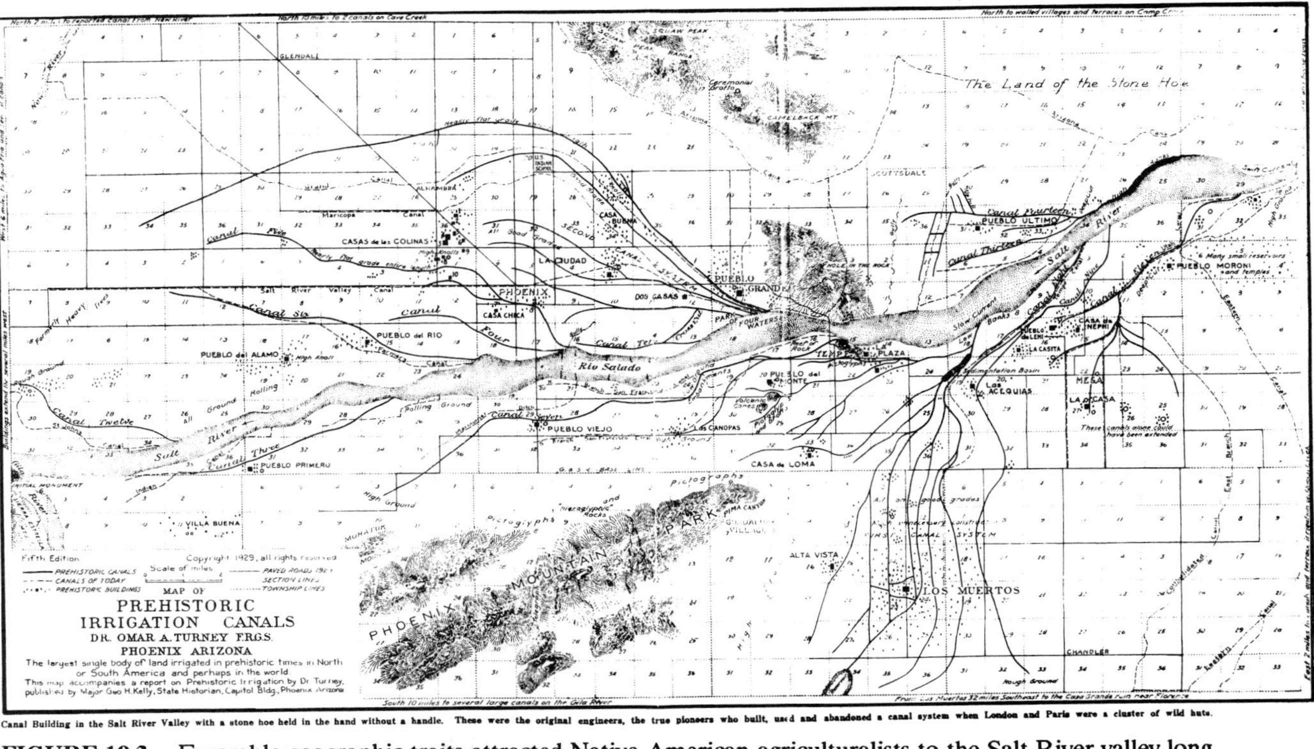

Canal Building in the Salt River Valley with a stone hoe held in the hand without a handle. These were the original engineers, the true pioneers who built, used and abandoned a canal system when London and Paris were a cluster of wild huts.

FIGURE 19.3 Favorable geographic traits attracted Native American agriculturalists to the Salt River valley long before Anglo settlers arrived. This map shows prehistoric irrigation canals in the valley, thought to be the largest single body of irrigated land in prehistoric times.

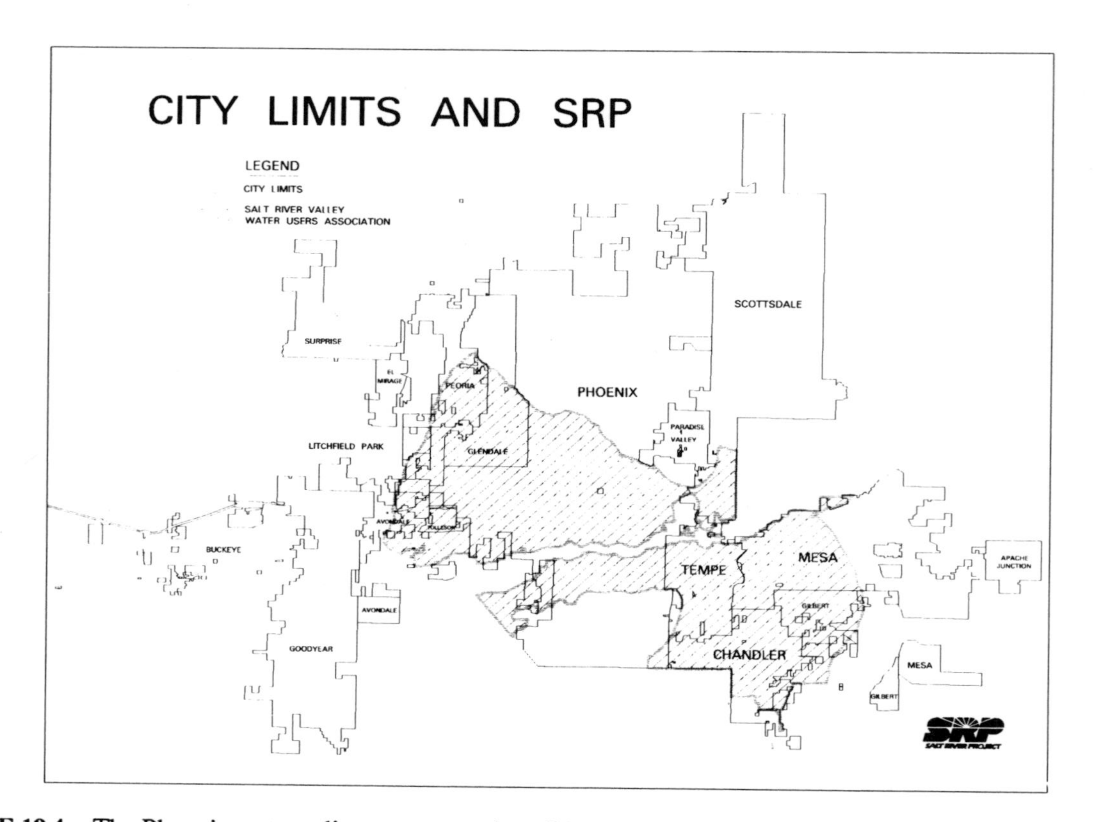

FIGURE 19.4 The Phoenix metropolitan area extends well beyond the lands holding water rights for the Salt and Verde rivers.

Dams

The SRP water supply system today has four dams along the Salt River and two on the Verde (Figure 19.5). All four dams on the Salt have hydroelectric capabilities while neither dam on the Verde does. The characteristics of the dams important for this discussion are:

1. Storage capacity for the system is not equally distributed between the Salt and Verde rivers. Most of the storage volume (nearly 85 percent) is on the Salt River. Total storage space on the Verde River is approximately equal to the Verde's annual flow, while storage space on the Salt is roughly three times the annual flow.
2. The total storage capacity of the entire system is less than double the average flow of the Salt and the Verde rivers combined. This is also about twice the annual demand placed on the system—a small storage capacity compared to the giant storage reservoirs on the nearby Colorado River.
3. The dams were built solely for water conservation. Thus, this system of dams is not truly a multipurpose reservoir system. The dams provide some incidental flood control benefits and the lakes behind them are heavily used for recreation, but in their origin, design, and legal obligations the dams are single-purpose structures.

Ground Water

Ground water often has been used to supplement surface water deliveries. Ground water provides anywhere from 5 percent (during times of ample surface water supplies) to 35 percent (during periodic droughts) of the million acre-feet per year of water demands on the Salt River Project. (Most of this demand is now urban rather than agricultural.)

SUPPLY VARIABILITY

The extraordinary variability of Salt River flows bedeviled Native American societies and drove farmers to seek federally-funded storage projects. Just how variable were these flows? Just how variable are they now? Tree-ring records show great variability in prehistoric times. The recent historical record gives an indication of what drove farmers to seek regulation of the Salt River (Figure 19.6).

The historical record begins with three wet years. The Salt and Verde together in 1889 produced 1.7 million acre-feet of water.

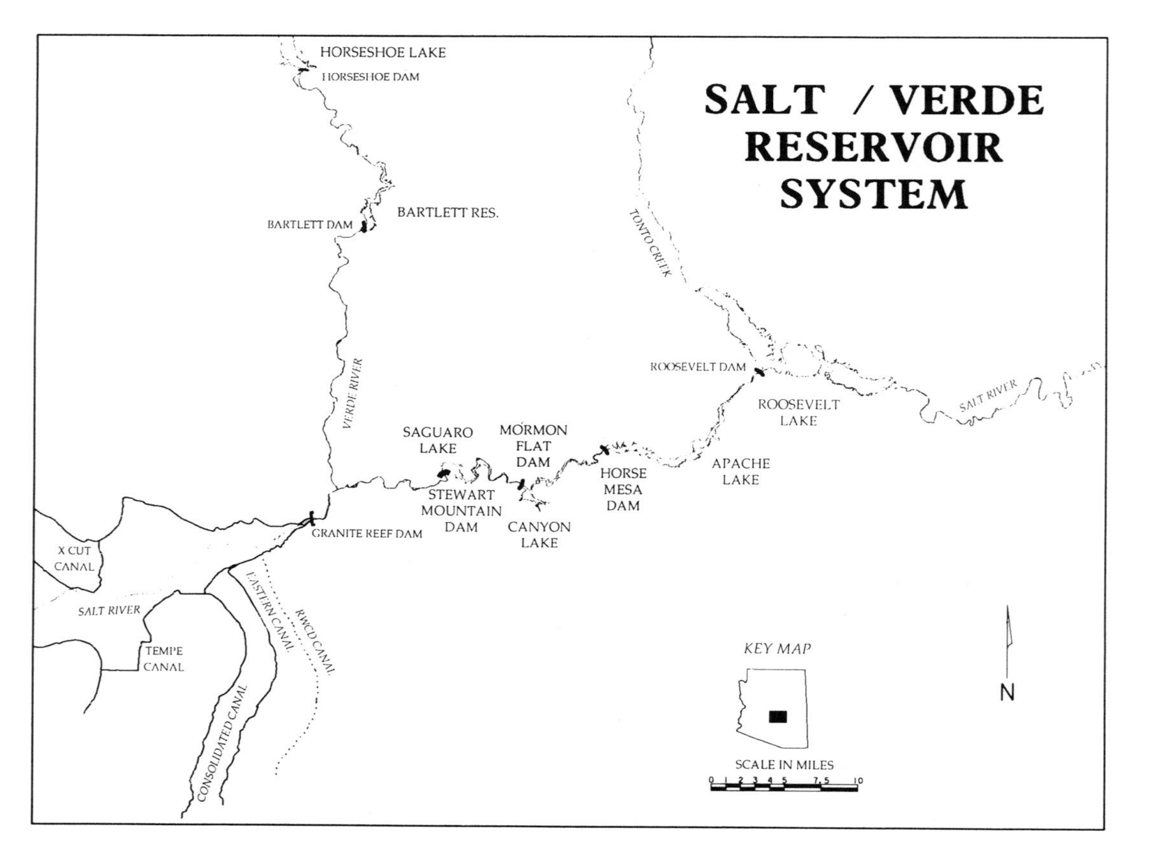

FIGURE 19.5 The Salt River Project manages four dams along the Salt River and two along the Verde River.

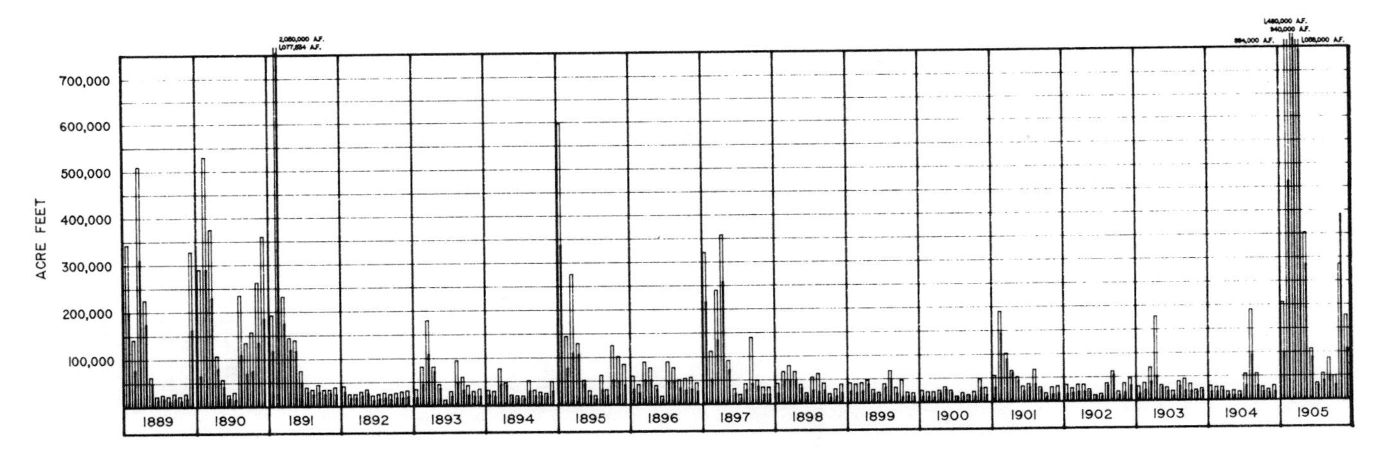

FIGURE 19.6 This historic record of streamflow on the Salt River shows the extreme variability of the flows.

In 1890, 2.6 million acre-feet of water flowed by, and in 1891 the figure rose to more than 3 million acre-feet. In that year, everything crossing the river was washed away; a flood several miles across inundated downtown Phoenix and a great deal of farm land. The next year, 1892, one of the driest years ever recorded was producing river flows of 299 thousand acre-feet. River flows remained well below average for several years thereafter. Several ups and downs of wet and dry years followed, and then a drought began in 1898 that lasted for 7 years. This drought remains the worst on record. Construction of Roosevelt Dam began in late 1904 and, as if on cue, in 1905, over 5 million acre-feet flowed down the rivers and washed out the early stages of work on the new dam. It remains the wettest year in the historical sequence.

The wild gyrations of weather at the turn of the century are not just a thing of the past; they have continued up to the present time even though their impact has been softened by storage reservoirs and ground water pumping (Figure 19.7). The Salt and the Verde watersheds suffered through the same dry conditions from 1974 through 1977 that were common throughout the western United States. The late 1970s and early 1980s then brought one of the wettest periods of central Arizona's history, with enormous flood damage. True to form, we are now back in drought conditions. The years 1989 and 1990 have been extraordinarily dry. Indeed, they represent one of the very driest 2-year periods on record, surpassed only by the drought at the turn of the century.

Average annual inflows to the reservoir system are around 1.2 million acre-feet, but this average is skewed by relatively rare, extremely large events. Median inflows are only around 950,000 acre-feet per year. The coefficient of variation for the river flows for the period 1889 through 1989 is very large, even when compared to other rivers in the western United States where variability is expected to be large (Figure 19.8). This historical record shows both prolonged wet periods and droughts lasting anywhere from 2 to 7 years; these appearances of dry and wet periods have no apparent pattern or sequence that is easily discernible or predictable (Figure 19.9).

Since most of the water in the Salt and Verde rivers originates from winter and spring precipitation, the variability of the winter season is the key to describing water supply variability in Arizona. One influence on Arizona water supply variability is a change in the pattern of Pacific storms that typically begins in late November or early December and continues until mid-March. Several very dry recent years resulted from a premature end to this pattern—a switch to dry, spring weather in early February. A second influence on

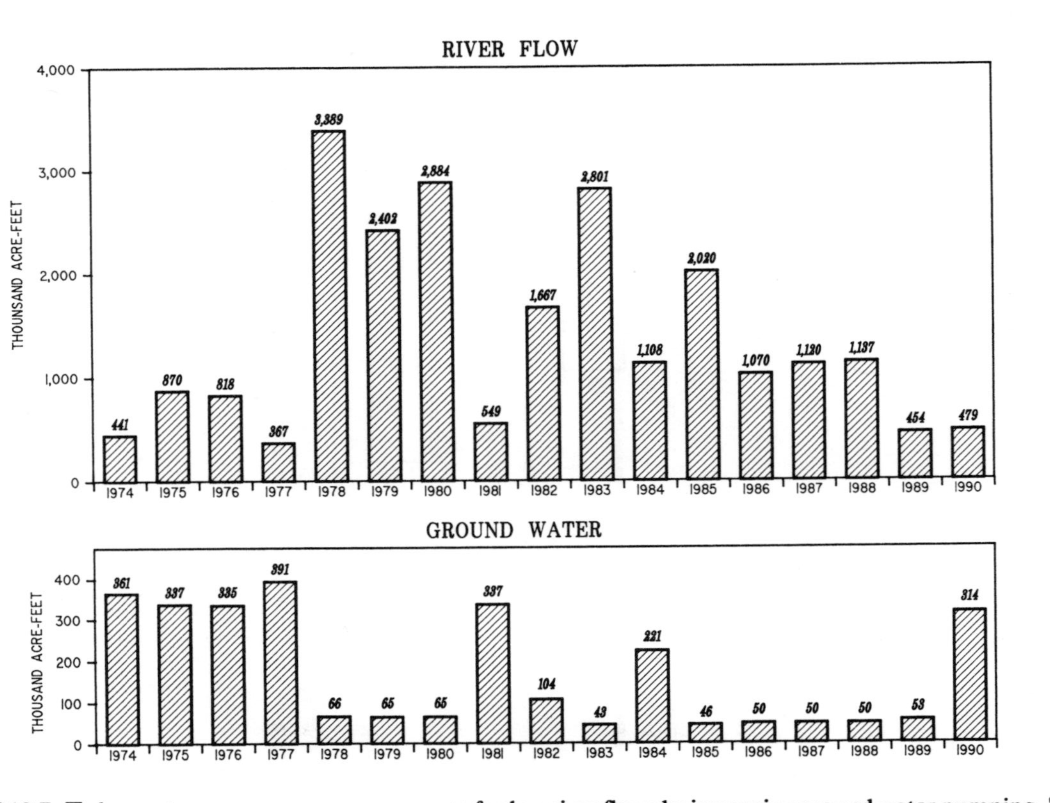

FIGURE 19.7 Today, water managers can compensate for low river flows by increasing ground water pumping. These figures show that pumping increases during low flow years.

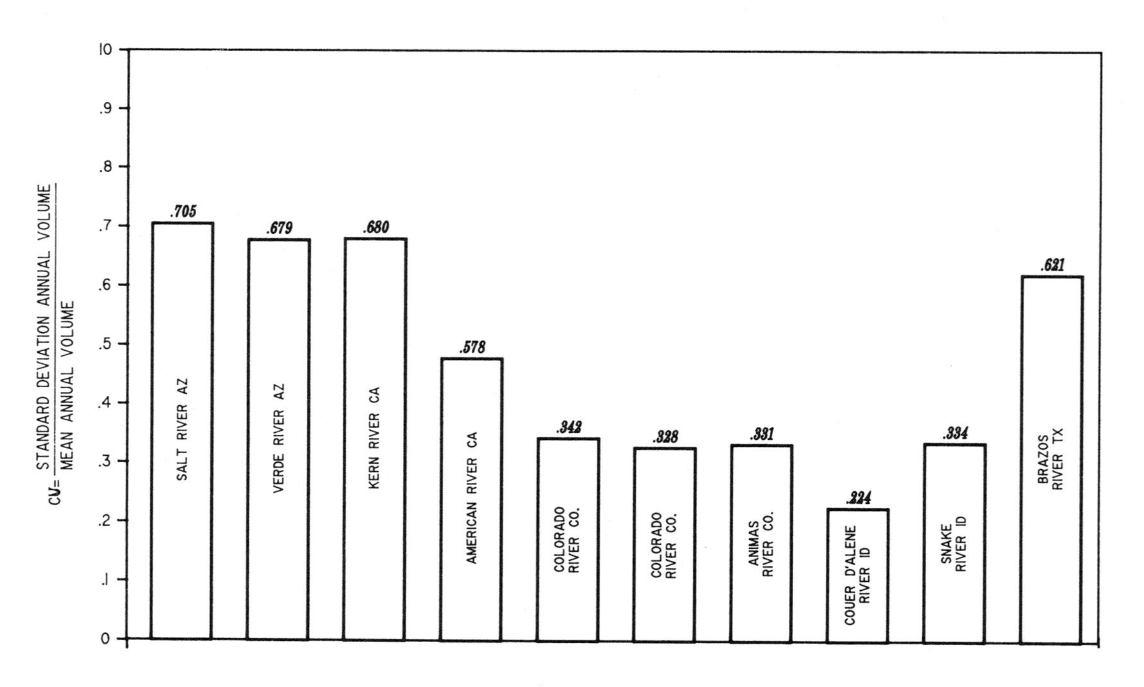

FIGURE 19.8 The coefficient of variation in annual Salt River flow is large even when compared to standard deviations from other rivers with highly variable flow.

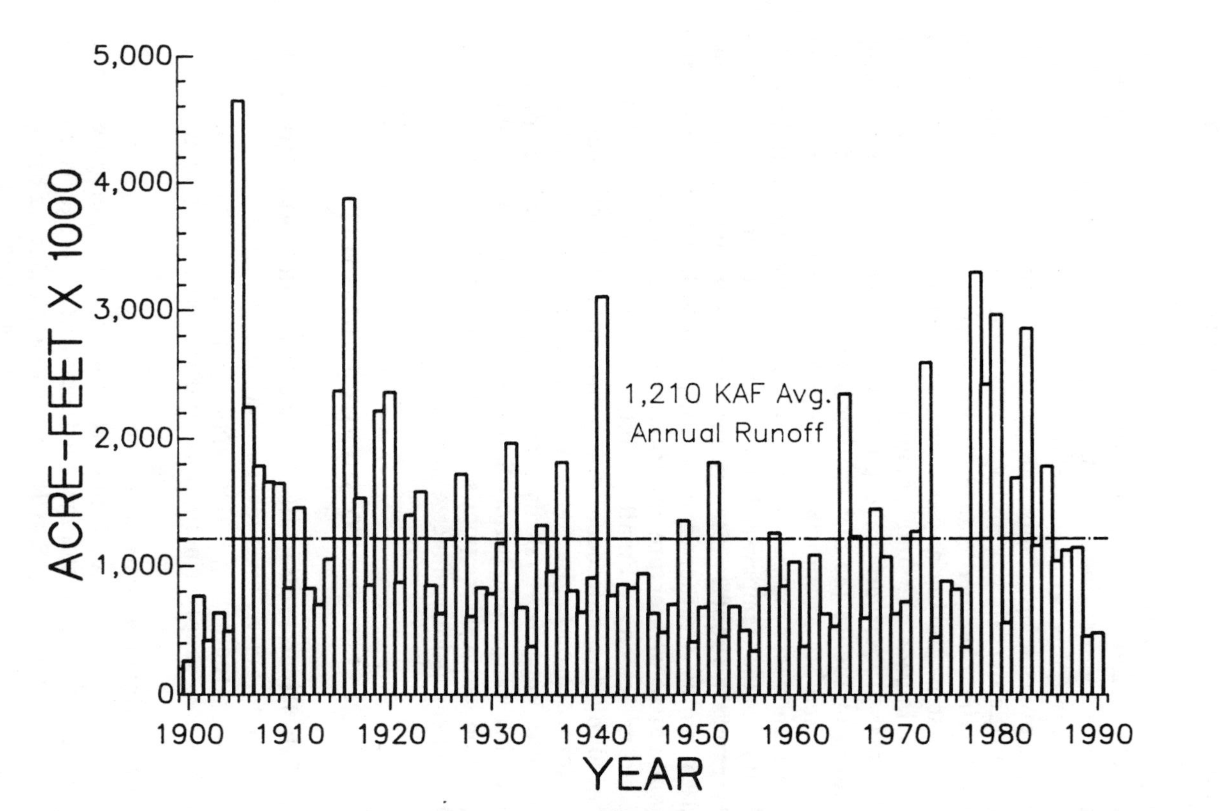

FIGURE 19.9 Annual runoff in the Salt, Tonto, and Verde river watersheds between 1900 and 1990. The frequency of wet and dry periods has no apparent pattern.

winter water supply variability are physical factors taking place on a gigantic scale. For example, the global effects of major El Niño episodes usually are associated with wet winters. A third influence on water supply variability is the temperature of the winter season. Because of Arizona's southerly latitudes and the relatively modest elevation of its mountainous areas, snowpack formation is very sensitive to winter temperatures. Relatively small temperature changes shift the snowpack up or down in elevation, and this causes drastic changes in the overall size and volume of the snowpack. Temperature also changes the percentage of the watershed that receives rain rather than snow during any winter storm event, which in turn can profoundly impact the total amount and distribution of winter runoff. Many observers have pointed out that a global warming trend could influence the distribution between rain and snow in the western United States, and the Salt and Verde watersheds are more sensitive than most areas. A fourth influence on winter water supply variability is the temporal distribution of winter storms, which affects rainfall runoff. Storms that are well spaced, with long periods of sunny weather in between, produce less runoff than the same amount of winter precipitation crowded into a few weeks of heavy storm activity. Finally, winter water supply variability is influenced by the antecedent soil moisture conditions before the winter storm season begins. Heavy rains in late summer and fall on the Salt and Verde watersheds can turn even average precipitation the following winter into better-than-average runoff. Average winter precipitation can produce poor seasonal runoff if dry conditions prevail during the prior summer and fall.

RESPONSES TO SUPPLY VARIABILITY

The Salt River Project, like most other water providers in the western United States, must meet a relatively constant water demand with a widely fluctuating water supply. Clearly, the SRP must plan for these water supply fluctuations and make changes to its operations.

The traditional western response to periodic excesses in surface water supplies is to reduce expensive ground water pumping and to increase surface water storage. Surface water storage is currently being increased along the Salt River. Roosevelt Dam is being raised, largely for flood control and dam safety considerations, but also to provide several hundred thousand acre-feet of additional conservation space. Increasing reservoir storage, however, is clearly no longer an easy or cheap solution. (In fact, the SRP did not acquire the additional storage at Roosevelt Dam; it was acquired by valley cities that need the water to supply the lands outside of SRP boundaries.) Additional storage has been suggested on the Verde River between the current Horseshoe and Bartlett reservoirs, and storage has been suggested at the confluence of the Salt and the Verde. Like new dams elsewhere in the West, these proposed structures ran into various environmental and social problems. By the time riparian habitats, bald eagles and Native American tribes were all considered in the decisionmaking process, the opportunities for additional surface water storage were thrown out. Few people expect to see them surface again any time soon. Even without social and environmental problems, such structures are very expensive, and Arizona's state government and the federal government are both wrestling with deficits of depressing proportions.

What other strategies can water managers adopt to make use of periodic high river flows? One often-mentioned possibility is to recharge excess surface water into the valley's alluvial basin. To a very significant extent this happens naturally; releases of excess water out of the lowest dams on the Salt and Verde into the dry river channel through Phoenix result in a good deal of natural recharge. There is limited opportunity for recharge projects to store more of these flood flows underground. However, these flows often come in quantities that are too great to capture and at time intervals that are too sporadic. Also, downstream water rights holders may want to benefit from these flows and therefore might oppose recharge projects.

What happens, on the other hand, during times when surface water supplies are shrinking and are not able to satisfy local demand? The SRP has responded in the same way as others all across southern Arizona have responded: by increasing ground water pumping. Southern and western Arizona host several large alluvial aquifers that have taken millennia to fill with water. In the last 70 years, pumping has significantly reduced some of these supplies. Pumping has long been crucial in the Phoenix area for handling supply variability. As water supplies and reservoir levels drop, pumping is increased. The customer simply pays a little more and otherwise notices little difference in the available water supply during surface water shortages.

Ground water is more costly to produce than Salt and Verde river water, not only because of pumping costs but also because of diminished water quality. Some of the ground water in southwestern alluvial basins was not of optimum quality to begin with, having a high total dissolved solids content. In many areas, man has made matters worse. Nitrates, insecticides, and industrial solvents exceed recommended levels in several SRP wells, reducing total pumping capacity. This problem is exaggerated because pumping capacity is not equally distributed throughout the system. Some municipal filter plants are getting far more of the poor quality ground water than others during dry periods, and this has effects on the taste and odor of the water and on treatment plant operations.

Water quality concerns and pumping costs aside, there is the longer-term problem of relying upon what in some ways is a finite ground water supply. Natural recharge in the Phoenix area is very small, especially because surface water flows are now captured behind the dams above the valley. Incidental recharge from both urban and agricultural water uses helps to replenish the aquifer to some extent, but at current levels of ground water pumping the aquifer under Phoenix could be largely dewatered in perhaps 200 years. Ten years ago, Arizona passed the Ground Water Management Act in an attempt to begin dealing with this problem. Currently, legislators are addressing the act's shortcomings to determine how to cut back on the use of the relatively cheap and dependable but finite supply that lies under our feet.

What else does the SRP do besides increasing pumping when it is faced with surface water shortages? As reservoir levels fall and as pumping approaches production capacity, the SRP reduces water allocations to its customers. In addition to straightforward supply cutbacks, there are, of course, any number of mechanisms that agricultural agencies and municipal governments can use to promote water conservation. Depending on the program involved, water use reductions may be imposed only temporarily, during drought periods, or they may aim for long-term demand reduction.

In drought periods, such as the one Arizona and California are now facing, reduced supply allocations bring up conservation issues, which frequently spill over into economic and social issues. Urban water users point to agricultural users as being selfish and wasteful and using too large a share of the water. Agriculturalists are frequently defensive and claim they cannot afford the investments in conservation technologies that urban dwellers feel they should be using. Similarly, some cities blame other cities for not trying hard enough to conserve, non-golfers point out how much water golf courses use, and so forth. There is plenty of finger pointing to go around.

Finally, as supplies shrink, interest grows in reusing sewage effluent, which is often touted as the only source of supply that is actually increasing. That water reuse will become a major force in the water future of the Southwest is clear. Effluent has its problems, however. Environmental and health regulations make it difficult and expensive to treat effluent for reuse or for recharge into aquifers. Some areas seem to be adding golf courses, parks, and artificial lakes just to put this resource to use. In the long run, it will make more sense to use effluent to displace ground water mining rather than to inflate demand to include effluent. That will happen, but it will take time.

So, the surface water supply of the Salt River Project is highly variable. Probably not much can be done to capture more of it during wet periods. Most of that water will continue to be lost in flood releases. Surface water supply shortages are going to occur sporadically under present conditions, and most climate change scenarios either leave this situation unchanged or make it worse. Turning to increased ground water pumping as we have done in the past cannot go on forever. It is not a very renewable resource, water quality is degenerating in many cases, and land subsidence can be a problem. Water conservation and reuse are clearly important and must become more so in the future. Conservation has its practical limits, however. Are there other sources that the Salt River Project could turn to for backup supplies when surface water supplies are low?

SEARCHING FOR BACKUP SUPPLIES

Population growth long ago outstripped renewable water supplies in central Arizona, although the Central Arizona Project (CAP) is trying to remedy this. This is true for the Salt River Project, and it is even more true for other areas in Arizona. Once an area has outstripped its local supplies, it goes looking for a backup either because it wants do deal with growth that has already occurred or because it wants backup supplies to allow further growth. What could the SRP do (as Los Angeles and Denver have done) by rerouting or transferring water to bring in new supplies? Where should we look for new supplies?

One idea that has been suggested time and time again over the last 50 years is simply to increase the flows of the Salt and Verde rivers themselves. Two techniques are usually suggested: watershed management and weather modification.

One form of watershed management is to decrease the water use by vegetation and, therefore, to eventually increase the streamflow. This involves either increased timber harvest, increased timber thinning, or chaparral brush reduction through fire or other means. It has been shown through research studies that runoff can be increased in these ways, although precise quantities of runoff increase are hard to pin down. In order to make a really significant change in river flows, however, there would have to be a concomitantly large change in the way the watersheds are managed. Most of the water-sheds in question (the middle and higher elevations) are U.S. Forest Service land or Native American reservations. In both cases, the lands are managed for a variety of products and a variety of uses. This web of multiple-use management, along with areas set aside for wilderness, sharply limits the kind of vegetation management that is possible. What is more, the public (probably rightly) feels that large-scale vegetation changes would be harmful. Even relatively small changes in land management—changes that may be ecologically sound and based on the ecological history of the Southwest—can be rejected by the public. Any kind of active management by Forest Service officials is often viewed as suspicious and dangerous. The end result is that changing land management in the future by the Forest Service and Native Americans may make subtle changes in river flows, but really significant increases in water supply cannot be expected from this type of activity.

Weather modification has been attempted in the West ever since the late 1940s. There are good scientific reasons why cloud seeding under certain conditions may produce an increase in winter orographic precipitation, and this in turn could increase snowpacks and runoff. These increases, however, are extremely difficult, if not impossible, to prove or quantify. It is difficult to expect governing boards and state legislatures to pay for weather modification programs when the benefit is uncertain and unmeasurable. Then too, the public may be skeptical of any attempt to "meddle with Mother Nature." As with watershed management, there may be opportunities here and there for small increases to river flows, but these will not be large enough or regular enough to be any kind of real solution to water supply variability problems.

For both of these flow augmentation techniques there is another problem. Increased flows from both weather modification and watershed management tend to come in the wetter years, when rivers and reservoir water levels are already high. Several simulation studies in Arizona have shown that a good portion of the increased water yield, from either watershed management or weather modification, would be spilled from already full reservoirs into the dry Salt River bed. In order to use the augmented flows of the Salt and Verde rivers, you would have to increase reservoir storage, and as I discussed earlier this is not easy to do.

If backup water supplies are not going to come from augmented flows of the Salt and Verde rivers, can the SRP turn to the Colorado River, that source that has been lusted after and tapped by all kinds of other thirsty water users? Southern California and Denver have been tapping the Colorado River in transbasin diversions for years. Las Vegas is attempting to use more Colorado River water. And of course the Central Arizona Project is now bringing substantial quantities of Colorado River water to supply parts of the valley not supplied by the SRP. Can the SRP tap into this source also? The SRP was offered a modest allocation of CAP agricultural water (30 thousand acre-feet) but turned it down. There were several reasons for this, and these reasons still pertain today. First, the use of this water would have subjected SRP agricultural water users to the Reclamation Reform Act, and no one undertakes that sort of burden lightly. Second, other water users in the state had no access to local surface water supplies like those of the Salt and Verde rivers. This is especially true of the cities in the valley that needed supplies for their non-SRP areas. For the SRP (already viewed as water rich) to try to corral a substantial share of Colorado River water would have been politically unpopular to say the least. Today the cities, developers, irrigation districts, and Native American tribes are all vying for the remaining unallocated shares of the Colorado River. The competition is intense and will become more so. Really large participation in the CAP by the SRP is not feasible at this point. The CAP may in the future bring some small relief to supply variability problems. Some cities are already beginning to use CAP water on SRP lands under certain circumstances. The future will no doubt bring various kinds of long-term or short-term exchanges between the two systems to help deal with droughts or system outages due to mechanical problems, but such exchanges will not provide a major new supply for the SRP.

Where else might the SRP go for backup water supplies? Water transfers from rural Arizona are one possibility. Non-CAP Colorado River water now being used for agriculture might be brought in through the CAP canal during times when excess canal capacity is available. Or, ground water from several basins in Arizona could be pumped and brought (perhaps through the CAP canal) to the Phoenix area. Water transfers have been a hot topic in recent years throughout the West and in Arizona. Several cities have bought water farms in western Arizona and in Pinal County, south of Phoenix, as backup supplies for the future. These water farms are a relatively expensive proposition, and if the CAP canal does not have excess capacity for transporting the water, an additional separate conveyance system would have to be built for some of these sources. The price would then become astronomical. (In 1991 the Arizona legislature banned ground water transfers from most rural areas, with some limited exceptions. The race for rural ground water supplies has stopped for now.) There could be both long- and short-term water leasing, as well. Leasing of agricultural water during dry periods holds some promise as a backup source for the thirsty urban areas. It is far from clear, however, whether the SRP will be involved in these kinds of transfers any time soon.

The water transfer issue, once raised, very quickly becomes a highly emotional one in which rural people are pitted against city dwellers. Rural people, even in areas that have little water that anyone wants, have organized to oppose potential water raids by the city dwellers. There are sometimes shrill and sometimes outrageously overdrawn depictions of urban octopi with their tentacles grabbing up the water supplies from the rural areas in order to suck them dry. These depictions raise important issues. Should Phoenix, Tucson, Los Angeles, and San Diego all continue to grow even though they far outstrip their supplies of water and clean air? Does it make sense to sacrifice the future of certain rural areas, however small and however bleak their future prospects may be, just to squeeze a few more people into already intolerably crowded cities? To the dismay of the city water planners, even many city citizens are beginning to sympathize with the rural point of view and find the notion of sacrificing rural futures for urban demands to be unacceptable. Water transfers are an old story in the West; they are an old story in Arizona. Some will continue, and some new ones will be undertaken. This is clearly not, however, a simple solution to the Salt River Project's supply variability problem.

CONCLUSIONS

Though future climate change patterns are anything but certain, some of the most likely scenarios now being painted suggest that the already extreme surface water supply variability in the Salt and Verde river watersheds may increase in the future. In the past, the supply variability problem in the Salt River basin has been met by pumping ground water. The ground water supply is still large; ground water will continue to be the Salt River Project's main backup in dealing with surface water variability for a long time to come. Problems with this ground water resource are beginning to surface, however. Water quality problems are multiplying, water levels in many parts of the SRP have declined significantly, and it is gradually becoming clearer that the ground water supply, which is renewed very slowly, cannot be mined indefinitely just because it is the cheapest alternative at hand. Dealing with the ground water over-draft is a little like dealing with our federal budget deficit: it is relatively cheap and painless in the short run just to let the deficit ride or even grow, but this course passes the burden to future generations. Slowly but surely, even in the conservative Phoenix basin, where government intervention is despised as much as any place else in the country, the necessity of government control of ground water mining is slowly becoming accepted. As taxes are levied to decrease consumption, ground water will become more and more expensive to pump.

Yet, the situation is not at a crisis point in the Phoenix area—at least not yet. There is ample room for conservation to expand supplies. Conservation by both agricultural and urban users is really just beginning. Reuse of sewage effluent will become much more important and will provide the valley with additional water for either future growth or for reducing the ground water overdraft. There is still a great deal of water use by agriculture, and this water will gradually, whether in an orderly or a disorderly fashion, be shifted to urban uses. So water will be argued over, traded, swapped and recirculated, but there will not be shortages severe enough to curtail urban growth for some time to come—several decades at least. Supply variability will be met by varying ground water pumping, by increasing conservation and effluent reuse, and by some water transfers.

The time will come in the next century when these various sources of water will no longer be sufficient to sustain Sun Belt growth rates. Planners might want this to occur relatively quickly. They might want to shift growth to other parts of the state or country where air quality is better, water is more plentiful, and the environment is less fragile. Given the history of the West and of the United States in general, however, it is highly unlikely that political restrictions on urban growth will ever be successful. Somewhere in the future, market-like mechanisms will begin to take hold. Either water will become very expensive or the quality of life will become so degraded (or both) that growth in areas like Phoenix and Tucson will slowly grind to a halt. By that time, we will have much larger urban populations, much less agriculture, and many more people expecting a steady supply of water. This, in turn, will make it more difficult to deal with increased supply variability that the greenhouse effect may bring about.

20

Public Involvement in Water Resources Decisionmaking in a Climate of Uncertainty

Helen Ingram
University of Arizona
Tucson, Arizona

Through much of this conference we have been considering uncertainty of the physical climate. This panel will consider another kind of climatic uncertainty—that of the political climate. As a political phenomenon, global climate change burst onto the public agenda in 1988. The conditions that allowed for the emergence of this issue to broad public consciousness have been discussed at length by myself and coauthors Hanna Cortner and Marc Landy (1990) in the book, Climate Change and U.S. Water Resources. Briefly summarized, the main components of the political atmosphere that brought public attention to this issue were:

1. The nature of the issue. It could be portrayed as serious and certain, consequences were perceived as happening soon, and sinners apart from ourselves (such as those who destroy tropical rain forests) could be blamed for the difficulty.

2. Science entrepreneurs. The issue was given a great deal of credibility by research scientists such as James Hansen and Stephen Schneider, who directed communications more to the public than to fellow scientists. Important to the effectiveness of science entrepreneurs were the science reporters who have come to hold positions on the nation's major newspapers and magazines and were poised to cover such a story.

3. Political entrepreneurs. Politicians, particularly those with national aspirations, are open to new issues with which they can make their mark. Senator Albert Gore and Representative Timothy Wirth helped to draw attention to the issue. For instance, Timothy Wirth told National Public Radio that he purposefully scheduled the testimony of Hansen before his committee on a day that the

National Weather Service predicted would be hot in order to maximize the public impact of the testimony.

4. Weather. The triggering event of the hottest, driest summer in decades gave people a graphic example of what might be in store because of global warming.

Getting an issue on the agenda is a different matter from keeping it on as a continuing agenda item or formulating and building a consensus for possible solutions to the problem. As a moderator to this panel, I wish to identify a few questions that I hope the panelists may choose to address:

- What happens if the weather is cool and wet during the next few summers?
- Will the disagreements within science increase, not so much about the correctness of the prediction but rather concerning the importance of the issue and what sort of research should be conducted? Will the public become disenchanted and skeptical about global climate change as it listens to conflicting scientific viewpoints?
- How will the policy debate change? Will alternatives in national energy policy continue to be the focal point of debate, or will other matters be considered? Will social science have a role in suggesting other responses, such as human adaptation to change through resettlement and changing lifestyles? Will we come to recognize the limits to the earth's capacity to sustain continued growth in human populations and increased human intervention into natural systems?
- Who are the winners and the losers in global climate change? Third World countries, which have contributed rather little to the creation of the climate change problem, may nonetheless carry the brunt of policies to limit development and to preserve forests. How can the burden of reducing the load of greenhouse gases be equitably shared among winners and losers?

I now turn to our speakers to address these and other issues.

REFERENCE

Ingram, H. M., H. J. Cortner, and M. K. Landy. 1990. The political agenda. Pp. 421-443 in P. E. Waggoner, ed., Climate Change and U.S. Water Resources. New York: John Wiley & Sons.

21

The First Rough Draft of History: How the Press Reports on Global Warming

Jim Carrier
The Denver Post
Denver, Colorado

Thomas Jefferson had some things to say about the newspaper business. Wise man that he was, he said that if he had a choice between a government without newspapers or newspapers without government, "I should not hesitate a moment to prefer the latter." And I would like to know how many agree with him. That's what I thought. He also said, "[t]he man that never looks into a newspaper is better informed than he who reads them because he who knows nothing is nearer the truth than he whose mind is filled with falsehoods and errors." Now, if you agree with that, you also might think that with 27 million trees killed for Sunday newspapers every week, illiteracy could save the world. My final note on Jefferson is his recommendation that "Editors should divide their papers into four chapters: (1) truths, (2) probabilities, (3) possibilities, and (4) lies." Which brings us to this conference.

What if the *National Enquirer* were here covering this? What would the headlines read and into what chapters would this information fall? I made some up. Our keynote speaker two nights ago: "Weiss Wants Sun Belt to Hold Its Water" and then a subhead, "Wishful Thinker Visits Scottsdale." Or how about our friend from Washington who gave us a talk yesterday at lunch: "Bureau-Rec Boss Brags 'All Things to All People'—Bureaucrat Bamboozles Public." Or, one of our scientist friends of yesterday, "Marshall Moss Mutilates Model: Egghead GCM No Match for Madonna."

On the global warming issue, the media generally is criticized in one of two ways. First, people say that we overplay it—and this I'm just going to dismiss with, "I disagree." It's a hell of a story; it's the Big One (to use the name of a show on NBC the other night); it's the Big One into the next century. And so what if we've overplayed it? What is the implication of an error on that side of the equation? The second criticism is that we ignore it, or that we badly report science, or that we speculate wildly. Stephen Schneider says we have to have news told four ways: drama, disaster, debate, or dichotomy. To that I plead, "Uncertainty."

The media is this great, imperfect information machine that puts out news hourly, daily, sometimes weekly. A newspaper like the *Denver Post*, for example, puts out the equivalent of two or three books in 24 hours; most of that news is written in an 8-hour shift. We are, what I like to call, the first rough draft of history. But like you, who spend your life, say, perfecting a model, we spend ours trying to improve on that machine—trying to produce better, clearer, more useful information. We have our problems. We get tired of stories, for instance. That's one of the problems with covering trends or long-term stories. But I think in that way we are a reflection of people. They get tired of stories, too; they get tired of bad news. We also are easily manipulated; I think a good example is the alar scare. We report too many football scores and too many stories like football games. We are not in the "truth" business; we are in the "perception" business. And we often act like a herd, or another analogy might be like bacteria in a petri dish: stories are reproduced to excess and then die off.

On the global warming issue, we have some specific problems. One is localizing, and I noticed in your talks that this is one you have problems with, too. How do you make that come home to the reader? "How is it going to affect me?" is a question we always ask of any story. Two, what's the evidence? What can a television camera take a picture of about global warming? And three is this question of probabilities. I don't think that's as much a problem as you think, because we report probabilities all the time in terms of health research about cancer risks and that kind of thing.

On the positive side, we put out another paper tomorrow. Competition keeps us honest and alert. We balance our views. We tend to be pro-environment. At the *Denver Post*, for example, we have three environmental reporters and we make it front-page news. And, we have raised this particular issue, as Helen Ingram mentioned, sufficiently to help generate a billion dollars for research. I think we do a reasonable job of science reporting—whether it's on cold fusion or the cholesterol issue or global warming. I think most readers are aware (and we may hear more about this because there's some research that has been done) of the "certainty of warming." We quite accurately reflect the "uncertainty" of the consequences and of the policy. We have a sense of what people will read and what will help make them read it. If they turn the page in a newspaper, then we've lost them and we've wasted our time. So, we need news hooks or articulate talking heads like Schneider, who has found a way to fit climate uncertainty into a sound bite.

I have to tell you that, frankly, the name of this conference is a tough one to deal with in our business. My editor would say, "What does that mean? How would that fit into a headline? How would that fit into a lead?" We try every day to make complicated stories interesting and intriguing and relevant enough to read. There is a misunderstanding of our role. We do, in fact, try for facts, or we try to carry factual accounts of someone's opinion. We rarely speculate. Although we have carried lots of speculation, it isn't our speculation; it's rare that we speculate on our own. As Helen mentioned, during the fires of 1988, the speculation came from somewhere else. And I'm also uncomfortable with this panel's title—that we should be part of building a public consensus. That's not our job. All sides expect us to be advocates for them. We carry water for a lot of people and if it's done right, with enough buckets of water, we reflect accurately the scientific and the policy debate.

Now, as you know, global warming became a media event in 1988. I happened to be camping in Yellowstone in July with my daughter that year. I left five months later with enough material to write a book. I raised some issues and formed some opinions about that fire and the way it was handled that the park people don't agree with. The first is that it was a triumph of science and a failure of political science. In Yellowstone in 1988, science became dogma. The "let burn" policy was so rigid that it actually ignored things: for example, the Palmer Index, which, when I began snooping around, I found rolled up in the fire cache. We in the press were criticized for pitting science against the emotion of fire: the scientist talking about the natural fire regime versus a cabin owner who lost his place to the fire or a person at Old Faithful when the fire storm blew over. But I think the biologists who consider Yellowstone a biome for their own personal study fail to see Yellowstone for what it is: a political stage on which our democracy plays out its environmental ethic. And global warming is such an issue as well: it is a political, value-packed debate—that's why it is acrimonious. You know, John Sununu says we don't have enough evidence of global warming. Well, I don't think that anyone is fooled by that; I think we should move beyond this issue of whether or not global warming is happening. I don't think that's an issue anymore. It's kind of like the cigarette industry saying they didn't have enough evidence to link smoking and cancer—ultimately it didn't make a lot of difference.

I think the implications of global warming are so horrendous that there is little dwelling on the pure science anymore. The mind immediately leaps to—what my mind leaps to—is, what about me? I think about where I'm going to live, and who's to blame, and what we can do, how much we must do, and what it will cost.

Your role as scientists in this debate is one of information. If you should, as Schneider has done, become a political scientist, then recognize that you're leaving the ivory tower and entering into the arena with the lions. And the history of scientists who have done that—the Einsteins, the Oppenheimers of World War II, the Craigheads of Yellowstone—is that they lost some skin in the process, despite the exoneration of history. The mass media in the real world—and this may be something that you don't agree with at all—is this great peer-review forum. It's the broad view, with experts, politicians, industrialists, minorities, and ordinary Joes all contributing to this great pile of information and opinion. And through the media we kind of lurch toward a consensus position that serves democracy best. Aldo Leopold had faith in this, and he could have been talking about global warming when he wrote: "[A] sufficiently enlightened society, by changing its wants and tolerances, can change the economic factors bearing on the land."

My plea to you is to join the chorus. Become part of the first rough draft of history. We depend on you to referee and validate what's going on. I wouldn't want to live in a world that is run by scientists, but I wouldn't want to live in one without them either. Don't pass up opportunities to speak to the press. There are ways to communicate that are better than others. One of the things that we use in our business all the time is to write as if, or talk as if you're speaking to your neighbor across the back fence. Metaphors help. I think Mr. Dickinson's metaphor of the "house warming" was beautifully done.

This, as I repeat, is a hell of a story, and thanks to you we know about it. Thanks to you and the much maligned press, we won't wake up one morning with Lake Powell dry or with the Atlantic Ocean running up Fifth Avenue.

22

Water Use Efficiency as a Response to Climate Uncertainty

Jim Dyer
Rocky Mountain Institute
Snowmass, Colorado

There is great uncertainty about the impact of the expected climatic change on western water supplies. On the other hand, it is highly likely that the cost of energy for pumping and heating water will increase, that environmental constraints will increase, and that periodic droughts will continue. It is a fact that we now are wasting energy, other resources, and money wherever water is used inefficiently. In other words, implementing water efficiency is a wise thing to do anyway, whatever the future may bring.

While researchers should continue their efforts to predict the impacts of climatic change on water supply, water planning—with a special emphasis on efficiency—should begin now. This will prove to be particularly valuable if climate turns out to limit water supplies, but, as mentioned above, it is worth doing anyway.

Water freed through residential efficiency can and should be recognized as an inexpensive and reliable new source of supply. It is relatively inexpensive because the saved water has already been delivered to the community, treated to drinking water quality, and often heated. It is reliable because any wise water efficiency program will be designed to provide the public not with brown lawns and dribbly showers but with the same or better water-related services at a lower cost. Satisfied consumers will gladly continue to use water-saving technologies that save them money.

Efficiency will be an integral part of any wise plan. Yet, efficiency is not an end in itself, but a tool. As such, it may compound problems or help solve them, depending largely on how the saved water is used. In the short term, it can also buy time for achieving better community planning, but it is no substitute for that planning. A comprehensive, long-range plan can reduce costs and future conflicts over competing uses of water and can contribute to a higher quality of life. However, such plans are merely words if the community does not participate in their design and support their implementation.

How can one achieve this involvement? The community should ask several fundamental questions:

- What lifestyle and quality of life are desired by those in the community?
- What land use, environmental quality, and settlement patterns would citizens like to see?
- What water-related services are required for that desired future?
- What is the cheapest way to provide those services that is consistent—economically, environmentally, and socially—with the desired community lifestyles?

To answer this last question, the answers to four more specific questions are required:

- What are the whole-system, avoidable costs of developing a new source of water supply (including the costs of construction, operation, water treatment, heating, and environmental mitigation)?
- What efficient technologies are available to provide the water-related services required by the community without diminishing the quality of service?
- How much will it cost to provide for those needs through efficiency?
- What are the most effective techniques available for implementing these efficiency measures?

The first steps toward building the required public consensus are to help the public recognize who will benefit—and in what ways—from using existing water supplies more efficiently and to get the comprehensive planning process under way.

Numerous studies have shown that efficiency programs can meet a community's water needs better and more cheaply than traditional supply programs. As the questions posed above are answered, this will tend to become clear to the community. Those who will benefit from a well-designed water efficiency program are to be found in all parts of the community. Consumers will enjoy lower water and energy bills and lower taxes as less money is needed to pay for water supply and treatment facilities. Utilities will avoid unnecessary expansions in supply and treatment facilities. Recreationists and the environment will benefit if at least a portion of the saved water reverts to streams, lakes, and wetlands. A wise program will reward the efficiency efforts of farmers and ranchers with reduced costs of production and possibly a market for their saved water. Future generations will be thankful for wise water decisions made today.

Thus, the whole community will be brought together to cooperate in building a long-term comprehensive water use plan. Such a plan, with efficiency at its center, will help cope with the water needs of today as well as with the future and its uncertain climate.

23

Climate Change, The Media, and Public Awareness

Roger E. Kasperson
Clark University
Worcester, Massachusetts

The wide-ranging discussion accompanying the emerging international response to global climate change and the upcoming 1992 United Nations Conference on Environment and Development in Brazil rest upon a series of assumptions of global public response to the issues associated with climate change. The assumptions are largely untested, and studies are only beginning to explore how public information, understanding, attitudes, and mobilization may affect the future course of these public processes and attempts at international initiatives.

In the United States, public awareness and concern about global environmental problems have appeared only recently. This is not surprising in light of the relatively recent scientific attention to these problems and the changes in media coverage referred to above. In a Cambridge Reports national poll in 1982, less than half of those interviewed had heard or read anything about the "greenhouse effect," and only one of every eight expressed a great deal of concern over it (Cambridge Reports, 1986). In 1986, a similar poll revealed rising awareness, but still only two of five persons interviewed had heard or read about it. Nonetheless, once it was explained, twice as many respondents in 1986 as in 1982 (24 percent as compared with 12 percent) felt that it was a very serious problem. In the public's volunteered list of environmental concerns solicited in 1983, 1984, and 1985, climate change, desertification, deforestation, and ozone depletion did not appear on the agenda of concerns. Air pollution topped the concerns, followed by water pollution and radioactive and toxic wastes.

In the second half of the 1980s, public awareness and concern over these problems, especially global warming, increased significantly. A national Roper poll conducted in December 1987 and

January 1988 showed relatively high rates of public concern: 47 percent of respondents judged ozone layer destruction to be very serious, 36 percent saw acid rain as very serious, and 33 percent viewed the greenhouse effect as very serious. Despite this, the greenhouse effect still ranked twenty-fourth out of twenty-eight environmental hazards viewed as very serious. Also noteworthy is the fact that public responses showed unusually high rates of "don't knows," even for acid rain, which has been extensively publicized. Although the data suggest increased public concern, potential climate change scored well down on the public's list of top environmental concerns, which continued to be dominated by issues associated with hazardous wastes, toxic materials, nuclear accidents, pesticides, and air and water pollution.

In 1987 and 1989, Cambridge Reports conducted national polls in the United States assessing the degree of threat to (1) personal health and safety and (2) the overall quality of the environment. The results showed dramatic increases in the perceived degree of threat associated with global environmental problems. Indeed, the largest increase in perceived "large threat" was for the greenhouse effect. These changes in opinion, however, must be placed in the context of the poll's finding that between 1987 and 1989, "Americans were not only growing vastly more concerned about environmental problems of all kinds, they were also increasingly likely to feel those problems posed a direct risk to their personal health and safety" (Cambridge Reports, 1990). It is also instructive that public perceptions of threats from climate change in the United States continued to follow well behind hazardous waste disposal, pollution, and pesticides.

Neither the public nor policymakers like large uncertainties. In our daily lives, we seek to remove them if the stakes involved matter greatly to us; or, finding this impossible, we either decide to forego the activity or make our uneasy peace. Which will it be for human-induced climate change? One possibility is that public responses to these problems will over time come to resemble those to nuclear war and nuclear power. Both conjure up futures of vast destruction and uncontrollability in the public mind (Slovic, Lichtenstein, and Fischhoff, 1979). Some, like Robert Lifton (1976), see the sources of public concern in a state of denial—what Lifton calls "psychic numbing"—in which, however uncertain, the catastrophic threatening of the earth itself is at stake. Others, such as Paul Slovic (1987) attribute it not only to the vast destructiveness and dread that it brings to mind but to a series of other qualitative properties—newness, involuntary nature, uncontrollability, unknowns of science—of the associated hazards. Some believe that the active scientific dispute over the uncertainties will only serve to convince publics around the globe that something is fundamentally amiss on a problem on which scientists have worked extensively but upon which they cannot agree. Will, as Gerard Blanc suggests, global change mobilize unconscious fears, producing *l'angoise planetarie*, and will it result in widespread and sustained public concern over this class of problems?

The ambiguities surrounding climate change will be no less pronounced than the uncertainties. Who exactly is responsible for the worldwide problems? What will the future be like at a time when we think that our descendants may experience harm? How should we reconcile our investments to protect both the worse off in this generation and the many unknown people in future generations? Will there really be winners as well as losers, and how will they be distributed? These and a host of other social and political considerations will add to the perplexity of the question, Who must respond, how, and when (O'Riordan and Rayner, 1990; Kasperson and Dow, 1991)?

The distinct possibility exists that the primary characteristic of media coverage and public responses to global environmental change will be oscillation. In a slowly changing global environment, the shifting averages may be imperceptible, and the periodic occurrence of extreme events may well drive the public agenda. In crisis years, the media will champion the global change cause. In more routine years, they will move on to more "newsworthy" topics (Schneider, 1988). Environmental groups also target particular problems for concerted effort, and these shift from year to year. And, of course, some policy outcomes (e.g., expansion of nuclear power) cut across traditional environmental interests (Ingram, Cortner, and Landy, 1990). Politicians, with characteristically short-term agenda horizons (the next election), may find it more convenient to wait out extreme events than to allocate costly expenditures designed to protect distant future generations or to subsidize preventive actions in far-off places. Then, too, it is a noteworthy characteristic of the most difficult types of environmental and technological controversies—those that can be described as "mysteries"—that the public debate oscillates between disputes over facts and disputes over values, with often growing confusion as to which are at stake or who is qualified to speak about them (Edwards and von Winterfeldt, 1986).

REFERENCES

Cambridge Reports, Inc. 1986. Emerging Environmental Concerns and Controversies. P. 20. Cambridge, Mass.: Cambridge Reports, Inc.

Cambridge Reports, Inc. 1990. Energy and Environment. Window on America. P. 4. Cambridge, Mass.: Cambridge Reports, Inc.

Edwards, W., and D. von Winterfeldt. 1986. Public disputes about risk of technologies: Stakeholders and arenas. Pp. 69-92 in V. T. Covello, J. Menkes, and J. Mumpower, eds., Risk evaluation and management. New York: Plenum.

Ingram, H, Cortner, H., and M. Landy. 1990. The political agenda. Pp. 421-443 (Chapter 18) in P. E. Waggoner, ed., climate change and U.S. water resources. New York: John Wiley & Sons.

Kasperson, R. E., and K. Dow. 1991. Developmental and geographical equity in global environmental change: a framework for analysis. Evaluation Review 15(1):147-169.

Lifton, R. 1976. Nuclear energy and the wisdom of the body. Bulletin of Atomic Scientists 32:16-20.

O'Riordan, T., and S. Rayner. 1990. Chasing a spectre: Risk management for global environmental change. Pp. 45-62 in R. E. Kasperson et al., eds., Understanding global environmental change: the contributions of risk analysis and management, Clark University. Worcester, Mass.: The Earth Transformed Program.

Schneider, S. H. 1988. The greenhouse effect and the U.S. summer of 1988: cause and effect or a media event? Climatic Change 13:113-115.

Slovic, P. 1987. Perception of risk. Science 236:280-289.

Slovic, P., S. Lichtenstein, and B. Fischhoff. 1979. Images of disaster: Perception and acceptance of risks from nuclear power. Pp. 223-245 in G. Goodman, and W. Rowe, eds., Energy risk management. London: Academic Press.

Appendix A

Biographical Sketches of Steering Committee Members

STEPHEN J. BURGES joined the faculty of The University of Washington in 1970 where he is a professor of civil engineering. He holds a Ph.D. from Stanford University and has major interests in science and practice of hydrology and water resources engineering. He is a former member of the Water Science and Technology Board and has been active in many WSTB activities. He is a former Co-Editor for Physical Sciences of Water Resources Research, published by the American Geophysical Union. His research interests include a broad range of topics in hydrology and hydrologic engineering, including streamflow production, and understanding and modeling hydrologic processes.

ROBERT E. DICKINSON holds a Ph.D. in meteorology (1966) from the Massachusetts Institute of Technology (MIT). After an early career at MIT, he moved to the National Center for Atmospheric Research in 1968 and remained there until his recent move to the University of Arizona's Institute for Atmospheric Physics. Dr. Dickinson's research interests include land surface processes and climate change. He is known for his work in global climate modeling. He is a member of the National Academy of Sciences and a member of the NRC's Committee on Global Change.

KENNETH D. FREDERICK is a Senior Fellow at Resources for the Future (RFF) in Washington, D.C. He has been a member of the research staff at RFF since 1971 and has served as director of the Renewable Resources Division from 1977 to 1988. Dr. Frederick received his Ph.D. in economics from the Massachusetts Institute of Technology in 1965 and his B.A. from Amherst College in 1961. He served as an economic advisor in Brazil for the U.S. Agency for International Development from 1965 to 1967 and an assistant professor of economics at the California Institute of Technology from 1967 to 1971. He is the author, coauthor or editor of six books and has authored more than 30 papers addressing the economic, environmental, and institutional aspects of water resource use and management.

ROGER E. KASPERSON received his Ph.D. in geography from the University of Chicago. He has been on the faculty at Clark University for most of his diverse career. His research has covered all aspects of water resources policy with emphasis on applications of risk assessment. He is currently conducting research on the contributions of risk analysis toward the understanding of global environmental change.

BRUCE A. KIMBALL received his Ph.D. in soils from Cornell University in 1970. He is presently a soil scientists at the USDA/Agricultural Research Service's Water Conservation Laboratory in Phoenix, Arizona. For most of his career, Dr. Kimball has been interested in ways to ameliorate the effects of plant stress on crop production. This has led to development of greenhouse computer models. He has also been investigating the effects of increasing carbon dioxide concentrations and changing climate on crop yield and water use.

DANIEL SHEER is president of Water Resources Management, Inc. (WRMI), a water resources consulting firm in Columbia, Maryland. Since founding WRMI in 1985, Dr. Sheer has worked for a variety of U.S. and international government agencies building simulation and optimization models to describe water resource systems for analysis, gaming, and computer-aided negotiating sessions. In 1980, Dr. Sheer became the first director of the ICPRB Section on Cooperative Water Supply Operations on the Potomac (CO-OP). Dr. Sheer has served as a member of the NRC's Water Science and Technology Board and participated in several intergovernmental assignments. He received his Ph.D. in environmental engineering from The Johns Hopkins University.

Appendix B

Biographical Sketches of Principal Contributors

LEON HARTWELL ALLEN, JR. is a soil scientist with the USDA's Agricultural Research Service at the University of Florida in Gainesville, where he also serves as Adjunct Professor in the Agronomy Department. He received his Ph.D. from Cornell University and has studied the effects of microclimate on plants for many years, including transport mechanisms for carbon dioxide and water vapor. Dr. Allen's current research is assessing the response of vegetation to rising levels of carbon dioxide and global climate change. He also has evaluated various carbon dioxide enrichment systems. Dr. Allen is a Fellow of the American Society of Agronomy.

EDITH BROWN WEISS is on leave from her position as professor of law at Georgetown University Law Center (where she has taught international law, international environmental law, water law, and environmental law) and is serving as Associate General Counsel for International Activities at the U.S. Environmental Protection Agency. Dr. Brown Weiss has served as chair of the Social Science Research Council's Committee on Research in Global Environmental Change and is former vice president of the American Society of International Law. Dr. Brown Weiss was a member of the Water Science and Technology Board from 1985-1988 and has served on several NRC committees. She is a member of the Board of Editors of the American Journal of International Law, International Legal Materials, and Climate Change Digest, and was elected to membership in the Council on Foreign Relations, American Law Institute, and the International Council on Environmental Law. Her book, *In Fairness to Future Generations*, received the Certificate of Merit from the American Society of International Law in 1990. She received her A.B. from Stanford University, LL.B. from Harvard Law School, and Ph.D. from the University of California at Berkeley.

DALE BUCKS is the National Program Leader for Water Quality and Water Management for the U.S. Department of Agriculture (USDA), Agricultural Research Service in Beltsville, Maryland. He is co-chair of the Research and Development Committee of the USDA Working Group on Water Quality. He began his career as a research agricultural engineer at the U.S. Water Conservation Laboratory in Phoenix, Arizona, where he published on the topics of irrigation practices, crop water requirements, and alternative agricultural management systems. Dr. Bucks holds a Ph.D. in soil and water science from the University of Arizona.

JIM CARRIER is a columnist who covers the West for *The Denver Post*. A journalist for 25 years, 15 of them in the West, Mr. Carrier has written five books on western issues, ranging from Yellowstone National Park to cowboys and Indians. His interest in water issues stems from an extensive series on the Colorado River, republished as "Down the Colorado." His report was updated in a June 1991 article in *National Geographic Magazine* titled, "The Colorado River in an Era of Limits."

ARNETT S. DENNIS has worked for the U.S. Department of the Interior's Bureau of Reclamation in Denver, Colorado, since 1981 on weather modification and climate change projects. He received his Ph.D. in Physics from McGill University in Montreal, Canada in 1955. He was vice president of the Weather Modification Company of San Jose, California, senior physicist at Stanford Research Institute, Menlo Park, California, and a consultant for the World Meteorological Organization in Geneva, Switzerland. He is a fellow of the American Meteorological Society and past president of the Weather Modification Association.

JOHN DRACUP is a professor in the Civil Engineering Department of the School of Engineering and Applied Science at the

University of California at Los Angeles. He received his Ph.D. from the University of California at Berkeley in civil engineering. His professional interests and expertise are in the fields of hydrology and water resource systems engineering. He has undertaken stochastic analysis of floods and droughts, investigation of the effects of climate stress on hydrologic processes, and development of computer models for the management of large-scale river basin systems.

JIM DYER is director of the Water, Agriculture, and Internship Programs at the Rocky Mountain Institute in Snowmass, Colorado. He earned a B.S. in meteorology from Saint Louis University and an M.A. in Natural Science from San Jose State University, with postgraduate study in renewable resources, meteorology, and geography. Mr. Dyer has a wide background in earth and soil science and in the environmental and physical sciences. He is a member of the American Meteorological Society.

MALCOLM K. HUGHES is director of the Laboratory of Tree-Ring Research and professor of dendrochronology at the University of Arizona. He received his Ph.D. from the University of Durham, England, and worked in Denmark and England before moving to Arizona in 1986. Dr. Hughes has been a member of the Council of the British Ecological Society, and he was Secretary of the British Ecological Society's Energy and Production Biology Group for several years. He is a member of the U.S. National Committee for the International Union for Quaternary Research and of the organizing committee for the 1989 Global Change Institute. He is a member of the advisory panel on paleoclimate formed by the National Oceanographic and Atmospheric Administration's National Geophysical Data Center, and he chairs the University of Arizona's Coordinating Committee on Global Change.

HELEN INGRAM is director of the Udall Center for Studies in Public Policy in Tucson, Arizona. She received her Ph.D. in political science from Columbia University, taught at the University of New Mexico, worked on the staff of the National Water Commission, and was a research fellow at Resources for the Future. She has been on the staff of the University of Arizona since 1972. Dr. Ingram has done extensive research on the public policy aspects of water resources. She is a former member of the Water Science and Technology Board, and was recently appointed to the NRC's new Commission on Geosciences, Environment, and Resources.

JOHN KEANE is a water policy analyst for the Salt River Project (SRP) in Phoenix, Arizona. Mr. Keane has held several technical and management positions with the SRP in the areas of supply forecasting, reservoir operations, water rights adjudications, state water regulations and legislative water issues. He holds an M.S. from the University of Arizona, focusing on watershed hydrology and resource policy.

DONALD R. KENDALL received his Ph.D. in Water Resources Hydrology from the University of California at Los Angeles. He is currently an assistant professor at Loyola Marymount University in Los Angeles. His professional interests include hydrology and water resource development. Dr. Kendall also serves as a consultant to the Metropolitan Water District of Southern California.

WAYNE MARCHANT is chief of the Research and Laboratory Services Division, U.S. Bureau of Reclamation, Denver, Colorado. He manages a diverse water resources research program that includes: water quality, water supply, hydroelectric power, and environmental components. He holds a Ph.D. in Chemistry from the University of California at Santa Barbara, and holds three U.S. patents based on his personal research. He has published on topics ranging from the irreversible chemical labeling of enzymes to industrial stack gas desulfurization. In 1987, he received the Department of the Interior's Meritorious Service Award and in 1988 the Secretary's Award for Exceptional Service.

DAVID MEKO is an associate scientist with the Laboratory of Tree-Ring Research at the University of Arizona. He holds degrees in meteorology and atmospheric sciences from Penn State University, and received his Ph.D. in hydrology from the University of Arizona. He served as a research hydrologist with the Tohono O'odham Nation in southwestern Arizona from 1986 to 1989. His interests include application of tree-ring data to infer the natural variability of runoff on time scales of decades to centuries.

MARSHALL MOSS is an hydrologist with the Water Resources Division of the U.S. Geological Survey (USGS). He has spent more than 25 years with USGS in varying capacities, collecting and interpreting hydrologic data, performing research, and serving as a senior executive. He holds a Ph.D. from Colorado State and is an active officer in the American Geophysical Union.

LINDA NASH is a research associate of the Pacific Institute for Studies in Development, Environment, and Security in Berkeley, California. Her current research is in the areas of water resources and water quality management. She is a member of the ASCE Committee on Climate Change and Water Resources. She previously worked as a policy analyst for the California Water Quality Control Board and as a technical analyst for the U.S. Environmental Protection Agency in the Superfund Program.

ROGER REVELLE received his Ph.D. from Scripps Institution of Oceanography at the University of California in 1936. The late Dr. Revelle, renowned for his many accomplishments and innovative thinking, was the recipient of numerous awards throughout his career, including the National Medal of Science from President Bush in 1990, the Tyler Award for Environmental Achievement, the Balzan Prize for Science, and the Agassiz Medal. Dr. Revelle was one of the founders of the Intergovernmental Oceanographic Commission of UNESCO. Dr. Revelle was a member of the National Academy of Sciences, and he served as a member of its Governing Board.

JOHN SCHAAKE, JR. is senior scientist, Office of Hydrology, National Oceanographic and Atmospheric Administration. He holds a Ph.D. in Environmental Engineering and Water Resources from The Johns Hopkins University, and did postdoctoral work with the Harvard Water Program. Dr. Schaake is author of a chapter in *Climate Change and U.S. Water Resources* dealing with the sensitivity of water resources to climate change. He is chair of the Science Panel for the GEWEX (Global Water and Energy Cycle Experiment) Continental-Scale International Project of the World Climate Research Program. His interests include the development of orographic precipitation models to analyze spatial distribution of precipitation in the western United States for input to hydrologic models and to assess information content of precipitation and snow measurement networks.

CHARLES W. STOCKTON is a professor of dendrochronology and has been on the staff of the Laboratory of Tree-Ring Research at the University of Arizona since 1970. He received his Ph.D. in hydrology from the University of Arizona in 1971. His research has been in hydrologic and climatological applications of tree-ring data, especially in the reconstruction of streamflow and the occurrence of drought.

A. DAN TARLOCK teaches at Chicago Kent College of Law. He has written extensively about water resources management and environmental law and policy. Mr. Tarlock has served on a variety of NRC activities; he is chair of the WSTB's Committee on Western Water Management and is vice chair of the Water Science and Technology Board. He received his law degree from Stanford University.

KEVIN TRENBERTH is head of the Climate Analysis Section at the National Center for Atmospheric Research (NCAR) in Boulder, Colorado. After completing a degree in mathematics at the University of Canterbury, New Zealand, he obtained his Sc.D. from Massachusetts Institute of Technology in meteorology in 1972. Following several years in the New Zealand Meteorological Service, he joined the Department of Atmospheric Sciences at the University of Illinois in 1977; he moved to NCAR in 1984. He has served on the National Research Council's Polar Research Board, several committees under the Board on Atmospheric Sciences and Climate, and the Space Science Board. He is a member of the NOAA Climate and Global Change advisory panel and has served as an editor of scientific journals. He is known for his work on global atmospheric circulation and climate change.

GILBERT WHITE is Distinguished Professor Emeritus of Geography in the Institute of Behavioral Science at the University of Colorado, Boulder. He received his Ph.D. from the University of Chicago, where he was on the faculty from 1956-1970. He has worked on water problems with the National Resources Planning Board and Bureau of the Budget from 1934-1942. His service includes: vice-chair of the President's Water Resources Policy Commission, 1950; chair of the Task Force on Federal Flood Policy, 1965-1966; and chair of the NRC's Commission on Natural Resources, 1977-1980. He is a member of the National Academy of Sciences and foreign member of the Soviet Academy of Sciences. He has served as chair for the NRC's Environmental Studies Board and as a member of the Technology Assessment Advisory Council of the United States Congress.